Learning and Analytics in Inte

C000064390

Volume 28

Series Editors

George A. Tsihrintzis, University of Piraeus, Piraeus, Greece

Maria Virvou, University of Piraeus, Piraeus, Greece

Lakhmi C. Jain, KES International, Shoreham-by-Sea, UK

The main aim of the series is to make available a publication of books in hard copy form and soft copy form on all aspects of learning, analytics and advanced intelligent systems and related technologies. The mentioned disciplines are strongly related and complement one another significantly. Thus, the series encourages cross-fertilization highlighting research and knowledge of common interest. The series allows a unified/integrated approach to themes and topics in these scientific disciplines which will result in significant cross-fertilization and research dissemination. To maximize dissemination of research results and knowledge in these disciplines, the series publishes edited books, monographs, handbooks, textbooks and conference proceedings.

More information about this series at https://link.springer.com/bookseries/16172

George A. Tsihrintzis · Maria Virvou ·
Anna Esposito · Lakhmi C. Jain

Editors

Advances in Assistive Technologies

Selected Papers in Honour of Professor
Nikolaos G. Bourbakis – Vol. 3

 Springer

Editors
George A. Tsihrintzis
Department of Informatics
University of Piraeus
Piraeus, Greece

Maria Virvou
Department of Informatics
University of Piraeus
Piraeus, Greece

Anna Esposito
Department of Psychology
University of Campania "Luigi Vanvitelli"
Caserta, Italy

Lakhmi C. Jain
KES International
Shoreham-by-Sea, UK

ISSN 2662-3447   ISSN 2662-3455 (electronic)
Learning and Analytics in Intelligent Systems
ISBN 978-3-030-87134-5   ISBN 978-3-030-87132-1 (eBook)
https://doi.org/10.1007/978-3-030-87132-1

This Springer imprint is published by the registered company Springer Nature Switzerland AG
The registered company address is: Gewerbestrasse 11, 6330 Cham, Switzerland

Foreword

Riddle of the Sphinx: "What is the creature that walks on four legs in the morning, two legs at noon, and three legs in the evening?"

The level of civilization of a human society is commensurate with the level of assistance it provides to its vulnerable people and its people in need. Indeed, since the very early days in History, humans have been inventing and using devices in order to assist vulnerable people, ease their disabilities, and support them in their daily routines. Today, life expectancy exceeds seven decades worldwide and eight decades in Europe, the USA, Australia, and Japan [1]. At the same time, the world population is expected to rise from the current 7.8 billion [2] to about 11 billion by the end of the 21st century [3, 4]. The rapid increase in both life expectancy and world population, along with the emergence of new diseases and the regrettable continuation of armed conflicts, is certain to boost the number of vulnerable people in need of assistance to over 2 billion within this century [5, 6]. Undoubtedly, if human societies wish to continue to assist those in need throughout the 21st century, it is mandatory that Academia, Industry, and Government develop and make broadly available ever more sophisticated Assistive Technologies.

Looking at researchers who have contributed significantly to Advances in the field of Assistive Technologies, Professor Nikolaos G. Bourbakis stands out as one of the main contributors. Indeed, Prof. Bourbakis has been enjoying a long and fruitful research career at various posts. His career spans almost five decades and has earned him broad international recognition. Currently, Prof. Bourbakis is a Distinguished Professor of Information & Technology and the Director of the Center of Assistive Research Technologies (CART) at Wright State University, Ohio, USA. He received a B.S. degree in Mathematics from the National and Kapodistrian University of Athens, Greece, a Certificate in Electrical Engineering from the University of Patras, Greece, and a Ph.D. degree in Computer Engineering and Informatics (awarded with excellence), from the Department of Computer Engineering & Informatics, University of Patras, Greece. In 2021, he was also awarded the title of Honorary Professor of the School of ECE of the Technical University of Crete, Greece. His many achieve-ments in various fields, including Assistive Technologies, are widely recognized via many distinctions and awards, including elevation to IEEE Fellow (1996); IEEE Computer Society Technical Research Achievement Award; Membership in the New York Academy of Sciences; Diploma of Honor in Artificial Intelligence, School of Engineering, University of Patras, Greece; ASC Outstanding Scientists & Engineers Research Award; Dr. F. Russ IEEE Biomedical Engineering Award, Dayton Ohio; Recognition Award for Outstanding Scholarly Achievements and Contributions in the field of Computer Science, University of Piraeus, Greece; IEEE EMBS-GR Award of Achievements; IEEE Computer Society 30 years ICTAI Outstanding Service & Leadership Recognition; Honorary Doctorate Degree of the University of Piraeus, Greece (2020).

Professors George A. Tsihrintzis, Maria Virvou, Anna Esposito, and Lakhmi C. Jain are honoring Professor Nikolaos G. Bourbakis with the edition of this special volume on Advances in Assistive Technologies. At the same time, they are presenting to researchers in this community and to a broader audience a high quality volume consisting of 13 chapters from many areas of the Assistive Technologies spectrum. All chapters have been authored by active and recognized researchers and report on the most recent findings in research and development. Despite the breadth and depth of the topics covered, the volume is well organized and easy to follow by both the expert and the general reader. An editorial/introductory chapter outlines both the discipline of Assistive Technologies (including related bibliography) and the book itself. The remainder of the volume is structured in four parts, which are devoted to *Advances in Assistive Technologies in Healthcare* (3 chapters), *Advances in Assistive Technologies in Medical Diagnosis* (3 chapters), *Advances in Assistive Technologies in Mobility and Navigation* (5 chapters), and *Advances in Privacy and Explainability in Assistive Technologies* (1 chapter).

I am confident this book will be very useful to researchers in Assistive Technolo-gies as well as to general readers interested in learning more about this field. Thus, I commend and I thank its editors for their great, detailed, and painstaking work in producing this volume and highly recommend it to both expert and general readers.

Apostolos Dollas
Professor of ECE
School of Electrical and Computer Engineering
Technical University of Crete
Chania, Crete, Greece

1. https://ourworldindata.org/life-expectancy
2. https://statisticstimes.com/demographics/world-population.php
3. https://ourworldindata.org/future-population-growth#global-population-growth
4. https://www.pewresearch.org/fact-tank/2019/06/17/worlds-population-is-pro jected-to-nearly-stop-growing-by-the-end-of-the-century/

5. https://www.disabled-world.com/disability/statistics/
6. https://www.cdc.gov/ncbddd/disabilityandhealth/infographic-disability-imp
   acts-all.html

Preface

A world-recognized researcher can be honored in a variety of ways, including elevation of his professional status or various prestigious awards and distinctions. When, additionally, the same researcher has served as advisor to generations of undergraduate, graduate, and doctoral students and as mentor to faculty and colleagues, the task of appropriately honoring him becomes even harder! Perhaps, the best way to honor this person is to ask former doctoral students, as well as colleagues and fellow researchers from around the world, to include some of their recent research results in one or more high quality volumes edited in his honor. Such an edition indicates that other researchers are pursuing and extending further what they have learned from him in research areas where he made outstanding contributions.

Professor Nikolaos G. Bourbakis has been serving the fields of Artificial Intelligence (including Machine Learning/Deep Learning) and Assistive Technologies from various posts for almost fifty years now. He received a BS in Mathematics from the National and Kapodistrian University of Athens, Greece, a Certificate in Electrical Engineering from the University of Patras, Greece, and a Ph.D. in Computer Engineering and Informatics (awarded with excellence), from the Department of Computer Engineering & Informatics, University of Patras, Greece.

Dr. Bourbakis (IEEE Fellow-1996) is currently a Distinguished Professor of Information & Technology and the Director of the Center of Assistive Research Technologies (CART) at Wright State University, Ohio, USA. He is the founder and Editor-in-Chief of the International Journal on Artificial Intelligence Tools, the International Journal on Monitoring and Surveillance Technology Research (IGI-Global, Publ.) and the EAI Transactions on Bioengineering & Bioinformatics. He is also the Founder and Steering Committee Chair of several International IEEE Computer Society Conferences (namely, ICTAI, ICBIBE, ICIISA), Symposia, and Workshops. He pursues research in Assistive Technologies, Applied Artificial Intelligence, Bioengineering, Information Security, and Parallel/Distributed Processing, which is funded by USA and European government and industry. He has published extensively in IEEE and International Journals and he has graduated, as the main advisor, several dozens of doctoral students. His research work has been internationally recognized and he has received several prestigious awards, including IEEE

Computer Society Technical Research Achievement Award; Member of the New York Academy of Sciences; Diploma of Honor in AI School of Engineering, University of Patras, Greece; ASC Outstanding Scientists & Engineers Research Award; Dr. F. Russ IEEE Biomedical Engineering award, Dayton Ohio; Most Cited Article in Pattern Recognition Journal; IEEE ICTAI and ICBIBE best paper Awards; Recognition Award for Outstanding Scholarly Achievements and Contributions in the field of Computer Science, University of Piraeus, Greece; IEEE EMBS-GR Award of Achievements; IEEE Computer Society 30 years ICTAI Outstanding Service & Leadership Recognition; Honorary Doctorate degree of the University of Piraeus, Greece.

We have been collaborating with Professor Nikolaos G. Bourbakis for very many years. Thus, we proposed and undertook with pleasure the task of editing a special book in his honor. The response from former mentees, colleagues, and fellow researchers of his has been great! Unfortunately, page limitations have forced us to limit the works to be included in the book and we apologize to those authors whose works were not included. Despite the decision not to include all proposed chapters in the book, it became apparent that not only one, but three volumes of the special book had to be developed, each of which would focus on different aspects of Dr. Nikolaos G. Bourbakis's research activities.

The book at hand constitutes the third volume and is devoted to *Advances in Assistive Technologies*. While honoring Professor Nikolaos G. Bourbakis, this book also serves the purpose of exposing its reader to some of the most significant advances in Assistive Technologies. As such, the book is directed toward professors, researchers, scientists, engineers, and students in computer science-related disciplines. It is also directed toward readers who come from other disciplines and are interested in becoming versed in some of the most recent Advances in Assistive Technologies. We hope that all of them will find it useful and inspiring in their works and researches.

We are grateful to the authors and reviewers for their excellent contributions and visionary ideas. We are also thankful to Springer for agreeing to publish this book in its *Learning and Analytics in Intelligent Systems* series. Last, but not least, we are grateful to the Springer staff for their excellent work in producing this book.

Piraeus, Greece   George A. Tsihrintzis
Piraeus, Greece   Maria Virvou
Lafayette, IN, USA   Lefteri Tsoukalas
Vietri, Italy   Anna Esposito
Shoreham-by-Sea, UK   Lakhmi C. Jain

Contents

1   **Introduction to Advances in Assistive Technologies**   1
    Nikolaos G. Bourbakis, Anna Esposito, George A. Tsihrintzis,
    Maria Virvou, and Lakhmi C. Jain
    1.1   Editorial Note   2
    1.2   Book Summary and Future Volumes   6
    References   6

Part I   Advances in Assistive Technologies in Healthcare

2   **Applications of AI in Healthcare and Assistive Technologies**   11
    Iosif Papadakis Ktistakis, Garrett Goodman,
    and Aikaterini Britzolaki
    2.1   Introduction   12
    2.2   Healthcare and Biomedical Research   13
          2.2.1   Controlled Monitoring Environment   13
          2.2.2   Evolving Healthcare Techniques   15
          2.2.3   Diagnosis   18
    2.3   Assistive Technologies   19
          2.3.1   Smart Homes and Cities   20
          2.3.2   Assistive Robotics   22
    2.4   Analysis and Forecasting   23
    2.5   Conclusions   26
    References   26

3   **A Research Agenda for Dementia Care: Prevention, Risk**
    Mitigation and Personalized Interventions   33
    Anna Esposito, Alessandro Vinciarelli, and Gennaro Cordasco
    3.1   Introduction   33
    3.2   Mild Behavioral Impairments (MBI) and Dementia   35
    3.3   Biometric Data   35
    3.4   Caring for Caregivers   36
    3.5   Tests   37

3.6   Conclusions 39
References 40

**4   Machine Learning and Finite Element Methods in Modeling
of COVID-19 Spread** 43
Nenad Filipovic
4.1   Introduction 43
       4.1.1   Physiology of Human Respiratory System 43
       4.1.2   Spreading of SARS-CoV-2 Virus Infection 45
       4.1.3   Machine Learning for SARS-CoV-2 46
4.2   Methods 49
       4.2.1   Finite Element Method for Airways and Lobes 49
       4.2.2   Machine Learning Method 50
4.3   Results 55
       4.3.1   Simulation of Virus Spreading by Finite Element
                Analysis 55
       4.3.2   Machine Learning Results 58
4.4   Conclusions 63
References 64

Part II   Advances in Assistive Technologies in Medical Diagnosis

**5   Towards Personalized Nutrition Applications with Nutritional
Biomarkers and Machine Learning** 73
Dimitrios P. Panagoulias, Dionisios N. Sotiropoulos,
and George A. Tsihrintzis
5.1   Introduction 73
       5.1.1   Summary 73
       5.1.2   Chapter Synopsis 74
       5.1.3   Goals and Perspective 76
5.2   Basic Concepts 76
       5.2.1   Personalized Medicine 76
       5.2.2   Next Generation Sequencing 87
       5.2.3   Obesity 90
       5.2.4   Nutritional Biomarkers 91
5.3   Neural Networks, Pattern Recognition and Datasets 94
       5.3.1   Neural Network 94
       5.3.2   Implementation Environment 100
       5.3.3   Proposed System 102
5.4   Implementation and Evaluation of the Proposed System 104
       5.4.1   Deep Back Propagation Neural Network 104
       5.4.2   Standard Biochemistry Profile Neural Network
                (SBPNN) 105
       5.4.3   Neural Network Dietary Profile 111

5.5	Conclusions and Future Research	....	113
	5.5.1	Prevention	....	113
	5.5.2	Modelling	....	113
	5.5.3	Automation	....	113
	5.5.4	Perspective	....	114
	5.5.5	Discussion on Feature Research	....	114
5.6	Appendix	....	114
References	....	119

**6 Inductive Machine Learning and Feature Selection
for Knowledge Extraction from Medical Data: Detection
of Breast Lesions in MRI** **123**
Evangelos Karampotsis, Evangelia Panourgias,
and Georgios Dounias

6.1	Introduction	....	123
6.2	Detailed Literature Review	....	124
6.3	Presentation of the Data	....	127
	6.3.1	Data Collection	....	128
	6.3.2	Description of Variables	....	129
	6.3.3	Dataset Preprocessing	....	133
6.4	Methodology	....	134
	6.4.1	Modeling Methodology	....	135
	6.4.2	Feature Selection Process	....	137
	6.4.3	Classification Method	....	138
	6.4.4	Validation Process	....	140
6.5	Modeling Approaches	....	143
	6.5.1	Experimental Process	....	143
	6.5.2	Experimental Results	....	144
6.6	Conclusions and Further Search	....	152
Annex 6.1—Abbreviations	....	154
Annex 6.2—Variables Frequency Charts (Original Dataset)	....	155
Annex 6.3—Variables' Values Range	....	157
Annex 6.4—Classification Tree (Benign or Malignant)	....	160
References	....	161

**7 Learning Paradigms for Neural Networks for Automated
Medical Diagnosis** **165**
Smaranda Belciug

7.1	Introduction	....	165
7.2	Classical Artificial Neural Networks	....	166
7.3	Learning Paradigms	....	170
	7.3.1	Evolutionary Computation Learning Paradigm	....	170
	7.3.2	Bayesian Learning Paradigm	....	173
	7.3.3	Markovian Stimulus-Sampling Learning Paradigm	....	176
	7.3.4	Logistic Regression Paradigm	....	176
	7.3.5	Ant Colony Optimization Learning Paradigm	....	177

7.4   Conclusions and Future Outlook 179
References 179

**Part III   Advances in Assistive Technologies in Mobility and
             Navigation**

8   Smart Shoes for Assisting People: A Short Survey 183
Nikolaos G. Bourbakis, Iosif Papadakis Ktistakis,
and Pulkit Khursija
8.1   Smart Shoes for People in Need 184
        8.1.1   Smart Shoes for Visually Impaired People [6] 184
        8.1.2   Smart Shoes for Blind Individuals [13] 185
        8.1.3   IoT Based Wireless Smart Shoes and Energy
                Harvesting System [7] 186
        8.1.4   Smart Shoes for Sensing Force [8] 187
        8.1.5   Smart Shoes for Temperature and Pressure [9] 187
        8.1.6   Smart Shoes in IoT [10] 188
        8.1.7   Smart Shoes for People with Walking Disorders
                [23] 189
8.2   Special Purpose Smart Shoes 190
        8.2.1   Smart Shoes with Triboelectric Nanogenerator [11] 190
        8.2.2   Smart Shoes Gait Analysis [12] 190
        8.2.3   Smart Shoes for Biomechanical Energy
                Harvesting [14] 191
        8.2.4   Smart Shoes with Embedded Piezoelectric Energy
                Harvesting [15] 192
        8.2.5   Pedestrian Navigation Using Smart Shoes
                with Markers [16] 193
        8.2.6   Smart Shoes with 3D Tracking Capabilities [17] 194
        8.2.7   Pedestrian's Safety with Smart Shoes Sensing [18] 194
        8.2.8   Smart Shoes Insole Tech for Injury Prevention [19] 195
8.3   Maturity Evaluation of the Smart Shoes 198
8.4   Conclusion 201
References 201

**9   Re-Examining the Optimal Routing Problem
      from the Perspective of Mobility Impaired Individuals** 203
K. Liagkouras and K. Metaxiotis
9.1   Introduction 203
9.2   Literature Review 205
        9.2.1   Mobility Aspects for People with Special Needs 205
9.3   Related Work 206
        9.3.1   Miller–Tucker–Zemlin Formulation
                of the Traveling Salesman Problem 206
        9.3.2   Dantzig–Fulkerson–Johnson Formulation
                of the Traveling Salesman Problem 207

9.4   The Optimal Routing Problem from the Perspective
        of Mobility Impaired Individuals 208
        9.4.1   Measuring Route Scores Based on the Degree
                 of Accessibility 208
        9.4.2   Problem Statement: The Optimal Routing
                 Problem from the Perspective of Mobility
                 Impaired Individuals 209
        9.4.3   The Proposed Solution Approach 211
9.5   The Experimental Results and Discussion 211
9.6   Conclusions 214
References 214

**10 Human Fall Detection in Depth-Videos Using Temporal
     Templates and Convolutional Neural Networks** 217
     Earnest Paul Ijjina
     10.1   Introduction 217
     10.2   Proposed Method 220
     10.3   Experiments, Results and Discussion 223
            10.3.1   SDU Fall Dataset 224
            10.3.2   UP-Fall Detection Dataset 229
            10.3.3   UR Fall Detection Dataset 230
            10.3.4   MIVIA Action Dataset 231
     10.4   Conclusions and Future Work 233
     10.5   Compliance with Ethical Standards 234
     References 234

**11 Challenges in Assistive Living Based on Tech Synergies: The
     Cooperation of a Wheelchair and A Wearable Device** 237
     Nikolaos G. Bourbakis
     11.1   Overall Description of the Challenges 238
     11.2   Background and Significance 239
     11.3   The Associated Research Challenges 241
            11.3.1   Main Innovative Tasks 241
     11.4   Discussion 253
     References 253

**12 Human–Machine Requirements' Convergence for the Design
     of Assistive Navigation Software: The Case of Blind
     or Visually Impaired People** 263
     P. Theodorou and A. Meliones
     12.1   Introduction 263
     12.2   Related Work 265
     12.3   Methodology 266
            12.3.1   Interviews with BVI People and Requirements
                     Classification 266

12.3.2   Description of the Participants 268
12.3.3   Requirements Classification 270
12.4   Analysis of the Elicited Requirements 271
12.4.1   Elicited Requirements of the BVI 271
12.5   Discussion 275
12.6   Conclusion 276
Appendix A 277
References 282

Part IV   Advances in Privacy and Explainability in Assistive Technologies

13   Privacy-Preserving Mechanisms with Explainability in Assistive AI Technologies

.... 287

Z. Müftüoğlu, M. A. Kızrak, and T. Yıldırım

13.1   Introduction 288
13.1.1   Data Ethics 289
13.1.2   Data Privacy 289
13.1.3   Data Security 290
13.2   AI Applications in Assistive Technologies 290
13.2.1   Explainable AI (XAI) 295
13.3   Data Privacy and Ethical Challenges for Assistive Technologies 296
13.3.1   Data Collection and Data Sharing 297
13.3.2   Secure and Responsible Data Sharing Framework 297
13.4   AI Assistive Technologies with Privacy Enhancing 299
13.4.1   Privacy-Preserving Mechanisms for AI Assistive Technologies 299
13.5   Discussions 304
References 305

Chapter 1
Introduction to Advances in Assistive Technologies

Nikolaos G. Bourbakis, Anna Esposito, George A. Tsihrintzis, Maria Virvou, and Lakhmi C. Jain

Abstract A fundamental priority of modern societies is to provide assistance to *vulnerable* people, including elderly people, people with cognitive disabilities, such as memory problems or dementia, and people with functional disabilities, such as those needing support in toileting, moving around, eating, bathing, dressing, grooming and taking personal care. *Assistive Technologies* is the term used to describe devices developed in order to assist vulnerable people. As life expectancy and the World population constantly increase, the need is becoming more and more pressing for development and availability of more advanced Assistive Technologies, which offer to vulnerable people the opportunity to live secure and controlled lives with active participation in society and at a reduced assistive cost. Such Assistive Technologies include prosthetics, exoskeletons, visual and hearing aids, cognitive aids and other devices possibly enhanced with Artificial Intelligence. The book at hand aims at exposing its readers to some of the most significant advances in Assistive Technologies and originated as an honour to a recognized researcher in Assistive Technologies, namely Professor Nikolaos G. Bourbakis. As such, the book is directed towards professors, researchers, scientists, engineers and students in all related disciplines. It is also directed towards readers who come from other disciplines and are interested in becoming versed in some of the most recent advances in Assistive Technologies. An extensive list of bibliographic references at the end of each chapter guides the readers to probe further into the application areas of interest to them.

N. G. Bourbakis
Wright State University, Dayton, OH, USA
e-mail: nikolaos.bourbakis@wright.edu

A. Esposito
Università delle Campania "Luigi Vanvitelli", Caserta, Italy
e-mail: iiass.annaesp@tin.it

G. A. Tsihrintzis (✉) · M. Virvou
University of Piraeus, Piraeus, Greece
e-mail: geoatsi@unipi.gr

L. C. Jain
Liverpool Hope University, Liverpool, UK

University of Technology Sydney, Ultimo, NSW, Australia

G. A. Tsihrintzis et al. (eds.), *Advances in Assistive Technologies*, Learning and Analytics in Intelligent Systems 28, https://doi.org/10.1007/978-3-030-87132-1_1

1.1 Editorial Note

Assistive Technologies are not new to humanity as, for thousands of years, humans have designed and developed various devices and systems to comfort disabilities and assist those with special needs [1–22]. Indeed, several of these Assistive Technologies tools have since the ancient times been in use by Sumerians, Egyptians, Greeks, Chinese and others and include the very old and very useful "cane" (i.e., walking stick) or prosthetic limbs. In Homer's "Odyssey", for example, Polyphemus (Πολύφημος in Greek), one of the mythical one-eyed Cyclops, after having been *blinded* by Ulysses, descends to the shore using a "lopped pine tree" as a walking aid [1]. Similarly and according to the Riddle of the Sphinx, "man is the creature that walks on four legs in the morning, two legs at noon, and three legs in the evening" [2], implying that a device like a walking stick has since ancient times been assisting people who are in need because of *old age.*

In the modern world, there has been significant progress in Assistive Technologies and there are many related inventions that continue to contribute to our lives in several different ways. Some of the most significant commonly-used Assistive Technologies include [6]:

- the first official text on sign language for the deaf-mutes, published in the 1620s;
- the world-famous binary writing system, invented in the 1700s;
- first attempts to develop hearing aids, in the 1820s;
- the first true hearing aid to embrace microphone and telephone technologies, developed in the 1890s;
- the invention of a stairs-chair for transporting up-and-down an individual with walking disabilities, in the 1920s;
- the first lightweight, collapsible wheelchair, presented in the 1930s;
- the light walker, presented in the late 1940s;
- a sip-and-puff device, designed in the 1960s; This device activated micro-switches that allowed an individual without limbs to operate electrical devices, namely lights and television, by sucking or blowing through a mouthpiece;
- the invention of the new wheelchair lift, in the late 1960s;
- speech-generating devices, firstly made available in the 1970s;
- major breakthroughs in prosthetic limbs which appeared after the 1980s and were due to technological advances;
- innovative/advanced software tools which were developed in the 1990s in universities and research centers and allowed accessibility to users with one or more disabilities.

In the 21st century, the number of people with special needs will reach 2 billion. Thus, advances of Information Technology embedded in intelligent systems, including Artificial Intelligence, Machine Learning and Deep Learning, are a necessary tool with potential to significantly boost Assistive Technologies and offer users with disabilities access to processing and evaluating valuable information. Moreover, the synergy between Information Technology and Artificial Intelligence-based

Technologies will create friendly intelligent interfaces to perform advanced functions using only signals directly from the brain for the operation of various devices or systems. In addition, satellite-navigation systems, intelligent Assistive Technologies devices, smart exoskeletons, intelligent robotics, autonomous vehicles and smart cities with access to global digitized maps can easily be used by people in need due to personalized intelligent software and internet services. This implies that the commercial "war" for developing and accessing Assistive Technologies products has already begun, with players that include Microsoft, IBM, General Motors, BMW, Google, Amazon, Apple, Mercedes-Benz, Toyota and others. In fact, these players are engaged in a "battle" to be the first to produce products for users with disabilities, and applications for smart houses and smart cities [9, 10, 14, 15].

Consequently, current Assistive Technologies form a multi-disciplinary research field enjoying worldwide interest, tremendous pace of growth, very high success rates and significant impact on science, technology and society. As Assistive Technologies are being developed to reduce differences between healthy people and individuals with physical and cognitive impairments and to assist vulnerable people and support their caregivers, their complexity raises ethical and acceptance issues [23]. Professor Nikolaos G. Bourbakis's research on Assistive Technologies was imperatively driven by both considering end-users' requirements and expectations, as well as ethical, legal, and social implications regarding their use. It seems worth in this editorial note to emphasize that such technologies must be easy to use, gain the acceptance of their users, and account for ethical and privacy issues associated with their use.

Any Assistive Technology developed for impaired individuals creates environmental and social modifications in their daily life. This issue is referred to as "*assistive technology and environmental interventions*" [24]. To this aim, Assistive Technologies are effective only if their implementation accounts for concepts that modulate the degree of "acceptance" by their users and avoid the violation of ethical principles governing social interaction policies [25]. To make sure that a technology is accepted, developers must consider users' level of technological experience and involve them in their design process. The acceptance of a technological device must satisfy the following constraints:

1. Highly motivate the user to take advantage of the proposed device
2. Be easy to use
3. Be designed in such a way as to inspire users' confidence
4. Be safe, comfortable in its use
5. Its design be based on "universal design concepts" to be environmentally and user adaptable.

Consequently, five ethical principles have been proposed to guide the development and application of Assistive Technologies: (i) beneficence, (ii) non-maleficence, (iii) justice, (iv) autonomy, and (v) fidelity [26]. *Beneficence* is associated with benefits that individuals can derive from technology use and trustworthiness in using it. *Non-maleficence* calls for preventing technology to cause pain or create harmful situations. The *justice* principle calls for fairness in the distribution of technology within individual, social, and organizational contexts. *Autonomy* relates to the individual's freedom of choice over the use of a given technology. In addition to the previous, a further ethical issue is *privacy*, which aims at preventing the diffusion of sensitive health data of the people involved. Future studies should focus on establishing guidelines to be followed in emergency situations, ensuring that individuals' wishes regarding their privacy are not violated.

The book at hand has originated as an honour to Professor Nikolaos G. Bourbakis, an outstanding researcher and educator in Assistive Technologies who has conducted leading research in this field for several decades and inspired, advised and mentored tens of students, fellow researchers and colleagues. However, the book also aims at exposing its readers, whether they are specializing in the Assistive Technologies field or they are general readers interested in becoming versed in some of the most recent Assistive Technologies, to some significant related advances, including:

I.   *Advances in Assistive Technologies in Healthcare,*
II. *Advances in Assistive Technologies in Medical Diagnosis,*
III. *Advances in Assistive Technologies in Mobility and Navigation,*
IV. *Advances in Privacy and Explainability in Assistive Technologies.*

More specifically, the book at hand consists of the current editorial chapter (In this chapter) and an additional twelve (12) chapters. All chapters in the book were invited from authors who work in the corresponding chapter theme and are recognized for their significant research contributions. In more detail, the chapters in the book are organized into four parts, as follows:

The *first part* of the book consists of three chapters devoted to *Advances in Assistive Technologies in Healthcare.*

Specifically, Chap. 2, by I. Papadakis-Ktistakis, G. Goodman, and A. Britzolaki, is entitled "*Applications of AI in Healthcare and Assistive Technologies.*" The authors describe the multiple applications of AI in healthcare and assistive technologies in the form of an extensive literature review, encompassing the overarching advances that AI has progressed, intelligent automation in assistive technologies, and the overall benefits from including AI in healthcare.

Chapter 3, by A. Esposito, A. Vinciarelli and G. Cordasco, is entitled "*A Research Agenda for Dementia Care: Prevention, Risk Mitigation and Personalized Interventions.*" The authors suggest a research agenda for dementia, connecting the latest clinical understanding of such neurodegenerative disorders, with social behavioral changes and social cognitive impairments.

Chapter 4, by N. Filipovic, is entitled "*Machine Learning and Finite Element Methods in Modeling of COVID-19 Spread.*" The author analyzes patient-specific machine learning and finite element approaches for modeling COVID-19 spread and for predicting the risk of mortality.

The *second part* of the book consists of three chapters devoted to *Advances in Assistive Technologies in Medical Diagnosis.*

Specifically, Chap. 5, by D. Panagoulias, D. Sotiropoulos and G.A. Tsihrintzis, is entitled "*Towards Personalized Nutrition Applications with Nutritional Biomarkers and Machine Learning.*" The authors formulate a general evaluation chart of nutritional biomarkers which is used to investigate how to best predict body mass index and to discover dietary patterns with the deployment of artificial neural networks.

Chapter 6, by E. Karampotsis, E. Panourgias and G. Dounias, is entitled "*Inductive Machine Learning and Feature Selection for Knowledge Extraction from Medical Data: Detection of Breast Lesions in MRI.*" The authors present an approach to the problem of breast cancer diagnosis through the data analysis of magnetic mammography observations and development of corresponding hybrid classification models of patient cases into benign or malign classes.

Chapter 7, by S. Belciug, is entitled "*Learning Paradigms for Neural Networks for Automated Medical Diagnosis.*" The author provides an overview of some learning paradigms for neural networks that can be applied to automated medical diagnosis and highlights their benefits in this crucial domain.

The *third part* of the book consists of five chapters devoted to *Advances in Assistive Technologies in Mobility and Navigation.*

Specifically, Chap. 8, by N. G. Bourbakis, I. Papadakis-Ktistakis and P. Khursija, is entitled "*Smart Shoes for Assisting People: A Short Survey.*" The authors offer a short survey on the state-of-the-art of technological accomplishments associated with smart shoes.

Chapter 9, by K. Liagkouras and K. Metaxiotis, is entitled "*Re-examining the Optimal Routing Problem from the Perspective of Mobility Impaired Individuals.*" The authors propose a routing recommendation algorithm, which takes into consideration the level of accessibility of the alternative routes, and report on numerical experiments to examine the performance of the proposed routing algorithm.

Chapter 10, by E. P. Ijjina, is entitled "*Human Fall Detection in Depth-Videos using Temporal Templates and Convolutional Neural Networks.*" The author uses motion information obtained from video to compute new temporal templates, which are in turn used by a convolutional neural network to recognize the human actions.

Chapter 11, by N. G. Bourbakis, is entitled "*Challenges in Assistive Living based on Tech Synergies: The Cooperation of a Wheelchair and a Wearable Device.*" The author discusses very important challenges associated with the synergy of the state-of-the-art technologies, like wearable devices and autonomous intelligent wheelchairs, for the elderly and the people in need.

Chapter 12, by P. Theodorou and A. Meliones, is entitled "*Human-Machine Requirements' Convergence for the Design of Assistive Navigation Software: The case of Blind or Visually Impaired People.*" The authors outline a subset of requirements identified with regard to assistive navigation software for the blind/visually impaired and moving devices equipped with Artificial Intelligence.

The *fourth part* of the book consists of one chapter devoted to *Advances in Privacy and Explainability in Assistive Technologies.*

Specifically, Chap. 13, by Z. Müftüoğlu, M. A. Kizrak and T. Yildirim, is entitled "*Privacy-Preserving Mechanisms with Explainability in Assistive AI Technologies.*" The authors tackle privacy-preserving solutions for assistive technologies and the relationship between explainability and privacy will be tackled.

1.2   Book Summary and Future Volumes

In this book, we have presented some very significant advances in Assistive Technologies, while honouring Professor Nikolaos G. Bourbakis who has been conducting leading relevant research for several decades and has inspired, advised and mentored tens of students, fellow researchers and colleagues. The book is directed towards professors, researchers, scientists, engineers and students in related disciplines. It is also directed towards readers who come from other disciplines and are interested in becoming versed in some of the most significant advances in Assistive Technologies. We hope that all of them will find it useful and inspiring in their works and researches.

On the other hand, societal demand continues to pose challenging problems, which require ever more advanced theories and ever more efficient tools, methodologies, systems and Assistive Technologies to be devised to address them. Thus, the readers may expect that additional related volumes will appear in the future.

References

1. https://www.sparknotes.com/lit/odyssey/full-text/book-ix/
2. *Oedipus and the Sphinx*, in M. Kallich, *Oedipus: Myth and Drama*, Bobbs-Merrill Co, June 1, 1968, ISBN-13: 978-0672630767
3. Assistive Technology Act of 1998, Retrieved from http://www.gpo.gov/fdsys/pkg/PLAW-108 publ364/html/PLAW-108publ364.htm
4. Assistive Technology from Ancient to Modern Times. (n.d.). Retrieved from http://bluebirdc are.ie/assistive-technology-ancient-modern-times/
5. S. Chang, *Disability in Smart Cities: Assessing Assistive Technologies and Urban Accessibility.* Oxford Urbanists, (January 11, 2019), https://www.oxfordurbanists.com/
6. L. Grant (Ed.), A Disability History Timeline -The struggle for equal rights through the ages. Retrieved from http://www.uhsm.nhs.uk/AboutUs/Equality and Diversity/DISABILITY Timeline–NHS.pdf
7. M. Scherer, *Assistive technology: Matching device and consumer for successful rehabilitation.* Washington, DC: American Psychological Association (2002)
8. R. Smith, K. Rust, A. Lauer, E. Boodey, Technical Report—History of Assistive Technology Outcomes (Version 1.0), Retrieved from http://www.r2d2.uwm.edu/atoms/archive/technical reports/fieldscans/tr-fs-history.html
9. N. Bourbakis, Research on Assistive Technologies, Proceedings of Int. IEEE-BAIS Symposium, Dayton, OH, USA, *IEEE Computer Society*, April 2007
10. A.R. Garcia-Ramirez, M.G. Gomes-Ferreira (Eds.), Assistive Technologies in Smart Cities, *Universidade do Vale do Itajaí,* Published: December 13th 2018, https://doi.org/10.5772/int echopen.72240
11. G.E. Lancioni, N.N. Singh (Eds.), Assistive Technologies for People with Diverse Abilities, Autism and Child Psychopathology Series, *Springer* (2014)
12. A.M. Cook, J. Miller Polgar, *Assistive Technologies: Principles and Practice.* Elsevier Publ. (2020)
13. N. Parikh, *A Brief History of Assistive Technology.* Kratu Inc., retrieved from http://test.kratu. org/2017/11/22/brief-history-of-assistive-technology-by-niraj-parikh/
14. N. Bourbakis, Smart Cities, CSE-7900, Lectures-notes (2017-2021), CART Center, *Wright State University*, Dayton Ohio, USA

15. C. Sik-Lanyi, Special issue on universal design and assistive technologies for accessibility in smart cities. Smart Cities **3**(4), 1293–1494 (2020). https://www.mdpi.com/search?journal=smartcities&special_issue=36007

16. A. Brooks, S. Brahnam, B. Kapralos, A. Nakajima, J. Tyerman, L. C. Jain (Eds.), *Recent Advances in Technologies for Inclusive Well-Being: Virtual patients, Gamification and Simulation.* Springer (2021)

17. A. Woodcock, L. Moody, D. McDonagh, A. Jain, A. and L.C. Jain (Eds.), Design of Assistive Technology for Ageing Populations, Springer-Verlag, Germany, 2020

18. A.L. Brooks, S. Brahnam, B. Kapralos, L.C. Jain (Eds.), *Technologies of Inclusive Well-Being: From Worn to Off-body Sensing, Virtual Worlds, and Games for Serious Applications.* Springer, Germany (2017)

19. A.L. Brooks, S. Brahnam, L.C. Jain, L.C. (Eds.), *Technologies of Inclusive Well-Being.* Springer, Germany (2014)

20. S. Brahnam, L.C. Jain (Eds.) *Advanced Computational Intelligence Paradigms in Healthcare 6: Virtual Reality in Psychotherapy, Rehabilitation, and Assessment.* Springer (2011)

21. N. Ichalkaranje, A. Ichalkaranje, L.C. Jain (Eds.), *Intelligent Paradigms for Assistive and Preventive Healthcare.* Springer (2006)

22. H.-N. Teodorescu, L.C. Jain (Eds.), *Intelligent Systems and Technologies in Rehabilitation Engineering.* CRC Press, USA (2001)

23. E. Tebbutt, R. Brodmann, J. Borg, M. MacLachlan, C. Khasnabis, R. Horvath, Assistive products and the sustainable development goals (SDGs). Globalization Health **12**, 79. https://doi.org/10.1186/s12992-016-0220-6 (2016)

24. J. Hammel, Technology and the environment: supportive resource or barrier for people with developmental disabilities?, Nurs. Clin. N Am. **38**, 331–349, Chicago, USA (2003)

25. M. Liegl, A. Boden, M. Büscher, R. Oliphant, X. Kerasidou, Designing for ethical innovation: a case study on ELSI co-design in emergency. Int. J. Human Comput. Stud. **95**, 80–95. https://doi.org/10.1016/ijhcs.2016.04.003 (2016)

26. A.M. Cook, Ethical issues related to the use/non-use of assistive technologies. Develop. Disab. Bull. **37**(1,2), 127–152 (2009)

Part I
Advances in Assistive Technologies in Healthcare

Chapter 2
Applications of AI in Healthcare and Assistive Technologies

Iosif Papadakis Ktistakis, Garrett Goodman, and Aikaterini Britzolaki

Abstract Artificial Intelligence (AI) has benefited the world greatly, since its popularity has grown in the recent decades. Specifically, it has affected manufacturing, education, transportation, healthcare, and many more areas of interest. One of the primary areas benefited from it, healthcare, has had the most interesting advances in the form of life saving predictions and faster prognosis. This is done with the use of complex algorithms to simulate human cognition in the analysis of data with and without human input. The biggest advantage originates from the ability to gather information, process it and provide a well-defined output to the user. Assistive technologies have also greatly been benefited as well from AI. People with disabilities and the elderly, commonly called as people in need, are being helped by researchers finding innovative ways to put artificial intelligence to work with their current conditions. With the advancements in technology and science, assistive technologies will continue to produce new and improved platforms to help create a better standard of living for those individuals. In this chapter, we describe the multiple applications of AI in healthcare and assistive technologies in the form of an extensive literature review. This review will encompass the overarching advances that AI has progressed, intelligent automation in assistive technologies, and the overall benefits from including AI in healthcare. Furthermore, we perform a maturity evaluation of the many areas within healthcare and assistive technologies in which AI is a prominent part of.

I. P. Ktistakis (✉)
ASML LLC US, Wilton, CT 06851, USA

G. Goodman
Center of Assistive Research Technologies (CART), Wright State University,
Dayton, OH 45435, USA
e-mail: garrett.goodman@wright.edu

A. Britzolaki
Department of Anesthesiology, Weill Cornell Medical College of Cornell University, 1300 York
Ave, New York, NY 10065, USA
e-mail: aib4001@med.cornell.edu

2.1 Introduction

Artificial Intelligence (AI) is emerging as one of the fastest growing fields in computer science and engineering with a wide range of applications. In recent years, AI has been vastly applied in healthcare and biomedical research, shaping the future of medicine. AI aims to mimic human behavior; a constellation of algorithms, applications, and technologies converge to recreate cognitive functions by acquiring, processing, and applying knowledge. When combined with powerful data analysis, AI can provide a deep understanding of the world around us [1]. Thus, emerging AI breakthroughs provide a paradigm shift in healthcare and biomedical research.

Notably, the role of AI in healthcare is comprised by three core functions: prognosis/diagnosis, decision making, and robotic assistive applications. Concurrent advancement in AI technologies to assist these functions have led to a dramatic increase of AI applications in the biomedical field and a robust interest in developing novel AI-controlled approaches. In that context, as extensively discussed in Jiang et al. [2], the evolution of AI in healthcare is advancing the medical field in great lengths through various applications in medical imaging, neurology, oncology, and cardiology. The importance of AI in medicine is further highlighted in a recent review by Yu et al. [3]. The authors discuss all current and potential applications in the biomedical field, including automated data collection, simulation of molecular dynamics, biomarker and drug discovery, gene annotation, diagnosis, selection of treatment, automated surgery, as well as patient monitoring [3]. Furthermore, it is well established that Machine Learning (ML) and Natural Language Processing (NLP) are essential elements in developing efficient AI technologies in the medical field. Notably, the successful application of AI in biomedicine is mainly attributed to the major evolution of Deep Learning (DL) in recent years, by employing training of complex artificial neural networks on big datasets (as reviewed in [2, 3]).

Moreover, assistive robotics and robotic surgery represent a major part of emerged AI technologies. AI-controlled robotic systems have been used in a range of environments including industrial assembly lines, preclinical, and clinical laboratory sites. Major focus has been put on developing autonomous robotic systems to perform automated surgeries, as well as assist in patient monitoring, care, rehabilitation, and performing activities of daily living (ADL) [3–7].

Most importantly, growing evidence supports the pivotal role of AI in future healthcare. Aging, neurological, and cardiovascular disorders are the major causes of disability. According to the World Health Organization (WHO) over 15% of the world's population have some form of disability, further highlighting the need for robust, efficient, and trustworthy assistive and diagnostic solutions. Herein, we are conducting a systematic review of key AI applications that have changed the landscape in healthcare, biomedical research, and assistive technologies in the past decade (2010–2020). In that context, we have categorized several studies into two main groups: (i) AI applications in healthcare and biomedical research, and (ii) AI applications in assistive technologies. Each group is further dissected into clusters of applications. To further assess the focus of research, we present a maturity analysis based on the number of citations within each cluster throughout the decade. Finally, we predict the research focus in the near future by forecasting the future trajectory of publications in the next 3 years.

2.2 Healthcare and Biomedical Research

In this section we will examine a large representative selection of the *state-of-the-art* technologies with respect to three categories within healthcare and biomedical research. That is, examining research in the controlled monitoring environment such as the Intensive Care Unit (ICU) of a hospital (Sect. 2.1), presenting key research outcomes on selected emerging healthcare techniques such as pain management (Sect. 2.2), as well as discussing how AI has benefited the realm of diagnosis in healthcare (Sect. 2.2).

2.2.1 Controlled Monitoring Environment

In this section we focus on representative research in the field of controlled monitoring environments and specific sub-categories. In that context, the first sub-category to be examined is the traditional image classification in hospital settings. Such an example is the classification of an Interventional X-Ray system into its Cardiac, Neurological, and Vascular X-Ray categories [8]. Specifically, Patil et al. [8] used Support Vector Machines (SVM), k-Nearest Neighbors (KNN), and Decision Trees (DT) to improve medical professional productivity. Furthermore, studies have attempted to increase multi-hospital collaboration with respect to congenital cataracts as specialized care is selective and leaves many people without a treatment location [9]. Long et al. [9] employed an AI agent that used Convolutional Neural Networks (CNN) to provide diagnostics, risk stratification, and treatment suggestions. In addition, extensive efforts by Ktistakis and Bourbakis [10] have focused on designing a novel robotic nurse to assist people in a controlled environment such as a smart home and rehabilitation centers [10]. More specifically, the designed robotic nurse, an intelligent robotic wheelchair, included two cameras, pressure sensors on the seat, a microphone, speakers, a computer for processing, and two robotic arms. More importantly, this intelligent wheelchair was designed to recognize a set of voice commands, body poses, as well as force from piezoelectric sensors to control the system. The ultimate end result was to assist an individual in need to complete getting up and sitting down tasks [10]. Other studies have focused on introducing AI in postoperative predictions. Specifically, Murff et al. [11] have investigated ways to utilize medical records to predict postoperative complications of surgery using NLP [11]. Interestingly, using NLP in Veterans Affairs medical centers have resulted in higher sensitivity and lower specificity as compared to traditional discharge coding. Additionally, Lee et al. [12] utilized DT, Random Forests (RF), Extreme Gradient Boosting (EGB), SVM, Neural

Networks (NN), and DL to examine postoperative acute kidney injury [12]. Notably, the proposed internet-based risk estimator successfully identified individuals more prone to postoperative acute kidney injury while using EGB in real time. Furthermore, wearable systems have been proposed for achieving controlled and continuous monitoring of people in need. Specifically, Pantelopoulos and Bourbakis [13] have designed a wearable system consisting of *off-the-shelf* Bluetooth sensors [13]. This system consisted of a smart phone acting as a sensor hub, a remote workstation for computing, a chest belt monitor for conducting 1-lead Electrocardiogram (ECG) and recording of heart rate, respiration rate, skin temperature, posture, and activity, as well as a PPG finger clip sensor and a blood pressure monitor. Intriguingly, the designed system was capable of greatly variable disease monitoring.

The second subcategory to review includes methods of unobtrusive monitoring in a controlled environment. Specifically, unobtrusive monitoring is defined as any system capable of monitoring an individual non-invasively and with minimal or no personal interaction. In that context, wearable devices are a prime focus towards this goal. Such an example by Goodman and Bourbakis [14] is a wearable ultrasound vest designed to generate a near-3D model of the heart in real time to continuously monitor the heart [14]. More specifically, the proposed device was designed for achieving continuous and autonomous monitoring of the external surface areas of the individual cardiac chambers. This allows the medical professional to focus on the results rather than the Ultrasonography task. In addition, a system collecting data from a wearable ECG has been created by Walinjkar and Woods [15] to classify arrhythmia and automatically update electronic health records [15]. Interestingly, the implementation of KNN and NN in this study showed a 97% accuracy. Further studies on wearable devices has yielded a physiological data fusion technique by Pantelopoulos and Bourbakis [13] using fuzzy regular formal language and fuzzy finite state machines to assist in prognosis via the multimodal input of sensor data [16]. This algorithm is generalizable in the sense that it can be easily adjusted to work with a variety of sensor setups in a wearable system. Apart from wearable systems, cameras in hospital rooms have been utilized for unobtrusive monitoring. Specifically, Rantz et al. [17] proposed an automated fall detection system designed to help reduce falls in hospital rooms [17]. The authors investigated how utilizing the Microsoft Kinect can successfully differentiate between genuine falls and other events such as sitting down, crouching, etc. Furthermore, when a fall event was captured, the camera recordings were used for medical professionals to learn and apply preventative measures to help prevent future occurrences. Finally, sleep monitoring has begun to take focus in the literature recently. Specifically, Romine et al. [18] began exploring unobtrusive sleep monitoring via a wearable wrist Electrodermal Activity (EDA) sensor [18]. Their work detailed a number of extracted EDA features which can be used in a ML framework for detecting subjective sleep quality.

The third subcategory regulates AI research to the ICU. An example of this is a ML approach by Eerikäinen et al. [19] to reduce arrhythmia false alarms in ICUs [19]. Their work used data from ECG and either Photoplethysmogram (PPG) or Arterial Blood Pressure (ABP) depending on the availability. The classifier utilized a traditional RF and showed an average of 82% false alarm detection rate to assist medical professionals. Further studies confined to the ICU involved research to continuously monitor, provide pre-diagnosis information, and alert medical professionals of issues using fuzzy logic [20]. This system by Leite et al. [20] takes in as input blood pressure, heart rate, respiratory rate, body temperature, and partial oxygen saturation and outputs a human readable message of the situation, a suggestion for the next step, and level or urgency corresponding to the situation. An important aspect of ICU is to determine the time sensitive patients as there is limited staff at hand. Therefore, Sadeghi et al. [21] studied early mortality predictions in the ICU using heart signals within the first hour of ICU admission [21]. The algorithms used were RF, Gaussian SVM, DT, Boosted Trees (BT), KNN, Logistic Regression (LR), Linear Discriminant Analysis (LDA), and linear SVM which interestingly showed a 0.97 F1-Score. Within the ICU, multiple medical conditions are handled, and AI classifications have been researched for multiple conditions. For Example,, Koyner et al. [22] researched Acute Kidney Injury (AKI) predictions using data from the ICU, wards, and emergency care [22]. The authors used a Gradient Boosting Machine (GBM), an extension of a tree algorithm, for this purpose. Their proposed algorithm is capable of predictions approximately 1–2 days faster than the standard methodology currently used in the ICU. Following, a ML approach was investigated by Moghadam et al. [23] to predict hypotensive events in the ICU [23]. Their method predicted hypotension using SVM and KNN up to 30 min in advance of the event. The last medical condition examined was sepsis, by Nemate et al. [24], that used the Weibull-Cox proportional hazard survival model [24]. Their results showed an ability to predict sepsis up to 12 h in advance for some cases.

We can see from these representative articles that there are a large variety of approaches to AI in healthcare and biomedical research with respect to controlled environments. These pathways include primarily AI in intensive care units, unobtrusive monitoring, and surgical outcomes. While we detailed a number of possible paths to pursue for future research in this area, there are a plethora of additional research potential according to the literature.

2.2.2 Evolving Healthcare Techniques

In this section, we examine a set of representative articles detailing evolving healthcare techniques. That is, research which assists people in need or has a unique approach to solving a problem that could be used within the hospital and specific sub-categories. The first sub-category analyzes pain assessment and management. For example, Gholami et al. [25] used a Relevance Vector Machine (RVM), a Bayesian extension of the SVM, for differentiating painful from non-painful facial expressions in neonates [25]. Pain differentiation is important as neonates (i.e., newborn children) are obviously incapable of effectively communicating that they are in pain. Similarly, the EDA signal was researched for its ability to identify pain as it can naturally distinguish stress and anxiety [26]. More specifically, Susam et al. [26] extracted salient features and used Time Scale Decomposition (TSD) with the standard deviation as the metric. The extracted features were fed into a Linear SVM for binary classification of pain or not pain and indicated an accuracy of 71.67%. In addition, Yang et al. [27] have created a methodology for predicting pain in patients suffering from Sickle Cell disease via physiological measures consisting of peripheral capillary oxygen saturation, systolic and diastolic blood pressure, heart rate, respiratory rate, and temperature [27]. Their methodology includes a method to impute missing medical data as a large portion of their dataset was incomplete. Interestingly, the results showed an ability to classify multiple pain scales using SVM, KNN, RF, and Multinomial Logistic Regression (MLR).

The second subcategory focuses on incorporating NLP, which is the interaction between computers and humans via natural language, for analyzing medical documentation. First, an NLP approach by Karakülah et al. [28] was created to automatically scan case documents related to congenital anomalies for newborn infants [28]. More specifically, the authors used the maximum entropy model for part-of-speech tagging to define lexical categories of the tokens. Their results showed that they could identify 89.20% of the features in documents related to phenotypic abnormalities. Furthermore, Grundmeier et al. [29] used NLP to assist medical professionals in improving healthcare quality with respect to identifying long bone fractures for pediatric emergencies [29]. That is, features were extracted from the text of radiology reports which were used as input to multiple ML algorithms consisting of Linear Decision Boundary (LDB), SVM with a Gaussian kernel, and RF for final classification. Finally, Karhade et al. [30] used NLP for automated detection of incidental durotomies from operative notes of patients undergoing lumbar spine surgery [30]. Notably, the authors created an EGB NLP algorithm for this problem. Impressively, their results showed an impressive AUC of 0.99 which can greatly assist in automatic tracking of incidental durotomies.

Continuing with the third sub-category focusing on event detection and explainability, Baig et al. [31] examined fuzzy logic for event detections during anesthesia [31]. More specifically, the methodology showed nearly continuous agreement between the fuzzy logic system and medical professionals for diagnosing hypovolemia events. As fuzzy logic is inherently an explainable AI approach, it co-exists well with healthcare as the prediction with a corresponding explanation can be more useful to a medical professional than just a prediction alone. Furthermore, explainable ML has been a researched topic since the conception of ML. To this end, Lundberg et al. [32] created an explainable ML algorithm was created for assisting anesthesiologists [32]. Specifically, the output of the algorithm is the prediction as well as a human readable explanation of said prediction. The authors hypothesized that if anesthesiologists utilize this system, it can double their ability to anticipate hypoxemia events. Finally, for explainability, Goodman et al. [33] created a preliminary methodology to detect family caregiver stress where the caregiver is of a family member with Alzheimer's, using a Fuzzy Inference System (FIS) [33]. By extracting features from a smart word scramble game, the FIS predicted word difficulty as a measure of task performance which is shown to be a measure of stress. As for event prediction, Nemati et al. [34] conducted research on using a wearable smart watch with a PPG sensor in conjunction with a ML model to detect Atrial Fibrillation events [34]. Specifically, the authors utilized an Elastic Net logistic model as their prediction scheme and achieved an accuracy of 95% on out-of-sample data. Finally, Ambale-Venkatesh et al. [35] used Random Survival Forests (RSF) to predict multiple cardiovascular outcomes of patients with atherosclerosis [35]. The authors found that including deep phenotyping in conjunction with the RSF increases prediction accuracy over the *state-of-the-art* in initially asymptomatic individuals.

With the final sub-category, we will examine articles that do not necessarily lie within any of the previous sub-categories of research discussed in this section. An example of this is an article by Narula et al. [36] discussing the usage of a ML algorithm for automation of morphological and functional 2D Echocardiography assessment [36]. Specifically, an ensemble algorithm using SVM, RF, and NN was created with a majority rule voting system for discrimination of hypertrophic cardiomyopathy. The author's results showed that including such an ensemble methodology can assist in discrimination of physiological versus pathological patterns of hypertrophic cardiomyopathy. Next, an article by Lee et al. [37] presented work towards an understandable DL algorithm for detecting acute intracranial hemorrhage [37]. More specifically, the authors used a dataset of only 904 training samples, substantially smaller than what most DL algorithms need, and achieved over 92% sensitivity and specificity. In addition, Gulshan et al. [38] used DL for detecting diabetic retinopathy in retinal fundus photographs [38]. Here, the authors utilized a much larger dataset as compared to [37] which is to be expected. Furthermore, their chosen DL algorithm was a standard CNN and showed impressive results of over 90% sensitivity and specificity. Following, Li et al. [39] used ML to classify multi-level ECG signal quality as it is an important and necessary method for the prognosis and diagnosis process of many diseases [39]. More specifically, the authors utilized a SVM as well as a novel methodology for providing ECG noise to different Signal-to-Noise Ratios (SNR). This methodology was used for creating simulated training and test sets and produced an impressive accuracy of 88.07%. Now, in the surgical case, Memarian et al. [40] used ML with multimodal data from clinical, demographical, imaging, and electrophysiological areas for predicting surgical outcomes of mesial temporal lobe epilepsy surgeries [40]. The supervised classifiers used were LDA, Naïve Bayes (NB), SVM with the radial basis function kernel, SVM with the multilayer perceptron kernel, and least-square SVM. Their methodology produced an impressive 95% prediction accuracy with the least-square SVM. Finally, Guidi et al. [41] detailed improvements for heart failure patient assistance using a ML clinical decision support system [41]. That is, their work compared NN, SVM, genetically produced fuzzy rules, and RF to assist heart failure patients.

We can see the multiple types of evolving healthcare techniques that include pain management, NLP, medical imaging, explainability, event detection, and more. While representative, there are certainly more possible research paths to choose for applying AI in general healthcare applications.

2.2.3 Diagnosis

In this section, we focus on diagnosis research and how AI has improved the field with respect to a set of sub-categories. Of which, the first sub-category examines cancer diagnosis AI systems. We begin with an article by Esteva et al. [42] that focused on skin cancer classification as it is a difficult task due to the variability of skin lesion appearances [42]. The authors used a single CNN trained with only pixels of skin lesions and disease labels as inputs. Their CNN performed on par as compared to dermatologist expert classifications. Another example by Bhardwaj and Tiwari [43] is of breast cancer diagnosis using a novel ML algorithm called a Genetically Optimized Neural Network (GONN) [43]. Their novel GONN was able to obtain an impressive 100% accuracy on a 70–30 training and test data split for the of Wisconsin Breast Cancer (WBC) dataset. Similarly, Asri et al. [44] examined breast cancer risk prediction and diagnosis [44]. The authors implemented SVM, DT, NB, and KNN on the WBC dataset and found that the SVM still provides a 97.13% accuracy and is far less complex than the previous article's GONN. Finally, research using a Particle Swarm Optimized Wavelet Neural Network (PSOWNN) has been performed by Dheeba et al. [45] for breast cancer classification [45]. The authors used digital mammograms as their input data and showed that the PSOWNN can reach an AUC value of 0.97.

The second sub-category focuses on articles discussing AI diagnosis assistance for cardiac diseases. The first example shows that hospitals have moved to creating electronic health records for ease of access and increasing productivity. This provides a large amount of data for ML or DL approaches. Therefore, research by Dilsizian and Siegel [46] has shown how big data approaches can be used on electronic health records and cardiac imaging to assist diagnosis and treatment [46]. Following, in an attempt to reduce arrhythmia false alarms, Antink and Leonhardt showed how using beat-to-beat heart intervals and average rhythmicity can help with this issue [47]. More specifically, after comparing Binary Classification Trees (BCT), Discriminant Analysis Classifiers (DAC), and SVM, the authors found that DAC performed the best with their novel feature selection process. Following, diagnosis efforts for heart disease have been made by Bhatla and Jyoti using fuzzy logic [48]. The authors wished to reduce the number of features needed for a diagnosis to consequently reduce the number of tests needed on a patient. Their work showed an astonishing 100% accuracy with only four features by using a combination of fuzzy logic, NB, and DT. Finally, Özçift [49] used an ensemble RF classifier to assist in cardiac arrhythmia diagnosis [49]. Specifically, the author used the RF along with a data resampling technique as the dataset was quite small and unbalanced. The resampling improved the diagnosis and showed an impressive 90% accuracy between 16 classes.

The third sub-category examines articles related to Alzheimer's disease (AD) diagnosis. In the first example, we examine a novel transfer learning technique by Hon and Khan [50] for AD classification [50]. Specifically, they used the VGG16 and Inception models and achieved 96.25% accuracy with a dataset that is 10 times smaller than the *state-of-the-art*. Following, Khedher et al. [51] used Magnetic Resonance Images (MRI) in conjunction with SVM as well as Principal Component

Analysis (PCA) and partial least squares for feature extraction for early diagnosis of AD. The authors showed the ability to distinguish between AD and Mild Cognitive Impairment (MCI), MCI and Elderly Normal Control (ENC), and ENC and AD, respectively. Similarly, Lie et al. [52] applied AI to early AD classification with a DL approach [52]. Specifically, the authors attempted to classify MCI using stacked auto-encoders and a SoftMax layer. Their results outperform previous iterations on the subject while needing a smaller amount of training samples.

The final sub-category focuses on general diagnosis articles and how AI has benefited the field. The first example covers an article by Lambrou et al. [53] which used Conformal Predictors (CP), ML algorithms that provide confidence measures along with their predictions, for medical diagnosis [53]. The article implemented a rule-based system that is improved via a Genetic Algorithm (GA). This allows for human readability (i.e., explainability) that shows improved results as well as a confidence measure per prediction for medical professionals to take into account. Ozcift and Gulten [54] implemented ensemble Rotational Forests of 30 classifiers to improve general medical diagnosis accuracy [54]. Specifically, the authors applied this method to diabetes, heart disease, and Parkinson's disease classification which showed better performance than the *state-of-the-art*. Similarly, another diabetes classifier that was created by Roychowdhury et al. [55] compared a Gaussian Mixture Model (GMM), KNN, SVM, and AdaBoost for classifying retinopathy lesions [55]. While these algorithms are less complex than a Rotational Forest, it still shows high performance of 0.904 Area Under the Receiver Operating Characteristic (AUROC). A third diabetes classifier by Zou et al. [56] focused on Diabetes Mellitus [56]. The authors implemented DT, RF, and NN for this purpose and showed an accuracy of 81%. As for heart beat classification, Pantelopoulos and Bourbakis [57] utilized ECG for Congestive Heart Failure victims using overcomplete dictionaries [57]. While the author's methodology does not outperform the *state-of-the-art* accuracies it showed real time capabilities which have its own purposes such as bedside classification. Finally, Michalopoulos and Bourbakis [58] used Electroencephalographic (EEG) data with Multiscale Entropy (MSE) to classify emotional states [58]. This unusual approach provides an impressive 0.80 AUROC.

These works detailing AI in diagnostic research has greatly and quickly moved the medical field forward. The sub-fields of AI in diagnosis covered here discuss cancer, cardiac diseases, Alzheimer's disease, and more. Though, there is still work to be done as our research community have yet to find a singular algorithm that can both outperform all others and is generally applicable to all use-cases.

2.3 Assistive Technologies

In this section we will examine another large representative selection of the *state-of-the-art* with respect to two categories within assistive technologies. In Sect. 3.1, we examine research in the smart cities and homes environment and in Sect. 3.2 we present research on different assistive robotics.

2.3.1 Smart Homes and Cities

AI has improved the technologies that can be used in a smart home/city environment to make people's lives easier, safer, and more robust. Improving the quality of life has generated an increase in life expectancy. Medjahed et al. [59] create a FIS for remote healthcare monitoring in a smart home environment by utilizing signal detection, speech classification, and speech recognition algorithms [59]. To be more specific, GMM algorithms and Linear Frequency Cepstral Coefficients (LFCC) acoustical parameters are used to distinguish the different sounds in a smart home and managed to achieve a high accuracy recognition rate prototype. The prototype was tested in a smart home with multiple people residing inside thus proving the high accuracy rate.

Several activity recognition research teams are working to improve the quality of life with another different approach which uses computer vision technologies. Some of the most representative ones can be found in the following survey paper by Schneider and Banerjee [60]. The authors not only describe the *state-of-art* on vision technologies for smart home monitoring but also try to answer questions such as what are the limitations in the real world and how AI could assist making them better. Diving further into the recognition area for monitoring, we can see several researchers using facial-expression recognition to detect the differences in the health of a person within a smart city environment. Muhammad et al. [61] used a system that applies a bandlet transform to a face image to extract sub-bands, leading to a produced feature vector. The vector is fed into two classifiers, GMM and SVM that end up with a 99.95% accuracy [61]. Following, human activity recognition could play a role into rehabilitation exercises along with the physical activities. Wearable sensors provide useful information and feature extraction algorithms use the raw data to select the desired features that are then post processed. Then, PCA and LDA are being used by Hassan et al. [62] to make the features robust which are then used to train a Deep Belief Network (DBN). When compared to more traditional algorithms such as SVM and NN, the DBN outperforms previous methods [62].

With smart cities being the future of the greatest technological changes, there are people with sensorial disabilities where autonomy in urban places is not yet achieved for them. An enhanced Internet of Things (IoT) that suits devices providing travel aid has been created by Ramirez et al. [63]. A haptic feedback architecture that acquires relevant environmental information was designed that when tested shows promising results. IoT technologies provide useful tools to enable people in need to interact with their surrounding environment. Next, an article detailed the use of Augmented Reality (AR) and Radio Frequency Identification (RFID) to help wheelchair users interact with items that are out of their reach by Rashid et al. [64]. By using fourteen wheelchair bound participants that performed autonomy and experience tests, the authors showed promising results towards the independence of wheelchair users. Following, personalizing the needs of the people in need in a smart city also requires personalized medicine on products. With such personalization, the need for tools to correctly predict allergens and nonallergens is growing, especially when they share sequences. Working towards that goal, the authors of this paper proposed the FuzzyApp, a system that utilizes FIS to enhance the predictive capabilities of post genomics diagnostics [65].

Continuing with IoT research and smart home monitoring, a system that communicates with smart devices in the healthcare environment supported by an IoT framework is proposed by Alhussein et al. [66]. They used EEG sensors to record and transmit EEG data of epileptic patients Furthermore, by utilizing this framework, they can make real time decisions while reaching an accuracy of 99%. Another way to improve transmission of data from sensors is a model proposed by Soraia et al. [67] that used an optimization algorithm, the Maximum Reward Algorithm (MRA), to enhance the delivery of healthcare resources [67]. After testing and simulations, the results showed that it can provide reliability and increase the performance by 50% to 77%. Continuing in the context of monitoring within a smart home, Hamid et al. [68] proposed an automatic in-home healthcare monitoring system that has a wide range of application [68]. Several sensors that can be installed in the house were used, thus providing a multimodal system that, with the use of FIS logic, ensured pervasive in-home health monitoring for older adults.

When it comes to older adults and assisting them, we need to consider ADL. Ktistakis et al. [69] proposed a case study of an autonomous intelligent robotic wheelchair (AIRW) scheme that could provide people in need independency in performing in-home actions and transferring to various rooms [69]. The case study showcased the synergistic collaboration between the AIRW and the user. Another system that is classified as a wearable device was proposed by Pantelopoulos et al. [70] with a systems functional scheme built on formal languages [70]. A simulation framework was built using Stochastic Petri Nets (SPNs) and used a voice recognition module to enhance the autonomous capabilities of the system. Another system that used SPNs for modeling and simulations was developed by Keefer et al. [71] that used studies containing blind participants to develop and refine the model [71]. Formal grammar was used to describe the system. Bourbakis et al. [72] provided an alternative mode to interact with the environment for visually impaired people [72]. The 3-D space sensation was studied and how it could be interpreted into a computer, leading to the design and creation of a prototype for the visually impaired with the results producing a useful assistive device.

Continuing on technologies for the visually impaired, Mostafa et al. [73] provided a study to assist in shopping tasks using mobile assistive technologies in a smart environment [73]. The main categories were presented showing advantages and weaknesses. Finally, Accessible Tourism is another area that AI is trying to provide assistance to people with disabilities. The United Nations state that almost 1/3 of the population is affected in a way by some kind of disability when it comes to travelling and which includes the visually impaired [74]. The article analyzed key requirements for an IoT application in a Smart City environment for this problem.

These works detailing AI in a smart home/city environment are some of the most representative in the last decade. It is important to note that there are more published works available and not all of them have been covered in this study. There is still plenty of room for work to be done as we as a research community have yet to optimize all the different algorithms and prototypes for a smart city.

2.3.2   Assistive Robotics

In a world where millions of people have some kind of disability and the elderly rely on caretakers, there are researchers who are working to develop prototypes that could assist users in a variety of activities [75, 76]. AI is playing an important role in the development of these prototypes. Assistive Robotics can help in ADL and one representative robot comes from Dr. Cooper's lab called PerMMA that stands for Personal Mobility and Manipulation Appliance [77]. The device enhances the user's capability of manipulating the arm and simulations showed its reachability and manipulability. Object manipulation is a critical task for the execution of activities and thus an alternative control method called ARoMA-V^2 was designed by Ka et al. [78] using computer vision and voice recognition to control the robotic manipulator [78]. Another alternative control system that was mentioned earlier [10] utilizes SPNs to model specific tasks of ADL and, more specifically, assists a user getting up from a sitting position with the help of a robotic wheelchair [79]. More tasks are also mentioned such as assisting a user to turn around and then sit down on the wheelchair [80]. Furthermore, while continuing with robotic wheelchairs. Polygerinos et al. [82] developed a soft robotic glove for assistance during ADL while the user is in a wheelchair [82]. The glove can detect muscle movement and accordingly activate the glove actuators using Electromyography (EMG) logic.

Assistive Robotics have entered the area of rehabilitation with the need coming from aging populations. Beckerle et al. [83] presented a human robot interaction perspective on the issues in the field and how ML can assist the future of rehabilitation robotics [83]. In a rehabilitation robot the wearer's motion estimation is a critical problem. J. Huang et al. [84] proposed an intention guided strategy for an upper limb power assisted exoskeleton with a human robot interface comprised of Force Sensing Resistors (FSRs) [84]. The proposed approaches were confirmed by experiments on a 3 Degree of Freedom (DOF) exoskeleton and proved the effectiveness of the system. Seth et al. [85] designed a soft exo-suit for arm rehabilitation that takes into consideration the limitations of the patients suffering from loss of the upper limbs while their validation is ongoing [85]. Gross et al. [86] described the implementation of a robot coach for walking and orientation training in clinical post stroke rehabilitation called ROREAS [86]. An overview of the training scenarios and its constraints are given as well its architecture and the required interaction skills. Finally, Peng et al. [87] presented a novel upper limb rehabilitation robot that can provide safe and compliant force feedbacks by using a virtual tunnel force field design [87].

Lastly, social robotics is another area within the assistive robotics domain that provides value to the user. Canamero et al. [88] presented Robin, a cognitively and motivationally autonomous affective robot toddler, to support emotional wellbeing to children that have diabetes [88]. The toddler robot has diabetes as well and its role is for the children to make new friends. Takeda et al. [89] provided an architecture on physical human support robots to work closely with the users, thus providing trust and creating a relationship [89]. Lepage et al. [90] presented a framework that is versatile enough to allow for custom designed integrated monitoring capabilities. The framework can handle remote patient monitoring, video visits, robot telepresence for the patients, and more [90]. Pandey et al. [91] presented Pepper, a mass produced sociable humanoid robot, to educational institutes in Japan to support the teaching of robot programming [91]. Arrichiello et al. [92] presented an architecture of operation of an assistive robot for users with motion impairment to assist them in ADL [92]. The robot receives commands from a Brain Computer Interface (BCI) showing the potential of the architecture.

2.4   Analysis and Forecasting

In this section, we provide an analysis on the maturity of the five subsections listed above. That is, of Controlled Monitoring Environments (CME), Evolving Health-care Techniques (EHT), Diagnosis (D), Smart Homes and Cities (SHC), and Assis-tive Robotics (AR), respectively. To do this, we will use the data kept by Google for paper citations and provide a statistical forecast of said citations. The data and corresponding forecasting will provide suggestions on possible future work domains for scientists to explore. The data were gathered by examining an author's Google Scholar page, finding the paper in question, and viewing the citations from publica-tion year to now and recording the data. This is done for all the discussed papers in the aforementioned five subsections.

The statistical forecasting method we will be using for this process is the Moving Average with a k (the number of years to include in the average calculation back-wards) equal to 3. This forecasting method simply takes the previous k number of data points, to predict the next data point. Therefore, for our forecasting, it will begin with an offset of 3 years and thus begin on 2013. We choose this method as it is suitable for univariate time series and we only have at most 10 data points per subsection [94]. Beginning, we gathered the raw data for citations of all papers in the subsections from Google Scholar and aggregated them by year as shown in Table 2.1. Subsequently, we calculate the moving average for the prediction with $k = 3$ as shown in Table 2.2. Following, we plotted both the original data with three years of forecasting as well as only the forecasted data with $k = 3$ as shown in Figs. 2.1 and 2.2, respectively. Finally, we present the Mean Absolute Deviation (MAD), which is simply the average of the absolute value between the forecasted and actual data, for each subsection. This is shown in Eq. 2.1 where n is the number of data points, x_i is the data point in question, and $\bar{x}$ is the forecasted data point. The MAD values are 48.88, 250.63, 444.75, 35.25, and 38 for CME, EHT, D, SHC, and AR, respectively.

Table 2.1 The data collected from the CME, EHT, D, SHC, and AR subsections aggregated by year

	2010	2011	2012	2013	2014	2015	2016	2017	2018	2019	2020
CME	5	7	48	55	47	69	81	83	139	283	208
EHT	5	4	7	8	11	21	31	261	697	1163	1006
D	0	10	20	41	53	110	186	580	1414	2194	1649
SHC	1	1	3	19	18	30	29	43	87	170	141
AR	0	2	2	3	0	5	34	57	129	185	112

Table 2.2 The forecasted data starting at 2013 and forcasting 3 years ahead to 2023 with respect to the current year of 2020

	2013	2014	2015	2016	2017	2018	2019	2020	2021	2022	2023
CME	20	36	50	57	65	77	101	168	210	245	208
EHT	5	6	8	13	21	104	329	707	955	1084	1006
D	10	23	38	68	116	292	726	1396	1752	1921	1649
SHC	1	7	13	22	25	34	53	100	132	155	141
AR	1	2	1	2	13	32	73	123	142	148	112

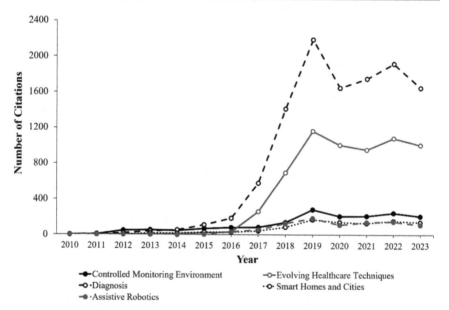

Fig. 2.1 Line graph showing the recorded data with three years of forecasting. Note that Diagnosis is the area within Healthcare that researchers are focusing on the most as compared to the rest of the sub-categories

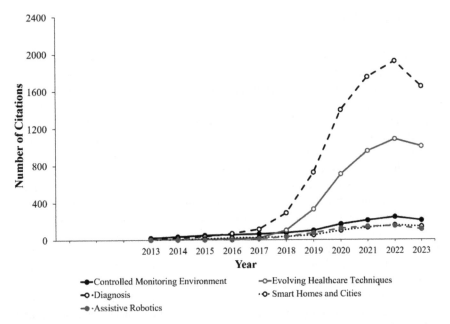

Fig. 2.2 Line graph showing only the forecasted data with k = 3. Note that after the forecasting, the Diagnosis sub-category is still the one that holds the attention of the researchers

$$\text{MAD} = \frac{\sum |x_i - \ddot{x}|}{n} \qquad (2.1)$$

The three years of forecasting from 2021 to 2023 shows that while CME, AR, and SHC will remain relatively consistent; D has a possibility to start a down trend. Though, there are 2 papers that skew D and EHT, one paper per subsection, respectively. From the data and anecdotal examination during the research for this article, there are significantly more papers related to diagnosis than the other sections. Therefore, the down trend could perhaps be attributed to the oversaturation of the field. Consequently, work could be redistributing to new areas in the near future as a theoretical wall could be reached. Furthermore, for EHT, the steady trend seems plausible as well due to the unique nature of the field. Articles placed here provide a novel approach to topics of healthcare with AI that have the potential of high citation performance given the correct journal or conference venue. Finally, from examining the MAD, the forecasting for D and EHT have a higher MAD due to the skewing from 2 specific papers. Namely, [31] for EHT and [35] for D. As for the remaining three sections, their MAD are relatively low which shows the forecasting has acceptable performance in these cases. Comparing Figs. 2.1 and 2.2, we can see the expected climb from the forecasting. Though, it does result in approximately the same number of citations value for both D and EHT, leaving us to be confident in these results moving forward.

2.5 Conclusions

Taken together, it is evident that AI can facilitate in the evolution of healthcare. In this review, we have focused on the presentation of some of the most representative AI applications in the fields of healthcare, biomedical research, and assistive technologies. We have further assessed the focal point of research and have predicted the future trajectory accordingly. Our maturity analysis has indicated that AI applications in healthcare and diagnosis have been in the hotspot the past decade, with a robust increase in most recent years. Notably, such applications are currently of higher interest due to the increased chance of pandemic breakouts, including the recent COVID-19 pandemic. Subsequently, AI applications have and will continue to play a pivotal role in improving the current healthcare system and shift the paradigm in clinical and preclinical medicine. To predict future trends, we have also forecasted that diagnostic and healthcare-related AI-controlled systems are more likely to continue to be of major interest in the near future. Thus, it is only expected that major efforts will continue to be put into developing novel and efficient AI applications in healthcare, especially after taking note of current circumstances. Taking into consideration the constant increase in prevalence of life-long debilitating disorders, and the subsequent high demand for more efficient healthcare applications and robust diagnostic tools, automated and intelligent solutions will be in the forefront of research in the years to follow. AI has already changed the landscape in the biomedical field, and it is only safe to say that it will comprise a key element in revolutionary future applications to assist professionals and patients.

References

1. S. Russell, P. Norvig, Artificial intelligence: a modern approach (2002)
2. F. Jiang et al., Artificial intelligence in healthcare: past, present and future. Stroke Vascular Neurol. 2(4), 230–243 (2017)
3. K.-H. Yu, A.L. Beam, I.S. Kohane, Artificial intelligence in healthcare. Nature Biomed. Eng. 2(10), 719–731 (2018)
4. H. Ashrafian, A. Darzi, T. Athanasiou, A novel modification of the Turing test for artificial intelligence and robotics in healthcare. Int. J. Med. Robot. Comput. Assisted Surg. 11(1), 38–43 (2015)
5. C. Di Napoli, S. Rossi, A layered architecture for socially assistive robotics as a service, in 2019 IEEE International Conference on Systems, Man and Cybernetics (SMC). IEEE (2019)
6. Ktistakis, I.P. and N.G. Bourbakis, Assistive intelligent robotic wheelchairs. IEEE Potentials, 2017. 36(1): p. 10–13
7. E.D. Oña et al., Robotics in health care: perspectives of robot-aided interventions in clinical practice for rehabilitation of upper limbs. Appl. Sci. 9(13), 2586 (2019)
8. M.A. Patil, R.B. Patil, P. Krishnamoorthy, J. John, A machine learning framework for auto classification of imaging system exams in hospital setting for utilization optimization, in 2016 38th Annual International Conference of the IEEE Engineering in Medicine and Biology Society (EMBC) (pp. 2423–2426). IEEE (2016)

9. E. Long, H. Lin, Z. Liu, X. Wu, L. Wang, J. Jiang, … J. Li, An artificial intelligence platform for the multihospital collaborative management of congenital cataracts. Nat. Biomed. Eng. **1**(2), 1–8 (2017)
10. I.P. Ktistakis, N. Bourbakis, A multimodal human-machine interaction scheme for an intelligent robotic nurse, in *2018 IEEE 30th International Conference on Tools with Artificial Intelligence (ICTAI)* (pp. 749–756). IEEE (2018)
11. H.J. Murff, F. FitzHenry, M.E. Matheny, N. Gentry, K.L. Kotter, K. Crimin, …T. Speroff, Automated identification of postoperative complications within an electronic medical record using natural language processing. JAMA **306**(8), 848–855 (2011)
12. H.C. Lee, H.K. Yoon, K. Nam, Y.J. Cho, T.K. Kim, W.H. Kim, J.H. Bahk, Derivation and validation of machine learning approaches to predict acute kidney injury after cardiac surgery. J. Clin. Med. **7**(10), 322 (2018)
13. A. Pantelopoulos, N. Bourbakis, A wearable platform utilizing off-the-shelf components and performing quality analysis of physiological data, in *Proceedings of the Fifth International Conference on Body Area Networks* (pp. 220–226) (2010)
14. G. Goodman, N. Bourbakis, A wearable ultrasound methodology for creating a real time near 3D model of the heart, in *2019 IEEE 19th International Conference on Bioinformatics and Bioengineering (BIBE)* (pp. 539–546). IEEE (2019)
15. A. Walinjkar, J. Woods, Personalized wearable systems for real-time ECG classification and healthcare interoperability: Real-time ECG classification and FHIR interoperability, in *2017 Internet Technologies and Applications (ITA)* (pp. 9–14). IEEE (2017)
16. A. Pantelopoulos, N.G. Bourbakis, Prognosis—a wearable health-monitoring system for people at risk: Methodology and modeling. IEEE Trans. Inform. Technol. Biomed. **14**(3), 613–621 (2010)
17. M.J. Rantz, T.S. Banerjee, E. Cattoor, S.D. Scott, M. Skubic, M. Popescu, Automated fall detection with quality improvement "rewind" to reduce falls in hospital rooms. J. Gerontolog. Nurs. **40**(1), 13–17 (2013)
18. W. Romine, T. Banerjee, G. Goodman, Toward sensor-based sleep monitoring with electrodermal activity measures. Sensors **19**(6), 1417 (2019)
19. L.M. Eerikäinen, J. Vanschoren, M.J. Rooijakkers, R. Vullings, R.M. Aarts, Decreasing the false alarm rate of arrhythmias in intensive care using a machine learning approach, in *2015 Computing in Cardiology Conference (CinC)* (pp. 293–296). IEEE (2015)
20. C.R. Leite, G.R. Sizilio, A.D. Neto, R.A. Valentim, A.M. Guerreiro, A fuzzy model for processing and monitoring vital signs in ICU patients. Biomed. Eng. Online **10**(1), 68 (2011)
21. R. Sadeghi, T. Banerjee, W. Romine, Early hospital mortality prediction using vital signals. Smart Health **9**, 265–274 (2018)
22. J.L. Koyner, K.A. Carey, D.P. Edelson, M.M. Churpek, The development of a machine learning inpatient acute kidney injury prediction model. Crit. Care Med. **46**(7), 1070–1077 (2018)
23. M.C. Moghadam, E.M.K. Abad, N. Bagherzadeh, D. Ramsingh, G.P. Li, Z.N. Kain, A machine-learning approach to predicting hypotensive events in ICU settings. Comput. Biol. Med. **118**, 103626 (2020)
24. S. Nemati, A. Holder, F. Razmi, M.D. Stanley, G.D. Clifford, T.G. Buchman, An interpretable machine learning model for accurate prediction of sepsis in the ICU. Crit. Care Med. **46**(4), 547 (2018)
25. B. Gholami, W.M. Haddad, A.R. Tannenbaum, Relevance vector machine learning for neonate pain intensity assessment using digital imaging. IEEE Trans. Biomed. Eng. **57**(6), 1457–1466 (2010)
26. B.T. Susam, M. Akcakaya, H. Nezamfar, D. Diaz, X. Xu, V.R. de Sa, … & M.S. Goodwin, Automated pain assessment using electrodermal activity data and machine learning, in *2018 40th Annual International Conference of the IEEE Engineering in Medicine and Biology Society (EMBC)* (pp. 372–375). IEEE (2018)
27. F. Yang, T. Banerjee, K. Narine, N. Shah, Improving pain management in patients with sickle cell disease from physiological measures using machine learning techniques. Smart Health **7**, 48–59 (2018)

28. G. Karakülah, O. Dicle, Ö. Kosaner, A. Suner, Ç.C. Birant, T. Berber, S. Canbek, Computer based extraction of phenoptypic features of human congenital anomalies from the digital literature with natural language processing techniques, in *MIE* (pp. 570–574) (2014)

29. R.W. Grundmeier, A.J. Masino, T.C. Casper, J.M. Dean, J. Bell, R. Enriquez, … N.T.P.E.C. Applied, Identification of long bone fractures in radiology reports using natural language processing to support healthcare quality improvement. Appl. Clin. Inform. **7**(4), 1051 (2016)

30. A.V. Karhade, M.E. Bongers, O.Q. Groot, E.R. Kazarian, T.D. Cha, H.A. Fogel, … J.D. Kang, Natural language processing for automated detection of incidental durotomy. Spine J. **20**(5), 695–700 (2020)

31. M.M. Baig, H. Gholamhosseini, M.J.Harrison, Fuzzy logic based smart anaesthesia monitoring system in the operation theatre. WSEAS Trans. Circuit. Syst. **11**(1), 21–32 (2012)

32. S.M. Lundberg, B. Nair, M.S. Vavilala, M. Horibe, M.J. Eisses, T. Adams, … S.I. Lee, Explainable machine-learning predictions for the prevention of hypoxaemia during surgery. Nat. Biomed. Eng. **2**(10), 749–760 (2018)

33. G. Goodman, T. Banerjee, W. Romine, C. Shimizu, J. Hughes, Caregiver assessment using smart gaming technology: a feasibility study, in *2019 IEEE International Conference on Fuzzy Systems (FUZZ-IEEE)* (pp. 1–6). IEEE (2019)

34. S. Nemati, M.M. Ghassemi, V. Ambai, N. Isakadze, O. Levantsevych, A. Shah, G.D. Clifford, Monitoring and detecting atrial fibrillation using wearable technology, in *2016 38th Annual International Conference of the IEEE Engineering in Medicine and Biology Society (EMBC)* (pp. 3394–3397). IEEE (2016)

35. B. Ambale-Venkatesh, X. Yang, C.O. Wu, K. Liu, W.G. Hundley, R. McClelland, … D.A. Bluemke, Cardiovascular event prediction by machine learning: the multi-ethnic study of atherosclerosis. Circulation Res. **121**(9), 1092–1101 (2017)

36. S. Narula, K. Shameer, A.M.S. Omar, J.T. Dudley, P.P. Sengupta, Machine-learning algorithms to automate morphological and functional assessments in 2D echocardiography. J. Amer. College Cardiol. **68**(21), 2287–2295 (2016)

37. H. Lee, S. Yune, M. Mansouri, M. Kim, S.H. Tajmir, C.E. Guerrier, … R.G. Gonzalez, An explainable deep-learning algorithm for the detection of acute intracranial haemorrhage from small datasets. Nat. Biomed. Eng. **3**(3), 173 (2019)

38. V. Gulshan, L. Peng, M. Coram, M.C. Stumpe, D. Wu, A. Narayanaswamy, … R. Kim, Development and validation of a deep learning algorithm for detection of diabetic retinopathy in retinal fundus photographs. JAMA **316**(22), 2402–2410 (2016)

39. Q. Li, C. Rajagopalan, G.D. Clifford, A machine learning approach to multi-level ECG signal quality classification. Comput. Methods Programs Biomed. **117**(3), 435–447 (2014)

40. N. Memarian, S. , Kim, S. Dewar, Jr, J. Engel , R.J. Staba, Multimodal data and machine learning for surgery outcome prediction in complicated cases of mesial temporal lobe epilepsy. Comput. Biol. Med. **64**, 67–78 (2015)

41. G. Guidi, M.C. Pettenati, P. Melillo, E. Iadanza, A machine learning system to improve heart failure patient assistance. IEEE J. Biomed. Health Inform. **18**(6), 1750–1756 (2014)

42. A. Esteva, B. Kuprel, R.A. Novoa, J. Ko, S.M. Swetter, H.M. Blau, S. Thrun, Dermatologist-level classification of skin cancer with deep neural networks. Nature **542**(7639), 115–118 (2017)

43. A. Bhardwaj, A. Tiwari, Breast cancer diagnosis using genetically optimized neural network model. Expert Syst. Appl. **42**(10), 4611–4620 (2015)

44. H. Asri, H. Mousannif, H. Al Moatassime, T. Noel, Using machine learning algorithms for breast cancer risk prediction and diagnosis. Procedia Comput. Sci. **83**, 1064–1069 (2016)

45. J. Dheeba, N.A. Singh, S.T. Selvi, Computer-aided detection of breast cancer on mammograms: a swarm intelligence optimized wavelet neural network approach. J. Biomed. Inform. **49**, 45–52 (2014)

46. S.E. Dilsizian, E.L. Siegel, Artificial intelligence in medicine and cardiac imaging: harnessing big data and advanced computing to provide personalized medical diagnosis and treatment. Curr. Cardiol. Reports, *16*(1), 441 (2014)

47. C.H. Antink, S. Leonhardt, Reducing false arrhythmia alarms using robust interval estimation and machine learning in *2015 Computing in Cardiology Conference (CinC)* (pp. 285–288). IEEE (2015)
48. N. Bhatla, K. Jyoti, A Novel Approach for heart disease diagnosis using Data Mining and Fuzzy logic. Int. J. Comput. Appl. **54**(17) (2012)
49. A. Özçift, Random forests ensemble classifier trained with data resampling strategy to improve cardiac arrhythmia diagnosis. Comput. Biol. Med. **41**(5), 265–271 (2011)
50. M. Hon, N.M. Khan, Towards Alzheimer's disease classification through transfer learning, in *2017 IEEE International conference on bioinformatics and biomedicine (BIBM)* (pp. 1166–1169). IEEE (2017)
51. L. Khedher, J. Ramírez, J.M. Górriz, A. Brahim, F. Segovia, S. Alzheimer, Disease Neuroimaging Initiative, Early diagnosis of Alzheimer's disease based on partial least squares, principal component analysis and support vector machine using segmented MRI images. Neurocomputing **151**, 139–150 (2015)
52. S. Liu, S. Liu, W. Cai, S. Pujol, R. Kikinis, D. Feng, Early diagnosis of Alzheimer's disease with deep learning, in *2014 IEEE 11th international symposium on biomedical imaging (ISBI)* (pp. 1015–1018). IEEE (2014)
53. A. Lambrou, H. Papadopoulos, A. Gammerman, Reliable confidence measures for medical diagnosis with evolutionary algorithms. IEEE Trans. Inform. Technol. Biomed. **15**(1), 93–99 (2010)
54. A. Ozcift, A. Gulten, Classifier ensemble construction with rotation forest to improve medical diagnosis performance of machine learning algorithms. Comput. Methods Programs Biomed. **104**(3), 443–451 (2011)
55. S. Roychowdhury, D.D. Koozekanani, K.K. Parhi, DREAM: diabetic retinopathy analysis using machine learning. IEEE J. Biomed. Health Inform. **18**(5), 1717–1728 (2013)
56. Q. Zou, K. Qu, Y. Luo, D. Yin, Y. Ju, H. Tang, Predicting diabetes mellitus with machine learning techniques. Front. Genet. **9**, 515 (2018)
57. A. Pantelopoulos, N. Bourbakis, ECG Beat Classification Using Optimal Projections in Over-complete Dictionaries, in *2011 IEEE 23rd International Conference on Tools with Artificial Intelligence* (pp. 1099–1105). IEEE (2011)
58. K. Michalopoulos, N. Bourbakis, Application of multiscale entropy on EEG signals for emotion detection, in *2017 IEEE EMBS International Conference on Biomedical & Health Informatics (BHI)* (pp. 341–344). IEEE (2017)
59. H. Medjahed, B. Dorizzi, D. Istrate, J.L. Baldinger, J. Boudy, L. Bougueroua, M.A. Dhouib, *A Fuzzy Logic Approach for Remote Healthcare Monitoring by Learning and Recognizing Human Activities of Daily Living*. INTECH Open Access Publisher (2012)
60. B. Schneider, T. Banerjee, Activity recognition using imagery for smart home monitoring, in *Advances in Soft Computing and Machine Learning in Image Processing* (pp. 355–371). Springer, Cham (2018)
61. G. Muhammad, M. Alsulaiman, S.U. Amin, A. Ghoneim, M.F. Alhamid, A facial-expression monitoring system for improved healthcare in smart cities. IEEE Access **5**, 10871–10881 (2017)
62. M.M. Hassan, M.Z. Uddin, A. Mohamed, A. Almogren, A robust human activity recognition system using smartphone sensors and deep learning. Future Gener. Comput. Syst. **81**, 307–313 (2018)
63. A.R.G. Ramirez, I. González-Carrasco, G.H. Jasper, A.L. Lopez, J.L. Lopez-Cuadrado, A. García-Crespo, Towards human smart cities: internet of things for sensory impaired individuals. Computing **99**(1), 107–126 (2017)
64. Z. Rashid, J. Melià-Seguí, R. Pous, E. Peig, Using Augmented Reality and Internet of Things to improve accessibility of people with motor disabilities in the context of Smart Cities. Future Gener. Comput. Syst. **76**, 248–261 (2017)
65. V. Saravanan, P.T.V. Lakshmi, Fuzzy logic for personalized healthcare and diagnostics: FuzzyApp—A fuzzy logic based allergen-protein predictor. Omics J. Integr. Biol. **18**(9), 570–581 (2014)

66. M. Alhussein, G. Muhammad, M.S. Hossain, S.U. Amin, Cognitive IoT-cloud integration for smart healthcare: case study for epileptic seizure detection and monitoring. Mobile Netw. Appl. **23**(6), 1624–1635 (2018)
67. S. Oueida, M. Aloqaily, S. Ionescu, A smart healthcare reward model for resource allocation in smart city. Multimedia Tools Appl. **78**(17), 24573–24594 (2019)
68. H. Medjahed, D. Istrate, J. Boudy, J.L. Baldinger, B. Dorizzi, A pervasive multi-sensor data fusion for smart home healthcare monitoring, in *2011 IEEE International Conference on Fuzzy Systems (FUZZ-IEEE 2011)* (pp. 1466–1473). IEEE (2011)
69. N. Bourbakis, I.P. Ktistakis, L. Tsoukalas, M. Alamaniotis, An autonomous intelligent wheelchair for assisting people at need in smart homes: A case study. In *2015 6th International Conference on Information, Intelligence, Systems and Applications (IISA)* (pp. 1–7). IEEE (2015)
70. A. Pantelopoulos, N. Bourbakis, Design of the new prognosis wearable system-prototype for health monitoring of people at risk, in *Advances in Biomedical Sensing, Measurements, Instrumentation and Systems* (pp. 29–42). Springer, Berlin, Heidelberg (2010)
71. R. Keefer, Y. Liu, N. Bourbakis, The development and evaluation of an eyes-free interaction model for mobile reading devices. IEEE Trans. Human Mach. Syst. **43**(1), 76–91 (2012)
72. N. Bourbakis, S.K. Makrogiannis, D. Dakopoulos, A system-prototype representing 3D space via alternative-sensing for visually impaired navigation. IEEE Sens. J. **13**(7), 2535–2547 (2013)
73. M. Elgendy, C. Sik-Lanyi, A. Kelemen, Making shopping easy for people with visual impairment using mobile assistive technologies. Appl. Sci. **9**(6), 1061 (2019)
74. M. Nitti, D. Giusto, S. Zanda, M. Di Francesco, C. Casari, M.L. Clemente, … V. Popescu, Using IoT for Accessible Tourism in Smart Cities, in *Assistive Technologies in Smart Cities*. IntechOpen (2018)
75. I.P. Ktistakis, N.G. Bourbakis, A survey on robotic wheelchairs mounted with robotic arms, in *2015 National Aerospace and Electronics Conference (NAECON)* (pp. 258–262). IEEE (2015)
76. O.R. Shishvan, D.S. Zois, T. Soyata, Machine intelligence in healthcare and medical cyber physical systems: a survey. IEEE Access **6**, 46419–46494 (2018)
77. J. Xu, G.G. Grindle, B. Salatin, J.J. Vazquez, H. Wang, D. Ding, R.A. Cooper, Enhanced bimanual manipulation assistance with the personal mobility and manipulation appliance (PerMMA), in *2010 IEEE/RSJ International Conference on Intelligent Robots and Systems* (pp. 5042–5047). IEEE (2010)
78. H. Ka, D. Ding, R.A. Cooper, ARoMA-V2: Assistive robotic manipulation assistance with computer vision and voice recognition, in *The 9th Conference on Rehabilitation Engineering and Assistive Technology Society of Korea*. RESKO (2015)
79. I.P. Ktistakis, N. Bourbakis, An SPN Modeling of the H-IRW Getting-Up Task, in *2016 IEEE 28th International Conference on Tools with Artificial Intelligence (ICTAI)* (pp. 766–771). IEEE (2016)
80. I. Papadakis Ktistakis, An Autonomous Intelligent Robotic Wheelchair to Assist People in Need: Standing-up, Turning-around and Sitting-down (2018)
81. T.T. Tran, T. Vaquero, G. Nejat, J.C. Beck, Robots in retirement homes: Applying off-the-shelf planning and scheduling to a team of assistive robots. J. Artif. Intell. Res. **58**, 523–590 (2017)
82. P. Polygerinos, K.C. Galloway, S. Sanan, M. Herman, C.J. Walsh, EMG controlled soft robotic glove for assistance during activities of daily living, in *2015 IEEE International Conference on Rehabilitation Robotics (ICORR)* (pp. 55–60). IEEE (2015)
83. P. Beckerle, G. Salvietti, R. Unal, D. Prattichizzo, S. Rossi, C. Castellini, … F. Mastrogiovanni, A human–robot interaction perspective on assistive and rehabilitation robotics. Front. Neurorobot. **11**, 24 (2017)
84. J. Huang, W. Huo, W. Xu, S. Mohammed, Y. Amirat, Control of upper-limb power-assist exoskeleton using a human-robot interface based on motion intention recognition. IEEE Trans. Autom. Sci. Eng. **12**(4), 1257–1270 (2015)
85. D. Seth, V.H.V. Varma, P. Anirudh, P. Kalyan, Preliminary Design of Soft Exo-Suit for Arm Rehabilitation, in *International Conference on Human-Computer Interaction* (pp. 284–294). Springer, Cham (2019)

86. H.M. Gross, A. Scheidig, K. Debes, E. Einhorn, M. Eisenbach, S. Mueller, ... A. Bley, ROREAS: robot coach for walking and orientation training in clinical post-stroke rehabilitation—prototype implementation and evaluation in field trials. Autonom. Robot. **41**(3), 679–698 (2017)
87. L. Peng, Z.G. Hou, L. Peng, L. Luo, W. Wang, Robot assisted rehabilitation of the arm after stroke: prototype design and clinical evaluation. Sci. China Inform. Sci. **60**(7), 073201 (2017)
88. L. Cañamero, M. Lewis, Making new "New AI" friends: designing a social robot for diabetic children from an embodied AI perspective. Int. J. Soc. Robot. **8**(4), 523–537 (2016)
89. M. Takeda, Y. Hirata, Y.H. Weng, T. Katayama, Y. Mizuta, A. Koujina, Accountable system design architecture for embodied AI: a focus on physical human support robots. Adv. Robot. **33**(23), 1248–1263 (2019)
90. P. Lepage, D. Létourneau, M. Hamel, S. Briere, H. Corriveau, M. Tousignant, F. Michaud, Telehomecare telecommunication framework—from remote patient monitoring to video visits and robot telepresence, in *2016 38th Annual International Conference of the IEEE Engineering in Medicine and Biology Society (EMBC)* (pp. 3269–3272). IEEE (2016)
91. A.K. Pandey, R. Gelin, A mass-produced sociable humanoid robot: Pepper: The first machine of its kind. IEEE Robot. Autom. Mag. **25**(3), 40–48 (2018)
92. F. Arrichiello, P. Di Lillo, D. Di Vito, G. Antonelli, S. Chiaverini, Assistive robot operated via P300-based brain computer interface, in *2017 IEEE International Conference on Robotics and Automation (ICRA)* (pp. 6032–6037). IEEE (2017)
93. N. Gandhi, M. Allard, S. Kim, P. Kazanzides, M.A.L. Bell, Photoacoustic-based approach to surgical guidance performed with and without a da Vinci robot. J. Biomed. Opt. **22**(12), 121606 (2017)
94. R.J. Hyndman, G. Athanasopoulos, *Forecasting: principles and practice.* OTexts (2018)

Iosif Papadakis Ktistakis is a Senior Mechatronics Engineer at ASML in Connecticut. He earned his Ph.D. degree in Computer Science and Engineering from Wright State University, USA. He also holds an integrated B.S./M.S. in Mechanical Engineering from the Technical University of Crete, Greece. He is a senior member of the Center of Assistive Research Technologies and an IEEE Member. His research interests lie in the intersection of Robotics, Assistive Technologies and Intelligent Systems.

Garrett Goodman earned his Ph.D. degree in Computer Science and Engineering from Wright State University, USA. He is a member of the Center of Assistive Research Technologies. He earned his B.S. and M.S. degrees in Computer Science at Wright State University as well. His work is focused on incorporating machine learning in health care to improve the lives and the wellbeing of people in need.

Aikaterini Britzolaki is a Postdoctoral Associate in Department of Anesthesiology, Weill Cornell Medical College of Cornell University, New York, USA. He is a senior Ph.D. candidate in Neuroscience at the Neuropsychopharmacology and Experimental Neuroscience Lab at the University of Dayton, Ohio. She earned her B.Sc. in Biology from the University of Crete and M.Sc. in Biomedicine from the University of Southern Denmark. She is an active member of the international Society for Neuroscience. Her work focuses on dissecting molecular mechanisms underlying complex human brain pathophysiology associated with neuropsychiatric and neurodevelopmental disorders utilizing animal models of disease.

Chapter 3
A Research Agenda for Dementia Care: Prevention, Risk Mitigation and Personalized Interventions

Anna Esposito, Alessandro Vinciarelli, and Gennaro Cordasco

Abstract This work aims to suggest a research agenda for dementia, connecting the latest clinical understanding of such neurodegenerative disorders, with social behavioral changes and social cognitive impairments. The proposed framework considers the manual and automatized analysis of behavioral and biometric data involving caregivers, patients, and healthcare professionals, and proposes a series of AI tools for extracting and identify, specific features of socio-emotional and functional impairments to develop personalized early risk prediction models, and personalized health care pathways empowering individuals with dementia and actively mitigating the effects of the disease.

Keywords Artificial Intelligence (AI) · Customer care · Daily social and functional activities · Biometric and behavioral data · Assistive technologies · Rehabilitation · Quality of life

3.1 Introduction

The number of individuals suffering of dementia is constantly increasing (from 47 million people living with dementia worldwide in 2017 to a projected number of 66 million by 2030 and 115 million by 2050, [1]). Given the rapidly aging of the worldwide population, the global cost of dementia is prospected to increase from the estimated $818 billions in 2015 to $1 trillion in 2018 and $2 trillion

A. Esposito (✉) · G. Cordasco
Dipartimento di Psicologia and IIASS, Università Della Campania "Luigi Vanvitelli", IT, Caserta, Italy

G. Cordasco
e-mail: gennaro.cordasco@unicampania.it

A. Vinciarelli
University of Glasgow (UK), Glasgow, Scotland
e-mail: Alessandro.Vinciarelli@glasgow.ac.uk

by 2030 (https://www.dementiastatistics.org/statistics/cost-and-projections-in-the-uk-and-globally/). There is a need to identify new dementia cares and develop tools for early diagnoses to reduce and delay the onset of the disease allowing as many elders as possible to reach the end of their life without developing the disease. The current understanding of neuro-generative disorders has notably increased. It has been learned that behavioral changes, such as lack of empathy, failure to reciprocate emotional signs, maladaptive emotional reactions, and repeated interruptions during conversations are symptomatic precursors of neuro-generative diseases. These behavioral changes may appear years before the disease is diagnosed [2].

The appropriate reciprocation of social signals discloses social cognition abilities which drive the success of human interactions. In such exchanges, verbal and nonverbal signals jointly cooperate in assigning semantic and pragmatic contents to the conveyed message, and cognitive and emotional feelings to their achievers tailoring with this information the meaning of the interactional process. Interactional exchanges are carried out through the multiple exploitation of speech signals, hand gestures, body movements, facial expressions, linguistic and paralinguistic information (such as speech pauses, turn takings, repairs, …) accounting of the multiple human's resources devoted to assist individuals in building meanings.

When it comes to dementia, the abovementioned interactional signals fail in their role to secure a successful and meaningful interaction. Recent research shows that some impairments of social cognition, i.e., the ability to understand and appropriately respond to others' mental states, emotions, and actions (i.e., social signals) are early precursors of mild cognitive impairments (MCI) and neurodegenerative diseases, among those Alzheimer's (AD) and vascular dementia (VD) [3–6]. Noticeably these signals of social inappropriateness exemplifying social cognition impairments (such as social withdrawal, loss of social etiquette, changes in personality and behavior) correlate with specific neurodegenerative disorders and occur years before neuronal impairments are detectable [2].

Considering this perspective, developing multimodal and multi-tasks approaches describing individuals' ability to exhibit and process social signals and initial manifestations of peculiar changes in social cognition and behaviors may help in the early stage's detection, the developing of personalized treatments and the implemention of autonomous complex systems for dementia's assistance. Automatic supervised/unsupervised artificial intelligent systems trained on comined features describing from one hand behavioral changes and the other hand clinical dementia assessment's procedures and MRI scans may offer personalized risk prediction, prevention, and intervention protocols to detect and discriminate among different neurogenerative diseases. Personalized cares may prevent negative consequences on the quality of life of patients thus resulting in better-quality healthcare services and healthcare outcomes [7].

3.2   Mild Behavioral Impairments (MBI) and Dementia

Mild behavioral impairments are sensed by carrying out, during social interactions, inappropriate verbal and nonverbal signals, such turning the face while others are approaching, incongruity in showing emotions, inability in noticing and understanding others' emotions, and responding to them with empathy and sympathy, as well as, inability to properly attribute mental states, beliefs, and desires, to oneself and others.

The engendering of these social abilities is subtle and requires possibly automatic analyses of how sufferers perceive, interpret and respond to such social cues.

In order to help doctors to delineate objective diagnoses of mild behavioral impairments (MBI) it is necessary to collect patients' intact perception, attention, and ability to perform executive functions. A not exhaustive list of MBIs [3] is the following:

(a)   Poor social perception (i.e., difficulties in identifying others' facial expressions, failure to recognize anger or boredom in conversational partners, failure to comprehend sarcasm or insincere speech).

(b)   Impaired meta-cognition (i.e., self-awareness deficits, failure to infer what others are thinking or feeling, indifference).

(c)   Reduced ability to show empathy and perform correct emotional processing (i.e., failure to share or maladaptive emotional reactions).

(d)   Anomalous social behavior (i.e., failure to follow elementary social rules, avoidance, poor conversational turn-taking, frequent conversational interruptions).

3.3   Biometric Data

In addition to behavioral information also biometrics data must be considered. In particular it has been shown that online drawing/writing measures, particularly pauses and hesitations while the pen is in the air (i.e., not on the pad surface) support the diagnosis of dementia [8, 9]. Drawing and writing measures also provide early warnings of motoric skills decline, inform of the deterioration status of already diagnosed dementia' patients, and can be used to detect patients' negative moods and emotional states through drawing and handwriting tasks such as two-pentagons copy, house drawings, clock's drawings, and capital and cursive handwritings of patients' native language phonetically complete sentences (see [10–12]). Biometric measures on body postures, meaningful hand gestures and emotion relevant activities, collected through lightweight wearable bracelets with built-in sensors (RSI information), and analyzed by AI systems may provide additional information on the stage of the disease [13]. Figure 3.1 summarizes a set of biometrics/behavioral data considered relevant for the diagnosis and early detection of neurogenerative disordes. Figure 3.2 shows changes in functional abilities such as handwriting and drawing.

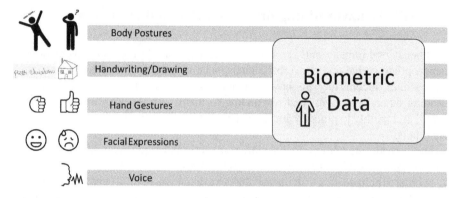

Fig. 3.1 Behavioral and Bionetric data relevant for the diagnosis and early detection of neurogenerative disorders

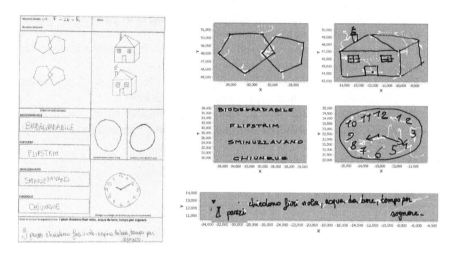

Fig. 3.2 (left) An A4 sheet with some handwriting tasks filled by a subject; (right) Writing and drawing samples. On paper and in air strokes are depicted in black and white respectively

3.4 Caring for Caregivers

The difficulty in being a caregiver of a person with dementia is that simple daily speech and associated gestures, memory of relatives, gratitude, and acknowledgements of received cares disappear at certain stages of the disease. Caring for people with dementia may be a source of high stress, to the extent that caregivers occasionally may lose their temper and immediately regret it, or adopt non-empathetic behaviors compromising the quality of care they are providing [14]. The burden on dementia's caregivers is the highest among of all caregiving groups with caregivers (especially family caregivers) exposed to negative physical and psychological aftereffects [15, 16]. Caregivers of people with dementia are at greater risk to develop anxiety and depressive symptoms (currently affecting up to 34% of dementia's family caregivers, [17]), are at greater risk of losing their temper and exhibit negative reactions, are vulnerable to (1) the amount of received formal and informal support, (2) type of bonds with the cared, (3) amount of care needed by the care-recipient, and (4) hours spent in providing care [18, 19].

Theoretical models must be considered explaining the negative emotional consequences of dementia's caregiving and how they are mediated by the way caregivers perceive, evaluate, and manage the caregiving process. Actions must be taken to educate/familiarize caregivers with dementia care processes to suggest more effective adoption of best practices for dementia care, enhance overall quality of care, decrease burdens, fear, and anxiety, increasing caregivers' knowledge and confidence in accomplishing their work.

Caregivers' burden is detected by collecting information on their strain and psychological wellbeing through questionnaires such as the Dementia Attitude Scale [20], the Modified Caregiver Strain Index [21] and the Carer's Needs Assessment for Dementia [22] scales. These tests are short enough (MCSI is composed by 12 questions, DAS has 20 items, CAN_D has 36 items) for not taking excessive time increasing caregivers' psychological and physical burdens. Caregivers must be almost continuously monitored to allow the timely identification of the magnitude of their burden and suggest fast, efficient, preventive, and supportive actions.

3.5 Tests

Data on patients with dementia status are collected through informant-based procedures such as the Everyday Cognition test (ECog, [23]). The choice of the Ecog questionnaire in alternative to patients' self-reports and patients' performance-based measures for assessing dementia status is motivated by the following reflections:

- Self-reports have been considered unreliable because of patients' difficulty to accurately describe their symptoms [24]. Patients' performance-based measures are more reliable but require the execution of specific tasks that need to be assessed by trained evaluators and are carried out under artificial conditions requiring extensive equipment, well defined environmental contexts, and time-consuming evaluation procedures [25].
- Others informant-based measures of everyday functions are hierarchical divided into activities of daily living (ADLs, which include basic self-care behaviors) and instrumental ADLs (IADL, which include planning and decision making). This subdivision is considered artificial since does not account of underlying cognitive abilities (such as experience) that may support patient's performances and not very sensitive to mild functional impairments and behavioral changes over time [26].

Ecog overcome these limitations and comprise multiple subscales measuring everyday/real-world functioning for the following specific neuropsychological domains: Everyday Memory, Everyday Language, Everyday Semantic Knowledge,

Everyday Visuospatial abilities, and three everyday executive domains including Everyday Planning, Everyday Organization, and Everyday Divided Attention. It can be administered to family caregivers from time to time and to monitor their perception of patient's health status. In addition to Ecog based measures, social and cognitive abilities of patients with dementia can be assessed through a series of selected tests proposed in a naturalistic formulation through structured interviews/interactions, among those:

- The Day Out Task model (DOT) proposed by Schmitter–Edgecombe et al. [27] which requires the programming of daily activities such as determine the moneys required for a bus ride and gather the changes needed, bring a magazine to read while waiting, gather the recipe for "Spaghetti, garlic, oil, and pepper" in a book of recipes, have a conversation with an old friend.
- The narration of stories from the Strange Story Test [28] and the Faux Pas Recognition Test [29, 30] appropriately adapted to the context and cultural background of involved subjects. The Strange Story and the Faux Pas Recognition Tests contain narratives where protagonists describe situations that must be metaphorically deciphered. For example, in a scene someone was going to a picnic, and while unpacking the food it starts to rain and it was commented that "it was a lovely day for a picnic". The patient will be asked to report the sentence's meaning. The correct response requires the exploitation of elements from the context. The stories in the Strange Story and Faux Pas Recognition Tests are an ecological and short (compared to long standardized tests) and serve to evaluate abilities of patients with dementia on the Theory of Mind (ToM) [31, 32]. The original stories must be modified to adapt and fit patients' cultural and social background.

Finally, it has been observed that when it comes to dementia the interaction between patients, caregivers, and healthcare-professional collide with different communicative needs of the triad. It requires a negotiation between the healthcare professional clarity versus sensitivity in delivering both the diagnosis and cares and it call for a compromise in minimizing the emotional impact of all parties, including doctors, who may need support in attending to the emotional reactions of patients and caregivers [33]. The analysis of the communication among this triad can provide guidance "to ensure clear communication and understanding in outpatient care whilst still maintaining a sensitive and empathic approach" [33]. These interactions, based on unstructured interviews leaving the triad freely express feelings, concerns and considerations can be recorded and automatically transcribed by specific automatic speech recognition systems. The texts can be analyzed through unsupervised sentence-level sentiment analysis techniques [34] to extract individual's sentiments and opinions experienced by participants (particularly the polarity of single sentiments expressed in single sentences). The reports generated by these analyses may providing guidance to healthcare professionals in delivering appropriate personalized cares and arouse personal awareness in patients and caregivers improving the quality of relationships among the triad.

The above discussed data collections of daily living (including basic self-care behaviors), instrumental (including planning and decision making, and handwriting) behavioral, biometric and social cognition activities aim to provide large behavioral, biometric, and clinical data sets to be mined for extracting and identifying specific socio-emotional impairments in MCI and dementia patients. Given the large amount of the collected data the mining must be largely automatized by exploiting supervised and unsupervised artificial intelligent techniques, such as deep learning [6] to reduce at the minimum manual annotation procedures.

3.6 Conclusions

A research agenda for dementia care should pursue the following objectives:

- Build a dataset of video recordings of emotional exchanges in healthy population, in order to create a reference sample for further investigation in individuals with mild cognitive impairment (MCI), Alzheimer (AD), vascular dementia (VD) and individuals with psychological disturbances (e.g., mood disorders, depression, burnout, apathy, anxiety, aggression);
- Investigate aspects of social cognition (i.e., empathy, emotional competence and regulation, theory of mind, speech prosody comprehension) comparing healthy and cognitively impaired individuals (MCI, AD and VD patients) with the aim to identify significant differences in social behavioral features exemplified by these groups and provide punctual descriptions of these behavioral changes describing differences between healthy/unhealthy individuals.
- Fuse the acquired (MBI, and biometric) knowledge on dementia with that deriving from classical clinical tests and ground truth (such as MRI scans) to explore multiple ways for advancing the instantiation of preventive measures and the implementation of automatic predictive models
- Investigate the negative emotional consequences of dementia's caregiving and implement actions empowering caregivers' knowledge and confidence in accomplishing dementia care reducing subjective and objective burdens
- Ascertain the existing supervised/unsupervised artificial intelligence and signal processing methods that best capture these changes from the collected data to automatize the detection and classification of dementia's categories

This methodology includes dyadic (or small group) interactions, involving healthy, cognitively impaired patients, and caregivers in naturalistic settled real-life, and computer-based scenarios, for providing (through multiple data analyses obtained with automatic supervised/unsupervised artificial intelligence data annotations and detection techniques) putative differences among stakeholders behavioral and biometric data and suggest appropriate actions in personalized health care pathways that actively empower health risk mitigation, and targeted dementia's prevention and intervention procedures. It provides a deep investigations on the relevant consequences that occur at the cognitive and emotional level of the final user.

Funding The research leading to these results has received funding from the EU H2020 research and innovation program under grant agreement N. 769872 (EMPATHIC) and N. 823907 (MENHIR), the project SIROBOTICS funded by Italian MIUR, PNR 2015–2020, D.D. 1735, 13/07/ 2017, and the project ANDROIDS funded by the program V:ALERE 2019 Università della Campania "Luigi Vanvitelli", D.R. 906 del 4/10/2019, prot. n. 157264,17/10/2019.The work of Alessandro Vinciarelli was supported by UKRI and EPSRC through grants EP/S02266X/1 and EP/N035305/1, respectively.

References

1. G. Livingston, A. Sommerlad, V. Orgeta, S.G. Costafreda, J. Huntley, D. Ames, C. Ballard, S. Banerjee, A. Burns, J. Cohen-Mansfield, C. Cooper, N. Fox, L.N. Gitlin, R. Howard, H.C. Kales, E.B. Larson, K. Ritchie, K. Rockwood, E.L. Sampson, Q. Samus, L.S. Schneider, G. Selbæk, L. Teri, N. Mukadam, Dementia prevention, intervention, and care. The Lancet **90**(10113), 1–129 (2017)
2. J.D. Henry, W. von Hippel, P. Molenberghs, T. Lee, P.S. Sachdev, Clinical assessment of social cognitive function in neurological disorders. Nat. Rev. Neurol. **12**(1), 28–39 (2016). https://doi.org/10.1038/nrneurol.2015.229
3. P. Desmarais, K.L. Lanctôt, M. Masellis, S.E. Black, N. Herrmann, Social inappropriateness in neurodegenerative disorders. Int. Psychogeriatr. **30**, 197–207 (2018). https://doi.org/10.1017/S1041610217001260
4. R. Hutchings, J.R. Hodges, O. Piguet, F. Kumfor, Why should I care? dimensions of socio-emotional cognition in younger-onset dementia. J. Alzheimers Dis. **48**(135–147), 2015 (2015)
5. Z. Ismail, L. Agüera-Ortiz, H. Brodaty, A. Cieslak et al., The Mild Behavioral impairment checklist (MBI-C): a rating scale for neuropsychiatric symptoms in pre-dementia populations. J. Alzheimer's Disease **2017**(56), 929–938 (2017). https://doi.org/10.3233/JAD-160979
6. Y. LeCun, Y. Bengio, G. Hinton, Deep learning. Nature, 521, 436–444Levenson R.W., Sturm V.E., Haase C.M. (2014). Emotional and behavioral symptoms in neurodegenerative disease: a model for studying the neural bases of psychopathology. Annu. Rev. Clin. Psychol. **28**, 581–606 (2015). https://doi.org/10.1146/annurev-clinpsy-032813-153653
7. M.J. Summers, T. Madl, A.E. Vercelli, G. Aumayr, D.M. Bleier, L. Ciferri, Deep machine learning application to the detection of preclinical neurodegenerative diseases of aging. DigitCult - Scientific J. Digital Cult. **2**, 9–24 (2017)
8. M.T. Angelillo, F. Balducci, D. Impedovo, G. Pirlo, G. Vessio, Attentional pattern classification for automatic dementia detection. IEEE Access **7**, 57706–57716 (2019)
9. D. Impedovo, G. Pirlo, G. Vessio, M.T. Angelillo, A handwriting-based protocol for assessing neurodegenerative dementia. Cogn. Comput. **11**, 576–586 (2019)
10. G. Cordasco, F. Scibelli, M. Faundez-Zanuy, L. Likforman-Sulem, A. Esposito, Handwriting and drawing features for detecting negative moods, in A. Esposito, M. Faudez-Zanuy, F. Mora-bito, E. Pasero (eds.)*Quantifying and Processing Biomedical and Behavioral Signals*. SIST 103, 71–86. Springer International Publishing (2019). https://doi.org/10.1007/978-3-319-950 95-2_7, https://doi.org/10.1007/978-3-319-95095-2_7
11. M. Faundez-Zanuy, E. Sesa-Nogueras, J. Roure-Alcobé, J. Mekyska, A. Esposito, K. Lopez-De-Ipiña, A preliminary study on aging examining online handwriting, in *Proceedings 5th IEEE international Conference on Cognitive InfoCommunications, Vietri sul Mare*, 5–7 Nov 2014, 221–224 (2014). https://doi.org/10.1109/coginfocom.2014.7020449
12. L. Likforman-Sulem, A. Esposito, M. Faundez-Zanuy, S. Clémençon, G. Cordasco, EMOTHAW: a novel database for emotional state recognition from handwriting and drawing. IEEE Trans. Human Mach. Syst. 47(2):273–284 (2017). https://doi.org/10.1109/THMS.2016. 2635441, http://ieeexplore.ieee.org/document/7807324/

13. R. Gravina, Q. Li, Emotion-relevant activity recognition based on smart cushion using multisensory fusion. Inform. Fusion **48**, 1–10 (2019)
14. R. Maskeliūnas, R. Damaševičius, C. Lethin, A. Paulauskas, A. Esposito, M. Catena, V. Aschettino, Serious Game iDO: towards better education in dementia care. Information **10**, 355 (2019)
15. B. Watson, G. Tatangelo, McCabeM, Depression and anxiety among partner and offspring carers of people with dementia: a systematic review. Gerontologist **59**(5), e597–e610 (2019). https://doi.org/10.1093/geront/gny049
16. Y. Kim, R. Schulz, Family caregivers' strains: comparative analysis of cancer caregiving with dementia, diabetes, and frail elderly caregiving. J. Aging. Health. **20**(5), 483–503 (2008)
17. N.S. Domingues, P. Verreault, C. Hudon, Reducing burden for caregivers of older adults with mild cognitive impairment: a systematic review. Am. J. Alzheimers Dis. Demen. **33**, 401–414 (2018). https://doi.org/10.1177/1533317518788151 PMID: 30041535
18. A.B. Sallim, A.A. Sayampanathan, A. Cuttilan, R. Chun-Man Ho, Prevalence of mental health disorders among caregivers of patients with alzheimer disease. J Am Med Dir Assoc **16**, 1034–1041 (2015). https://doi.org/10.1016/j.jamda.2015.09.007 (PMID: 26593303)
19. J. van der Lee, T.J. Bakker, H.J. Duivenvoorden, R.M. Droes, Multivariate models of subjective caregiver burden in dementia; a systematic review. Ageing Res. Rev. (2014). https://doi.org/10.1016/j.arr.2014.03.003 (PMID: 24675045)
20. M.L. O'Connor, S.H. McFadden, Development and psychometric validation of the dementia attitudes scale. Int. J. Alzheimer Dis. Article ID **454218**, 1–10 (2010)
21. Onega, L.L., The modified caregiver strain index (MCSI). TryThis, 10 (2018). https://consultgeri.org/try-this/general-assessment/issue-14.pdf
22. T. Novais, V. Dauphinot, P. Krolak-Salmon, C. Mouchoux, How to explore the needs of informal caregivers of individuals with cognitive impairment in Alzheimer's disease or related diseases? A systematic review of quantitative and qualitative studies. BMC Geriatr. **17**(1), 86 (2017). https://doi.org/10.1186/s12877-017-0481-9.PMID:28415968;PMCID:PMC5393006
23. S.T. Farias, D. Mungas, B.R. Reed, D. Cahn-Weiner, W. Jagust, K. Baynes, C. DeCarli, The measurement of everyday cognition (ECog): scale development and psychometric properties. Neuropsychology **22**(4), 531–544 (2008). https://doi.org/10.1037/0894-4105.22.4.531
24. T. Buckley, E.B. Fauth, A. Morrison, J.A. Tschanz, P.V. Rabins, K.W. Piercy, M. Norton, C.G. Lyketsos, Predictors of quality of life ratings for persons with dementia simultaneously reported by patients and their caregivers: the Cache County (Utah) Study. Int. Psychogeriatr. **24**(7), 1094–1102 (2012)
25. T. Giovannetti, K.S. Schmidt, J.L. Gallo, N. Sestito, D.J. Libon, Everyday action in dementia: Evidence for differential deficits in Alzheimer's disease versus subcortical vascular dementia. J. Int. Neuropsychol. Soc. **12**, 45–53 (2006)
26. S.T. Farias, D. Mungas, W. Jagust, Degree of discrepancy between self and other-reported everyday functioning by cognitive status: Dementia, mild cognitive impairment, and healthy elders. Int. J. Geriatr. Psychiatry **20**, 1–8 (2005)
27. M. Schmitter-Edgecombe, C. McAlister, A. Weakley, Naturalistic assessment of everyday functioning in individuals with mild cognitive impairment: The Day-Out Task. Neuropsychology **26**(5), 631–641 (2012)
28. F.G. Happé, An advanced test of theory of mind: understanding of story characters' thoughts and feelings by able autistic, mentally handicapped, and normal children and adults. J. Autism Dev. Disord. **24**, 129–154 (1994)
29. C. Gregory, S. Lough, V.E. Stone, S. Erzinclioglu, L. Martin, S. Baron-Cohen, odges J, , Theory of mind in frontotemporal dementia and Alzheimer's disease: Theoretical and practical implications. Brain **125**, 752–764 (2002)
30. V.E. Stone, S. Baron-Cohen, R.T. Knight, Frontal lobe contributions to theory of mind. J. Cogn. Neurosci. **10**, 640–656 (1998)

31. F.G. Happé, E. Loth, Theory of mind' and tracking speakers' intentions. Mind Lang. **17**, 24–36 (2002)
32. A.D. Hutchinson, J.L. Mathias, Neuropsychological deficits in frontotemporal dementia and Alzheimer's disease: a meta-analytic review. J. Neurol. Neurosurg. Psychiatry **78**, 917–928 (2007). https://doi.org/10.1136/jnnp.2006.100669
33. J. Dooley, C. Bailey, R. McCabe, Communication in healthcare interactions in dementia: a systematic review of observational studies. Int. Psychogeriatr. **27**(8), 1277–1300 (2015)
34. M.F. Mowlaei, M.S. Abadeh, H. Keshavarz, Aspect-based sentiment analysis using adaptive aspect-based lexicons. Expert Syst. Appl.**148**, Article 113234 (2020)

Chapter 4
Machine Learning and Finite Element Methods in Modeling of COVID-19 Spread

Nenad Filipovic

Abstract In this chapter we analyze patient-specific machine learning and finite element approach for modeling COVID-19 spread. Many factors are influencing COVID-19 development in humans, including genetics, blood markers, geographical position and medical Roentgen images. Our unique machine learning (ML) tool provides a set of solutions for COVID-19 development in the patient-specific model. This tool uses the following data: demographic data, clinical picture, blood test data and imaging data. The result of the model is prediction of the risk of mortality. A unique feature of COVID-19 interstitial pneumonia is an abrupt progression to respiratory failure. Our patient-specific finite element lung model focuses on the spread of virus-laden bioaerosols to many regions of the lungs from the initial site of infection. Both tools can help medical experts to decide whether the patent should be subjected to further analysis and prescribe adequate therapy. Predictive models based on ML can provide useful data in terms of prediction of epidemiological events, which can save time for timely and optimal response of both the health system and the society. The finite element tool can contribute to better treatment of COVID-19 patients by using mechanical ventilation.

4.1 Introduction

4.1.1 Physiology of Human Respiratory System

The human respiratory system consists of the upper respiratory tract and the lower respiratory tract. The upper respiratory tract is made up of mouth, nose and nasal cavity, throat (pharynx) and voice box (larynx). The lower respiratory tract consists of two lungs, trachea (windpipe), bronchi, bronchioles and alveoli. The lungs are the

N. Filipovic (✉)
Faculty of Engineering, University of Kragujevac, Sestre Janjić 6, 34000 Kragujevac, Serbia
e-mail: fica@kg.ac.rs

Bioengineering Research and Development Centre (BioIRC), Prvoslava Stojanovića 6, 34000 Kragujevac, Serbia

G. A. Tsihrintzis et al. (eds.), *Advances in Assistive Technologies*, Learning and Analytics in Intelligent Systems 28, https://doi.org/10.1007/978-3-030-87132-1_4

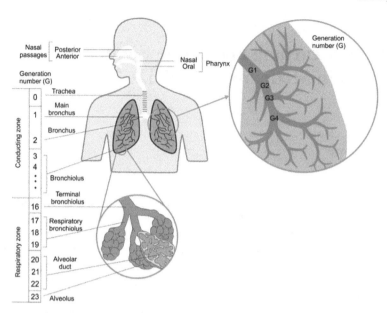

Fig. 4.1 Schematic representation of the human respiratory system, from extra-thoracic components to alveoli, and cast of human airways

main part of the respiratory system. They are irrigated with fresh air through a dyadic structure called the tracheobronchial tree. The respiratory muscles form a complex arrengement around the lungs (Fig. 4.1). These muscles contract during inspiration, causing the lungs to expand, and relax at expiration, resulting in lung deflation. Lung physiology and mathematical modeling of ventilation are given in [9, 79, 82]. The inhaled fresh air rich with dioxygen (O_2) enters the lungs and reaches the alveoli where gas exchange takes place. The oxygen is loaded into the bloodstream and is further carried out through the body. In the opposite direction carbon dioxide passes from the blood into the alveoli and is then exhaled. The tracheobronchial tree in humans has a non-symmetric dyadic branching structure. It is a complex system that transports gases from the trachea, which divides into two airways irrigating the right and left lung, down to the acini. The tracheobronchial tree contains approximately 24 generations of dichotomous branching, from upper airways with centimetric diameter down to 15 networks of millimeter size diameter [82]. In the first 17 generations, the airflow is convective. In lower branches of the tree, Reynolds numbers are low and the flow regime is diffusive. There are about 30000 acini, which are dyadic terminal sub-trees. They end in alveolar sacs where gas exchange takes place. Three hundred millions alveoli represent an exchange surface of about 100 m^2 [9].

4.1.2 Spreading of SARS-CoV-2 Virus Infection

There are three ways in which virus can spread in tissues: virus transport, virus multiplication in host cells and virus-induced immune response. It is well known that cytotoxic T cells remove infected cells with a rate determined by the infection level. The mathematical model consists of reaction-diffusion equations which describe the different regimes of infection spreading. A low level infection, a high level infection or a transition between both are determined by the initial virus load and by the intensity of the immune response. Viruses are non-cellular organisms that need cells to replicate their genomes and produce progeny. They will expand locally around the entry site of a newly infected organism, depending on the mode of transmission. The virus cannot divide itself automatically [94]. As an organism consisting of a nucleic acid genome, surrounded by a protein coat, the denaturation of virus can be regarded as a temperature-dependent rate problem, just as any natural substances do. It is very important to understand SARS-CoV-2 replication cycle (Fig. 4.2) and its interactions with the immune system. This cycle contains four stages. The first stage is the binding to receptors of the host cell where the virus RNA is uncoated in the cytoplasm. The second stage is transcription/translation processes which generate new viral RNA material and proteins. The third stage is virus assembly within vesicles followed by virus release. The fourth stage is infection of other cells.

In certain cases, histological findings of the lungs infected with SARS-CoV-2 showed bilateral diffuse alveolar damage with cellular fibromyxoid exudates, desquamation of pneumocytes, pulmonary edema and hyaline membrane formation. There is also some evidence of direct injury to the lung tissue from a viral infection [90]. According to some studies, progression to Acute Respiratory Distress Syndrome (ARDS) related to COVID-19 largely depends on the time from the manifestation of symptoms to hospitalization and mechanical ventilation. Moreover, it has been

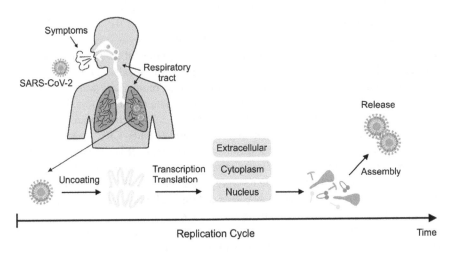

Fig. 4.2 SARS-CoV-2 replication cycle

reported that patients with COVID-19 lung disease have significantly higher compliance than it is typical for their shunt fraction, indicating this may be a very different phenotype than typical ARDS. The explanation remains unclear, with pulmonary perfusion dysregulation posited as one possible explanation.

4.1.3 Machine Learning for SARS-CoV-2

It is still not known how SARS-CoV-2 triggers a wide spectrum of heterogeneous clinical manifestations, from asymptomatic cases through ARDS to multiple organ failure and death [5, 34, 26, 54]. Therefore, researchers all over the world have been looking for the ways to define and stratify predictors of the severity of COVID-19 disease in an attempt to properly guide medical management. Basic knowledge of pathogenesis of diseases and methods of discerning and assessing COVID-19 infection have been established. Common blood hematology and clinical biochemistry tests are inexpensive, simple and widely-accessible biomarkers. As such, they have become the preferred methods of tracking and predicting disease effects [4]. Recognizing the variation and phenotype of certain biomarkers as a result of multiple outcomes of COVID-19 will help establish a risk-stratified strategy for the treatment of patients with this disease. Among other issues, it is very important to predict the unfavorable progression of the illness easily, reliably and timely. The possibility to detect cases that are at imminent risk of death has become an urgent, but necessary task [91].

Numerous researchers have studied predictors of disease severity in COVID-19 patients. Some studies indicate that serious or lethal cases of COVID-19 disease have been linked with elevated white blood cell count, creatinine, blood urea nitrogen, markers of liver and kidney function, interleukin-6 (IL-6), C-reactive protein (CRP), lower lymphocyte ($<1 \times 10^9$/L) and platelet counts ($<100 \times 10^9$/L), as well as albumin levels compared with milder cases of patients who survived the disease [23, 26, 65]. These research results provide an initial insight into the effects of the infection by SARS-CoV-2. However, due to geographical restrictions, specific experience of the clinical centers and small cohorts, the results cannot be generalized [49]. Malik and associates found that particular biomarkers were correlated with poor outcome results in COVID-19 hospitalized patients in meta-analysis of 32 trials reflecting 10,491 reported patients with COVID-19. The biomarkers included a low lymphocyte count, a lower count of platelets and elevated CRP, creatine kinase (CK), procalcitonin (PCT), D-dimer, lactate dehydrogenase (LDH), alanine aminotransferase (ALT), aspartate aminotransferase (AST) and creatinine [49]. Assandri et al. [5] found that certain laboratory tests were characteristically altered in COVID-19 cases and proposed new tests as sensitive and rapid alternatives for identification of possible COVID-19 cases-specific biomarkers were found to be associated with the clinical outcome [10]. Some of the best prognostic markers were found to be neutrophil/lymphocyte ratio (in patients with a more serious illness) [46], CRP and inflammatory cytokines such as IL-6, IL-1 or *tumor necrosis factor alpha* (TNF-α) [66]. Other studies also found that CRPO7 mg/dL can identify subjects that would develop critical condition [5, 10]. After an extensive research, it can be concluded that the results obtained so far show inconsistency. Biomarkers shown to be the predictors of COVID-19 disease progression differ from research to research and are sometimes even contradictory (Cheng et al. [54]. Therefore, there is a strong need for new tools and strategies such as ML methods to obtain a further insight into prognostic biomarkers.

The ML algorithms have been used in COVID-19 research for many purposes including epidemiological and clinical issues such as timely detection of disease outbreaks, fast diagnosis, classification and stratification of radiological images, risk factors analysis, as well as prediction of final clinical outcomes [15, 23]. For example, Yan et al. [91] conducted a research that used a blood sample database of 404 infected patients in the region of Wuhan, China, to classify important biomarkers of the disease seriousness in support of decision-making and logistics planners of health systems. To achieve this, three biomarkers were selected using an ML method which had an accuracy of more than 90%: LDH, lymphocyte and high sensitivity CRP (hs-CRP). In fact, comparatively high LDH levels alone are critical in distinguishing the vast majority of cases developing critical state. This result is consistent with current medical knowledge that elevated levels of LDH are related to deterioration of tissues, which appear in pneumonia and other infectious and inflammatory diseases [36]. The main advantage of this chapter is that it introduces a clear and operational formula to easily forecast and assess clinical outcome—death/survival, allowing critical patients to be prioritized, and potentially reduce the mortality rate. The main drawback of this study is that the classification is binary—survival/death, which may not be the best type of classification in situations where the healthcare systems are overloaded. In these situations, it is better to have a more distinct classification into clinical conditions, in order to determine which patients will develop critical condition and, therefore, should stay in hospital, and which ones could be discharged and treated at home, having only mild symptoms. The same group of researchers lead by Yan studied the samples of blood of 485 patients from the area of Wuhan, China, retrospectively to recognize robust and relevant markers of risk mortality [92]. The goal was to identify the most discriminatory biomarkers of patient mortality by using state-of-the-art interpretable ML algorithms. Inputs contained the specific details, symptoms, blood sample and laboratory results on the liver, renal functions, coagulation activity, electrolytes and inflammatory factors, collected from patients at different stages of the disease, and related results for death or survival. The research employed a supervised XGBoost model classification. A tree-based ML model was used to predict the outcomes of individual patients (death/survival) using a sample blood test database. The same biomarkers (features) as in the previous study were experimentally chosen with strong predictive degradation values or fatality of disease, also matching other studies found in the literature [34], Wang et al. [54], Li, [90], Tan et al. [73]. The classification was binary and, except predicting survival, it did not help in reducing the burden put on the healthcare system. Huang and associates say that although the studies by Yan and colleagues reveal that the mortality outcomes in COVID-19 patients are associated with three biomarkers using a single-tree XGboost model, which is a recognizable result, and the forecast results are over-optimistic. The results of the forecast remain ambiguous as they indicate high variability. Hence, Yan and colleagues' assertion on the successful prediction dates are insufficiently strong and not fully validated by the data provided. The prediction results may be unreliable. As also discussed by the scientists, the efficiency and stability of the proposed mortality forecast model should be tested further in future by a broader, representative population of COVID-19 patients [34].

Furthermore, Gao et al. [21] presented a COVID-19 Mortality Risk Prediction (MRPMC) model that uses hospital admission data to streamline patients via risk of mortality, allowing the prediction of physiological decline and death up to 20 days in advance. They implemented four methods of machine learning: Logistic Regression, Support Vector Machine, Gradient Boosted Decision Tree and Neural Network. The MRPMC was tested on internal and external evaluation cohorts and achieved the average area under the curve (AUC) of 95% on internal and 97% and 92% on two different external cohorts, showing that it can have possible high utility in future clinical practice [21].

Although there are many studies that deal with the implementation of ML in proposing valid prognostic biomarkers and predictors of survival several days in advance, unsatisfied needs still remain. Moreover, there is limited research in the field focused on classification of patients in more subtle severity-of-illness categories (e.g. mild, moderate, severe and critical) during hospital stay. Such knowledge could help physicians and hospital managers in decision-making process aiming to avoid not only patients' final unfavorable outcome but also to improve other important secondary treatment endpoints and institutional performances which are deteriorated by inappropriate measures such as unnecessary prescription of adjunctive drugs (e.g. wide-spectrum antibiotics and immune-modulatory biologics), over-utilization of sophisticated and invasive diagnostics and inappropriate allocation of intensive care beds. Therefore, in this chapter we propose a methodology based on ML to classify patients into several categories and predict the outcome in advance (change of severity of clinical condition). The main objectives and contributions of this chapter are:

- Examine rule-based algorithms that are more suitable for implementation in clinical practice rather than black box models (i.e. neural networks).
- Classify patients into 4 distinct categories (mild, moderate, severe and critical) of COVID-19 disease.
- Predict disease progression (mild to moderate, moderate to severe, severe to critical clinical condition).
- Work with a limited dataset, but implement different methods to overcome the drawbacks of small datasets.

4.2   Methods

4.2.1   Finite Element Method for Airways and Lobes

We developed a finite element model of human lungs airway, using optimized material and friction parameters, based on CT images using functional residual capacity (FRC) to total lung capacity (TLC). In addition, we developed a model that predicts a displacement field for lobe sliding and continuum-based using average landmark error and correlation with the lobe-by-lobe deformable image registration results [19]. On a macroscopic level, the airflow into and out of the lungs is driven by pressure differences between the alveoli and the outside environment. The activity of the muscles surrounding the parenchyma induces lung volume variations during breathing. The lungs are divided into units called lobes that are not mechanically attached and that can slide with respect to one another. The human left lung contains two and the right lung contains three lobes. Our model is the airway coupled to a parenchyma model (Fig. 4.3). The upper airways geometry is segmented from a CT scan. This framework is used to mimic virus spread from the alveoli to the other airway geometry. We used around 500,000 finite elements to model both the airway and the lobes [18, 19, 76, 77] (Fig. 4.3).

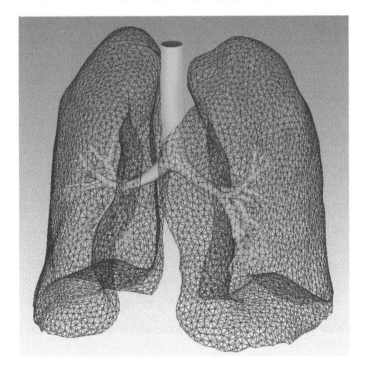

Fig. 4.3 Finite element mesh of the upper airways with the lung mesh and all lobes

4.2.2   Machine Learning Method

This part of the chapter explains the dataset first, and after that the proposed ML methodology. Since the methods can be divided into unsupervised (clustering) and supervised methods (regression and classification), and independent feature extraction, separate sections are created for each implemented sub-method.

4.2.2.1   Dataset

Blood biomarkers of the patients from two hospitals were used—Clinical Center of Kragujevac, Serbia (45 patients) and Clinical Center of Rijeka, Croatia (60 patients). In total, the results of blood analyses of 105 COVID-19 positive patients were collected (42% women, 58% men) and the age distribution of the patients in the form mean $\pm$ standard deviation was 52.77 $\pm$ 16.63. For all patients, fever was the most common symptom (83%), followed by cough (74.6%) and fatigue (45.7%). The described dataset is shown in Fig. 4.4.

We divide the clinical data into three subgroups:

- Demographic data (gender and age).
- Symptoms (fever, cough, fatigue, chest pain, muscle pain, headache, dyspnea, loss of taste or smell).
- Blood analysis:

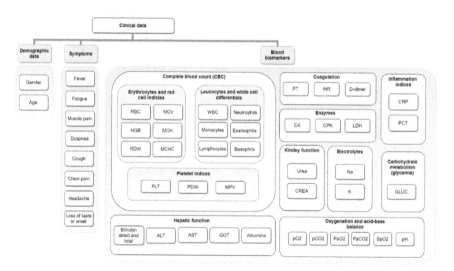

Fig. 4.4 Graphical representation of clinical data which consist of demographic data, symptoms and blood analysis

- Complete blood count (CBC): *erythrocytes* (red blood cells (RBC))—red cell indices: hemoglobin (HGB), mean corpuscular volume (MCV), mean corpuscular hemoglobin (MCH), mean corpuscular hemoglobin concentration (MCHC), red cell distribution width (RDW); *leucocytes (white blood cells (WBC))*—white cell differentials: neutrophils, lymphocytes, monocytes, eosinophils (EOS) and basophils (BASO), *platelet indices*: platelets (PLT) platelet distribution width (PDW), mean platelet volume (MPV).
- Coagulation: prothrombin time (PT), international normalized ratio (INR), D-dimer.
- Kidney function: urea, creatinine (CREA).
- Hepatic function: bilirubin—direct and total, alanine transaminase (ALT), aspartate transaminase (AST), gamma-glutamyltransferase (also γ-glutamyltransferase, GGT), albumin.
- Enzymes: creatine kinase (CK), also known as creatine phosphokinase (CPK) or phosphocreatine kinase; lactate dehydrogenase (LDH).
- Electrolytes: Sodium (Na), potassium (K).
- Oxygenation and acid-base balance: arterial blood gas (ABG) analyses/tests: partial pressure of oxygen (pO2), arterial partial pressure of oxygen (PaO2) partial pressure of carbon dioxide (pCO2), arterial partial pressure of carbon dioxide (PaCO2), SpO2 (peripheral oxygen saturation s. oxygen saturation as measured by pulse oximetry), pH.
- Inflammation indices: C-reactive protein (CRP), procalcitonin (PCT).
- Carbohydrate metabolism (glycemia): GLUC—glucose.

Most of the patients had multiple blood samples taken during their hospital stay. We took into consideration the day of admission to the hospital and days 2, 5, 7, 9, 11 and 14 after hospital admission. In order to deal with missing data, we used data imputation method. To fill in the missing values, we used the mean value of four different values (values from 2 days before and 2 days after). The aim of this type of data imputation is to include the values of analyses of adjacent days that are not directly included in the prediction.

4.2.2.2 ML Model Development

The ML system consists of two main tasks. The first task is prediction of blood analysis in advance (in days) and based on that prediction, the second task is to determine the severity of clinical condition (mild, moderate, severe and critical). For the blood analysis prediction, we used ML supervised learning, more concretely, regression. Feature selection method and application of both unsupervised (clustering) and supervised (classification) ML algorithms are required to determine the severity of clinical condition. The role of the mentioned approaches that are part of this ML system is shown in Fig. 4.5.

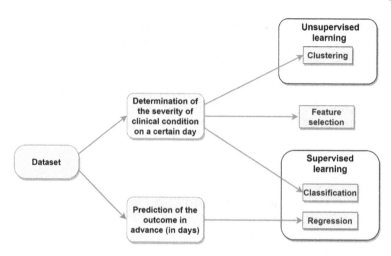

Fig. 4.5 Schematic representation of ML methodology for determination of the severity of clinical condition

4.2.2.3 Feature Selection

To evaluate which biomarkers are crucial for determination of the severity of clinical condition, through a feature selection process we selected ten features that had the greatest influence in creating an effective model. In addition, another important aspect of considering 10 features instead of the 44 original ones is reduction of the computational complexity.

Unsupervised methods

In this section, we propose the use of clustering methods to estimate the missing output labels. This section is particularly important, since without enough output labels no ML method will be able to perform classification into different categories.

Clustering

In addition to the missing values in the blood tests, there are also some missing values in the output labels. To solve that problem, unsupervised clustering analysis was used. Based on the outcomes of the clustering method, missing labels were filled in. Since the application of different clustering methods obtained small differences between clustering solutions [2], we applied the simplest and most commonly used k-means algorithm, which requires a predefined number of clusters, and for determining the optimal number of clusters the Elbow method was used. The method will be examined starting from two clusters and increasing the number of clusters in each step by one until the optimal number is determined. For each given number of clusters, the sum of intra-cluster distances is computed. The variance of the intra-cluster distances between two consecutive numbers of clusters is computed. Then, the Elbow method looks into the percentage of variance explained as a function of the number of clusters.

One should choose a number of clusters so that adding another cluster in the analysis does not give much better explanation of the variance. The optimal number of clusters that was found in this chapter was 4, which is in accordance with the manually marked target outputs (doctor's categorization of the patients into 4 categories (mild, moderate, severe and critical).). This confirms the right use of such ML methodology.

As a result, the dataset was divided in four clusters and, as we expected, cluster distribution is the following:

- 31.80% of total data belongs to the cluster of mild clinical condition,
- 50.90% belongs to the cluster of moderate clinical condition,
- 13.88% belongs to the cluster of severe clinical condition, and
- 3.42% belongs to the cluster of critical clinical condition.

Supervised methods

In this section, we implement regression analysis to predict the values of important features (biomarkers) in the following days, in order to be able to further perform the classification of the patients. Such coupled methodology will enable us to estimate the patients' condition in time.

Regression

Following the previous findings on the importance of biomarkers such as CRP, Creatinine, LDH, etc. the next step is to predict the change of biomarkers values in time. The aim is to predict the patient's clinical condition two weeks after hospital admission based on blood analysis from previous days, starting from the day of admission and ending on the 11th day. The main limitation was the lack of data for biomarkers in time (full blood analysis on the admission day was available for all 105 patients, blood analyses from 104 patients were available for the second day, on the 5th day data for 67 patients were available, on the 7th day data for 68 patients were available, on the 9th day data for 59 patients were available, on the 11th day data for 51 patients were available and on the 14th day only data for 44 patients were available). It should be emphasized that not always were the data for one patient available at each and every time point, but rather spread across some days (i.e. for patient 1, blood analysis was available on days 0, 5, 11, while for patient 2 for days 0, 7, 9, 14 etc.). Due to the small amount of patients' data available in time, in order to predict blood biomarkers values, we decided to select 34 patients with a full blood analysis for all days. In this case, blood biomarkers were used as labels, and for our regression model, we created new features such as the same blood biomarker in the previous moment (the previous day observed) and the difference between values of the biomarker in the current and previous recorded moment. In this way, we created time dependencies between a patient's blood analysis throughout time and expanded the dataset for the regression problem. This means that for day 5, the analyses were taken on day 2 as t-1 and day 5 as t, in order to adapt the dataset for regression. Figure 4.6 shows a schematic representation for the described methodology.

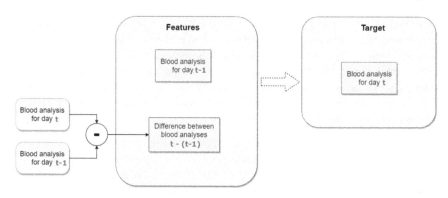

Fig. 4.6 Schematic representation of principles used for database organization

After establishing the database, we divided the data in training and test sets, where the training set consists of data from days 2, 5, 7, 9 and 11. Therefore, day 14 belongs to the test set.

For prediction of each blood biomarker, we used the boosting ensemble ML approach—Gradient boosting regressor (GBR). Boosting is a powerful strategy of learning firstly designed for classification problems, but it has been extended to regression as well. The motivation for boosting is to combine the output of many "weak" methods such as decision trees into a "powerful" structure. Gradient boosting regressor sequentially constructs many decision trees in each sequence, and the GBR model based on its performances in the previous sequence reweights the training data (instances with a more substantial error term get higher weights) [34].

This model was trained with optimal hyperparameters settings based on the grid search method:

- number of tree estimators set to 500,
- max depth equals to 4,
- min samples split equals to 5, and
- learning rate equals to 0.05.

Classification

The main task of this chapter is to assess how COVID-19 develops over time in patients, or more precisely, to determine a patient's condition 14 days after hospital admission. Firstly, it is necessary to assess the value of biomarkers and then classify them into one of 4 classes (mild, moderate, severe and critical). For classification purposes, the original dataset was modified and all the patients with blood analysis results on a particular day were included. This would mean that within this dataset, the same patients were repeated multiple times without any dependences between different days. This modification is justifiable as in this case it is not important to observe the same patient over time, but to create as many different instances as possible in order to expand the dataset and prepare for the classification task. As a result, the dataset included 497 instances (158 instances belonged to the cluster of mild condition, 253 to moderate, 69 to severe and 17 instances belonged to the cluster of critical condition).

For the described classification task, the aim was to construct a simplified, rule-based decision model, and for that purpose, the model of extreme gradient boosting (XGBoost) was used. XGBoost is a supervised ML algorithm based on the recursive construction of a decision tree, and those trees that most influence the decision of the predictive model can be identified. At each decision step in the XGBoost trees, the significance of each feature is determined by its accumulated use [12, 92]. This is an advantage of rule-based algorithms because internal model strategies are easy to interpret. Also, the importance of these algorithms in clinical prediction is reflected in the fact that the feature values for decision-making are known, unlike black box models whose rules and strategies are difficult to interpret.

XGBoost was trained with optimal hyperparameters obtained based on the grid search method:

- number of tree estimators set to 100,
- max depth equals to 5,
- learning rate equals to 0.3,
- gamma equals to 1,
- 'subsample' and 'colsample bytree' are both set to 1.

4.3 Results

4.3.1 Simulation of Virus Spreading by Finite Element Analysis

Our initial simulation results with the dynamics infection model defined by (3–5) after 5 and 10 days are presented in Fig. 4.7. The red object simulates the virus spreading inside the lobes and airways. When the virus propagates through the airway system towards the alveoli, it binds to their wall and cause high angiotensin converting enzyme 2 (ACE2) expression. It induces innate inflammation by pro-inflammatory macrophages and granulocytes and causes accumulation of liquid in the alveoli space and bronchi. It seems that the liquid can be seen in CT images as GGO. Our simulation mimics this severe process in the period of several days during which the liquid can accumulate causing the growth of the space it occupies inside the airway system. We compared our finite element simulation of SARS-CoV-2 virus infection spreading with CT findings as GGO, particularly (Fig. 4.8).

We hypothesize that the abrupt deterioration several days or several hours after the initial infection in some patients is due to induction of cytokine storms in multiple regions of the lungs, caused by the spread of virus-laden bioaerosols through the airways to many different regions from the initial site of lung infection. Clinically,

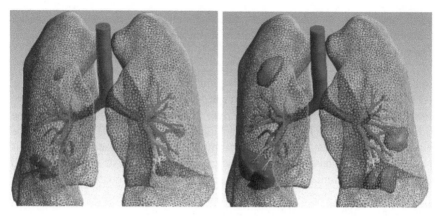

Fig. 4.7 Simulation of SARS-CoV-2 infection spreading 5 and 10 days from the start of infection. The red object simulates virus spreading inside the lobes and airways

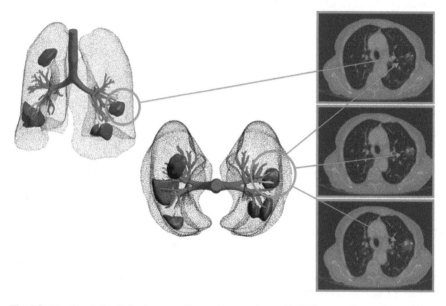

Fig. 4.8 Results of virus infection spreading and comparison with CT images. Comparison of finite element simulation of virus spreading and CT findings (GGO particularly)

severe SARS-CoV-2 infections exhibit multi-region patchy patterns of ground-glass opacity (GGO) throughout the lungs detected by computed tomography (CT) [51, 95]. This presentation is a unique characteristic of COVID-19 pneumonia.

To explore this possibility, we created a computer simulation demonstrating the spread of virus infection between regions of the lungs. Initially, some part of the lung is infected by inhaled SARS-CoV-2 virus (Fig. 4.9a). As a result, pulmonary

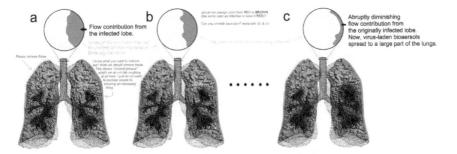

Fig. 4.9 SARS-CoV-2 virus spreading in the lungs. **a** Suppose some part of the lungs is initially infected by inhaled SARS-CoV-2 virus (in this example, we assume that the left upper lobe is initially infected). As a result, near the lesion, virus-laden meniscus may be formed in the alveolar ducts and a rupture of those menisci may form a number of virus-laden small bioaerosols (shown as red dots). **b** Those bioaerosols move toward the airway opening with the expired air. While most of them exit the body, some of them may remain suspended in the airways. Some of the particles, which have exited the body, may reenter the lungs in the next inhalation and move around the lungs, but deposition of those reentered particles may not deposit much due to their low diffusivity. Some of the particles, which remain suspended in the airways, can go back to the original position due to the negligible (but not zero) diffusivity. **c** However, as the disease progresses, the production of virus-laden bioaerosols increases, the tissues available for gas exchange become smaller and smaller as the infected lesion become larger and larger. This may dramatically alter the distribution of airflow in the lungs, resulting in a substantial spread of the infection to other parts of the lungs. This may cause a cytokine storm. (Adapted from [20]

edema develops locally and virus-laden fluid fills the alveoli and surrounding alveolar ducts and meniscus may be formed. During breathing, the menisci travel back and forth in the duct, diminishing in size. A rupture of those menisci, however, can form a number of small droplets called bioaerosols [48]. The bioaerosols move toward the airway opening with the expired air in a piggy-back fashion; while most of the bioaerosols exit the body, some of them, especially near the airway walls where the airflow velocity is small, remain suspended in the airways. In the next inhalation, some of the particles which exit the body, may reenter the lungs and move around the lungs, but deposition may not be appreciable due to their small amount of diffusion, or at least do not cause a cytokine storm. The particles, which remain in the lungs, are likely to retrace their path back to the original position due to the small magnitude of diffusion (Fig. 4.9b).

As the disease progresses, however, the distribution of airflow in the lungs changes (see the cross-section maps of flow distribution shown in Fig. 4.9). The infected part of the lungs contributes a smaller volume of the airflow, and, at the same time, the production of virus-laden bioaersols increases. Therefore, the produced virus-laden bioaersols are more likely to spread to other parts of the lungs (Fig. 4.9c). If multiple areas of the lungs are simultaneously infected, an overwhelming innate immune response may occur resulting in a cytokine storm. Considering the negligible (but not zero) diffusion of bioaerosols, the observation that the deterioration occurs several days after the initial infection is consistent with this hypothesis.

Table 4.1 Ten blood analyses that had the greatest influence on COVID-19 patients' clinical condition assessment

Blood biomarker	Mean ± standard deviation	Normal values	Units
White blood cell count	8.25 ± 4.72	3.70–10.00	10^9/L
Lymphocyte count	22.15 ± 11.06	20.00–46.00	%
MCHC	335.7 ± 8.16	320–360	g/L
RDW	13.69 ± 1.07	11.6–14.5	%
Hgb	130.08 ± 21.05	138–175	g/L
Urea	7.35 ± 5.76	3.0–8.0	mmol/L
Creatinine	92.46 ± 69.62	49–106	μmol/L
Albumins	33.57 ± 6.23	35–52	g/L
LDH	424.86 ± 239.26	220–450	U/L
CRP	69.11 ± 93.55	0.0–5.0	mg/L

4.3.2 Machine Learning Results

A total of ten blood biomarkers were selected as the best features that had the greatest contribution in the classification task and most reliably described the progression of COVID-19 disease in patients, according to the previously described methodology. These blood biomarkers and their values are presented in Table 4.1. In addition, there are reference values of biomarkers and their values from our database in the form of mean ± standard deviation.

For the evaluation of the importance of features, correlation of each two features was used, after which the importance scores were computed. The importance of all ten features is shown in Fig. 4.4. Lactate dehydrogenase (LDH) has the highest importance score, which may be contributed to the fact that this enzyme is widely distributed in tissues and its elevated serum levels could be caused by systemic hypoxemia [57]. It was found that high serum activity of LDH in earlier stages of the disease is a good predictor of a lung injury and poor risk outcomes [92]. Besides LDH, which was shown to be by far the most important biomarker in predicting severity of clinical condition, other parameters which have the most influence on the model are urea (2nd most important), creatinine (3rd most important), C-reactive protein (CRP) (4th most important), white blood cell (WBC) (5th most important), etc. (Fig. 4.10). The explanation for the importance of urea and creatinine biomarkers could be found in the fact that both biomarkers are related to kidney function. Namely, in one research study that included the analysis of 701 COVID-19 patients' conditions, the conclusions were that incidence of acute kidney injury and death was significantly higher in patients which had elevated baseline serum creatinine levels in comparison to the patients with normal baseline values [54]. The explanation for this was found in hematogenous spread and accumulation of the virus in the kidney, which causes renal cell necrosis. Further, CRP is used as a biomarker for different inflammatory and infectious conditions for clinical purposes. Elevated CRP is directly correlated

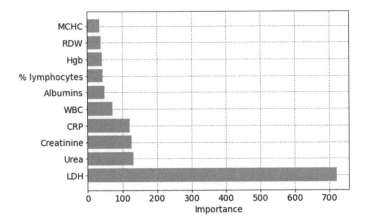

Fig. 4.10 Importance scores of ten best features

with the level of inflammation and therefore could be correlated with severity of clinical condition. Generally, several studies have found that serious or lethal effects of COVID-19 disease are linked to liver and kidney function markers, C reactive protein (CRP), interleukin-6 (IL-6), lower lymphocytes and albumin blood levels relative to milder form of the disease in the survivors [49].

This feature selection method reduces the amount of blood tests that need to be assessed. Values of all ten selected blood biomarkers are assessed for 34 patients on day 14 after the admission day by gradient boost regressor model. Due to the small amount of patients' data available on time, we decided to select these 34 patients with a full blood analysis for all days. Table 4.2 shows the root mean square error between predicted and actual values of blood analysis.

It can be seen that normal values of biomarkers such as white blood cell (WBC), % lymphocytes, MCHC, RDW, urea and albumins fall under the narrower range as

Table 4.2 Root mean squared error between predicted and actual values of blood biomarkers

Blood biomarker	RMSE
White blood cell (WBC)	3.03
% lymphocytes	1.57
MCHC	2.77
RDW	0.35
Hgb	10.12
Urea	2.54
Creatinine	55.23
Albumins	1.63
LDH	35.51
CRP	41.68

it is shown in Table 4.1, so we can expect smaller RMSE. On the other hand, Hgb, creatinine, LDH and CRP have greater deviation between real and predicted values. Results for each of these four biomarkers will be discussed individually.

In Fig. 4.11, a comparison between the actual and predicted values of Hgb is shown. RMSE is 10.12, which may be due to the fact that this biomarker had a wider range in patients in our study (from around 80 to around 150). Although RMSE for Hgb is larger in comparison to the other biomarkers' RMSE, greater deviation from the actual value is expected, as the range of actual values (80–150) is wider compared to the ranges of the other biomarkers (i.e. range of actual values of RDW is 11–18). For this reason, the value of RMSE does not affect the model's decision in assessment of clinical condition of COVID-19 patients.

In Fig. 4.12, the comparison between the actual and predicted values of CRP is given. RMSE for this analysis is 41.68, and explanation of this RMSE values is similar to the RMSE value of Hgb, where CRP has even wider range of values (0–400) than Hgb.

Figure 4.13 shows the comparison between the actual and predicted values of creatinine. This blood biomarker has the highest value of RMSE, and the cause may be the existence of some values that are drastically higher compared to the rest of the analyses, meaning that some outliers could be present in the dataset.

The same reason why we see greater RMSE for LDH analysis where comparison between predicted and actual values is shown in Fig. 4.14.

After the evaluation of the patients' hematology results and clinical biochemistry analyses, it is possible to predict the patients' clinical condition in advance, and this was achieved by the Xgboost classification algorithm. The model was tested on 34 patients and achieved an accuracy of 94% in predicting the patients' condition on

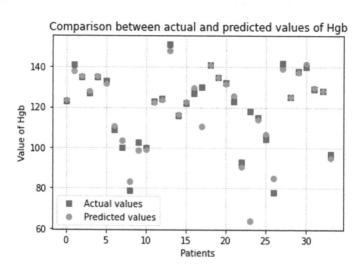

Fig. 4.11 Comparison between actual and predicted values of Hgb

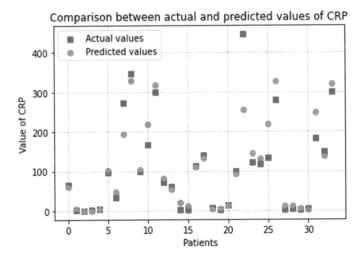

Fig. 4.12 Comparison between actual and predicted values of CRP

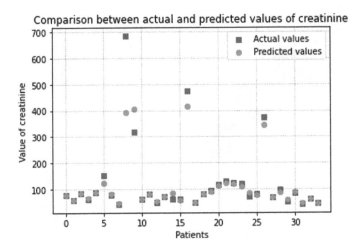

Fig. 4.13 Comparison between actual and predicted values of creatinine

day 14 after hospital admission. We computed a confusion matrix for this test set, which is shown in Fig. 4.15, both for normalized and regular number of patients.

Since the dataset is unbalanced, we considered other metrics such as precision, recall and F1-score in addition to accuracy. In Table 4.3, all of these metrics for each class individually are shown.

Our validated XGboost model is a rule-based model, consisting of hundreds of trees with known rules are known and stand as a reason behind the final decision. Figure 4.16 shows one such tree that is part of the XGboost classifier, the rules of which are understandable and based on if-then-else. This type of model is suitable

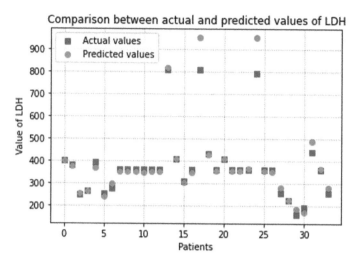

Fig. 4.14 Comparison between actual and predicted values of LDH

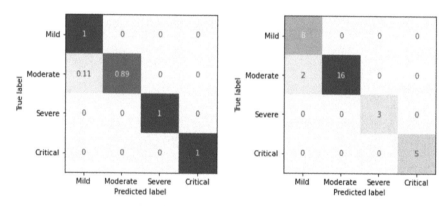

Fig. 4.15 Confusion matrix with normalized (left) and regular (right) values of patients

Table 4.3 Overview of classification metrics for each class on test data

Class	Accuracy	Precision	Recall	F1-score
Mild	1.00	0.80	1.00	0.89
Moderate	0.89	1.00	0.89	0.94
Severe	1.00	1.00	1.00	1.00
Critical	1.00	1.00	1.00	1.00

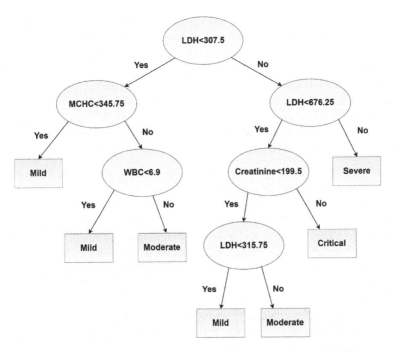

Fig. 4.16 An example of a tree which is a part of the XGBoost classification model

for application in clinical practice due to its comprehensibility. In this example of a tree, it can be seen that decisions about determination of a class of patient, indicating severity of clinical decision, are based on a single tree. Although one tree is not enough for final decision, when several trees (i.e. hundreds) are investigated and analyzed, high accuracy will be achieved for the problem of classification. It should be emphasized that this analysis is automated and, from a clinical point of view, the final decision is met within seconds.

4.4 Conclusions

The main task of this chapter was to assess how COVID-19 develops over time in patients. A unique feature of COVID-19 interstitial pneumonia is an abrupt progression to respiratory failure. Our patient-specific finite element lung model focuses on the spread of virus-laden bioaerosols to many regions of the lungs from the initial site of infection. The airway coupled to a parenchyma model was modeled using the finite element method. The results show virus spread starting from the alveoli to the other parts of the airway. It is shown that the finite element tool can enhance COVID-19 patient treatment with the use of mechanical ventilation.

In this chapter, we proposed the use of automatic ML methods for classification of COVID-19 patients into several categories and prediction of the outcome of the disease in advance (change of severity of clinical condition). Our research analyzes the possibility for prediction of the change of category in advance, which means that the model predicts the patient's clinical course during hospital stay until discharge or death. It represents a proof of concept that the ML model is an efficient and informative method to gain insight into the COVID-19 disease process. From extensively implemented algorithms and hyper-parameter optimization, the following conclusions can be drawn:

- Ten most important variables (features) from blood analysis are extracted and they are strongly associated with patient outcomes.
- Gradient boost regressor proved to be the best in predicting blood biomarkers, achieving root mean square error for WBC, lymphocytes, MCHC, RDW, urea, albumins less than 3, and some larger RMSE for Hgb, creatinine, LDH and CRP, which can be explained with wider ranges of those biomarkers.
- The proposed methodology using XGboost classifier is adequate for classification of patients into 4 distinct categories of clinical conditions (mild, moderate, severe and critical) with 94% of accuracy.
- Since the proposed methodology is rule-based (XGboost can be described using IF-than rules) rather than black box, it is more suitable for implementation in real clinical practice.
- Coupled unsupervised and supervised algorithms are able to predict disease progression (mild to moderate, moderate to severe, severe to critical clinical condition) in advance.

This hybrid approach, fusing finite element and ML methods, can open a new strategy for prediction of virus spreading in specific patients. Further analyses are necessary to go in this direction.

Acknowledgements This research is funded by Serbian Ministry of Education, Science, and Technological Development [451-03-68/2020-14/200107 (Faculty of Engineering, University of Kragujevac)]. This research is also supported by the project that have received funding from the European Union's Horizon 2020 research and innovation programme under grant agreement No 952603 (SGABU project). This article reflects only the author's view. The Commission is not responsible for any use that may be made of the information it contains.

References

1. T. Ai, Z. Yang, H. Hou, C. Zhan, C. Chen, W. Lv, Q. Tao, Z. Sun, L. Xia, Correlation of chest CT and RT-PCR testing for coronavirus disease 2019 (COVID-19) in China: a report of 1014 cases. Radiology **296**(2), E32–E40 (2020). https://doi.org/10.1148/radiol.2020200642
2. N. Altman, M. Krzywinski, Points of significance: clustering. Nat. Methods **14**, 545–546 (2017)

3. R.M. Anderson, H. Heesterbeek, D. Klinkenberg, T.D. Hollingsworth, How will country-based mitigation measures influence the course of the COVID-19 epidemic? The Lancet (2020). https://doi.org/10.1016/S0140-6736(20)30567-5

4. J.K. Aronson, R.E. Ferner, Biomarkers—a general review. Curr. Protoc. Pharmacol. **76**(1), 9–23 (2017)

5. R. Assandri, E. Buscarini, C. Canetta, A. Scartabellati, G. Viganò, A. Montanelli, Laboratory Biomarkers predicting COVID-19 severity in the Emergency room. Arch. Med. Res. **51**(6), 598–599 (2020)

6. P. Baccam, C. Beauchemin, M. Ca, F.G. Hayden, A.S. Perelson, Kinetics of influenza A virus infection in humans. J. Virol. **80**(15), 7590–7599 (2006)

7. H.X. Bai et al., AI augmentation of radiologist performance in distinguishing COVID-19 from pneumonia of other etiology on chest CT. Radiology **296**(3):201491 (2020)

8. H. Banchner, (chief editor of JAMA) interview with Dr. M. Gong https://www.youtube.com/watch?v=TH9skp5R9F4. Accessed 1 Apr 2021

9. J. Bates, *Lung Mechanics: An Inverse Modeling Approach* (Cambridge University Press, Cambridge, 2009). https://doi.org/10.1017/CBO9780511627156

10. G. Benelli et al., SARS-COV-2 comorbidity network and outcome in hospitalized patients in Crema. Italy. medRxiv (2020). https://doi.org/10.1101/2020.04.14.20053090

11. N. Chen, M. Zhou, X. Dong et al., Epidemiological and clinical characteristics of 99 cases of 2019 novel coronavirus pneumonia in Wuhan, China: a descriptive study. Lancet **395**(10223), 507–513 (2020)

12. T. Chen, C. Guestrin, Xgboost: a scalable tree boosting system. Proceedings of the 22nd ACM SIGKDD international conference on knowledge discovery and data mining, California, San Francisco, USA, August 2016. Association for Computing Machinery New York, NY, United State, p 785–794.

13. A. Cheng et al., Diagnostic performance of initial blood urea nitrogen combined with D-dimer levels for predicting in-hospital mortality in COVID-19 patients. Int. J. Antimicrobial Agents **56**(3), 106110 (2020)

14. Y. Cheng et al., Kidney disease is associated with in-hospital death of patients with COVID-19. Kidney Int. **97**(5), 829–838 (2020). https://doi.org/10.1016/j.kint.2020.03.005

15. A. Cho, AI systems aim to sniff out coronavirus outbreaks. Science **368**(6493), 810–811 (2020)

16. S.M. Ciupe, J.M. Heffernan, In-host modeling. Infectious Disease Model. **2**(2), 188–202 (2017). https://doi.org/10.1016/j.idm.2017.04.002

17. Y. Fang, H. Zhang, J. Xie, M. Lin, L. Ying, P. Pang, W. Ji, Sensitivity of chest CT for COVID-19: comparison to RT-PCR. Radiology **296**(2), 200432 (2020)

18. N. Filipovic, B. Gibney, M. Kojic, D. Nikolic, V. Isailovic, Y. Alexandra, M. Konerding, S. Mentzer, A. Tsuda, Mapping cyclic stretch in the postpneumonectomy murine lung. J. Appl. Physiol. **115**(9), 1370–1378 (2013). https://doi.org/10.1152/japplphysiol.00635.2013

19. N. Filipovic, B. Gibney, D. Nikolic, M. Konerding, S. Mentzer, A. Tsuda, Computational analysis of lung deformation after murine pneumonectomy. Comput. Methods Biomech. Biomed. Engin. **17**(8), 838–844 (2012). https://doi.org/10.1080/10255842.2012.719606

20. N. Filipovic, I. Saveljic, K. Hamada, A. Tsuda, Abrupt Deterioration of COVID-19 patients and spreading of SARS COV-2 virions in the lungs. Ann. Biomed. Eng. **48**, 2705–2706 (2020). https://doi.org/10.1007/s10439-020-02676-w

21. Y. Gao et al., Machine learning based early warning system enables accurate mortality risk prediction for COVID-19. Nat. Commun. **11**(1), 1–10 (2020). https://doi.org/10.1038/s41467-020-18684-2

22. GitHub "COVID X ray-dataset," 2020. (Online). Available: https://github.com/ieee8023/covid-chestxray-dataset. Accessed 19 Apr 2020

23. P. Goyal et al., Clinical characteristics of Covid-19 in New York city. N. Engl. J. Med. **382**, 2372–2374 (2020). https://doi.org/10.1056/nejmc2010419

24. F. Graw, A.S. Perelson, Modeling viral spread. Annu. Rev. Virol. **3**, 555–572 (2016). https://doi.org/10.1146/annurev-virology-110615-042249

25. W. Guan et al. Comorbidity and its impact on 1590 patients with Covid-19 in China: a Nation-wide analysis. Europ. Respiratory J. **55**(5), 2000547 (2020). https://doi.org/10.1183%2F1399 3003.00547-2020

26. W.J. Guan, Z.Y. Ni, Y. Hu, W.H. Liang, C.Q. Ou, J.X. He et al., Clinical characteristics of Coronavirus Disease 2019 in China. N. Engl. J. Med. **382**, 1708–1720 (2020). https://doi.org/ 10.1056/NEJMoa2002032

27. A. Handel, I.M. Longini, R. Antia, Neuraminidase inhibitor resistance in influenza: assessing the danger of its generation and spread. PLoS Comput. Biol. **3**(12), 2456–2464 (2007)

28. D.M. Hansell, A.A. Bankier, H. MacMahon, T.C. McLoud, N.L. Müller, J. Remy, Fleischner society: glossary of terms for thoracic imaging. Radiology **246**, 697–722 (2008)

29. T. Hastie, R. Tibshirani, J. Friedman, *The Elements of Statistical Learning: Data Mining, Inference, and Prediction*. Springer Science & Business Media (2009)

30. E.A. Hernandez-Vargas, Modeling and Control of Infectious Diseases: with MATLAB and R. 1st ed. Elsevier Academic Press (2019)

31. E.A. Hernandez-Vargas, R.H. Middleton, Modeling the three stages in HIV infection. J. Theor. Biol. **320**, 33–40 (2013)

32. E.A. Hernandez-Vargas, E. Wilk, L. Canini, F.R. Toapanta, S.C. Binder, A. Uvarovskii et al., Effects of aging on influenza virus infection dynamics. J. Virol. **88**(8), 4123–4131 (2014)

33. C. Huang et al. Model stability of COVID-19 mortality prediction with biomarkers (2020). medRxiv https://doi.org/10.1101/2020.07.29.20161323

34. C. Huang, Y. Wang, X. Li et al., Clinical features of patients infected with 2019 novel coronavirus in Wuhan China. Lancet **395**(10223), 497–506 (2020). https://doi.org/10.1016/S0140-6736(20)30183-5

35. R.J. Jose, A. Manuel, COVID-19 cytokine storm: the interplay between inflammation and coagulation. Lancet Respir Med. **8**(6), e46–e47 (2020). https://doi.org/10.1016/S2213-2600(20)302 16-2

36. V. Jurisic, S. Radenkovic, G. Konjevic, The actual role of LDH as tumor marker, biochemical and clinical aspects. Adv. Cancer Biomarkers **867**, 115–124 (2015). https://doi.org/10.1007/ 978-94-017-7215-0_8

37. Kaggle "COVID CT," Kaggle, 2020. (Online). Available: https://www.kaggle.com/luisblanche/ covidct. Accessed 19 Apr 2020

38. Kaggle "COVID-19 Open Research Dataset Challenge," 2020. (Online). Available: https:// www.kaggle.com/allen-institute-for-ai/CORD-19-research-challenge. Accessed 19 Apr 2020

39. Kaggle, "Chest X ray pneumonia," 2019. (Online). Available: https://www.kaggle.com/paulti mothymooney/chest-xray-pneumonia. Accessed 19 Apr 2020

40. H.J. Koo, S. Lim, J. Choe, S.H. Choi, H. Sung, K.H. Do, Radiographic and CT features of viral pneumonia. Radiographics **38**, 719–739 (2018)

41. J. Lei, J. Li, X. Li, X. Qi, CT imaging of the 2019 novel coronavirus (2019-nCoV) pneumonia. Radiology (2020). https://doi.org/10.1148/radiol.2020200236:200236

42. X. Li et al., Clinical characteristics of 25 death cases with COVID-19: a retrospective review of medical records in a single medical center, Wuhan, China. Int. J. Infect. Dis. **94**, 128–132 (2020). https://doi.org/10.1016/j.ijid.2020.03.053

43. X. Li et al., Risk factors for severity and mortality in adult COVID-19 inpatients in Wuhan. J. Allergy Clin. Immunol. **146**, 110–118 (2020). https://doi.org/10.1016/j.jaci.2020.04.006

44. N. Lin, W. Yu, J. Duncan, Combinative multi-scale level set framework for echocardiographic image segmentation. Med. Image Anal. **7**, 529–537 (2003)

45. Q. Liu, Y.H. Zhou, Z.Q. Yang, The cytokine storm of severe influenza and development of immunomodulatory therapy. Cell. Mol. Immunol. **13**, 3–10 (2016)

46. Y. Liu et al., Neutrophil-to-lymphocyte ratio as an independent risk factor for mortality in hospitalized patients with COVID-19. J. Infect. **81**(1), e6–e12 (2020). https://doi.org/10.1016/ j.jinf.2020.04.002

47. R. Lu, X. Zhao, J. Li et al., Genomic characterisation and epidemiology of 2019 novel coronavirus: implications for virus origins and receptor binding. Lancet **395**(10224), 565–574 (2020)

48. A. Malashenko, A. Tsuda, S. Haber, Propagation and breakup of liquid menisci and aerosol generation in small airways. J. Aerosol Med. Pulm Drug Deliv. **22**(4), 341–353 (2009)
49. P. Malik, U. Patel, D. Mehta et al., Biomarkers and outcomes of COVID-19 hospitalisations: systematic review and meta-analysis. BMJ Evidence-Based Med. (2020). https://doi.org/10.1136/bmjebm-2020-111536
50. X. Mei et al., Artificial intelligence-enabled rapid diagnosis of patients with COVID-19 (2020). medRxiv. https://doi.org/10.1101/2020.04.12.20062661
51. H. Meng, R. Xiong, R. He, W. Lin, B. Hao, L. Zhang, Z. Lu, X. Shen, T. Fan, W. Jiang, W. Yang, T. Li, J. Chen, G. Qing, CT imaging and clinical course of asymptomatic cases with COVID-19 pneumonia at admission in Wuhan China. J Infect **81**(1), e33–e39 (2020). https://doi.org/10.1016/j.jinf.2020.04.004
52. D.R. Milovanovic, S.M. Jankovic, D. Ruzic Zecevic, M. Folic, N. Rosic, D. Jovanovic, D. Baskic, R. Vojinovic, Z. Mijailovic, P. Sazdanovic, LEČENJE KORONAVIRUSNE BOLESTI (COVID-19). Medicinski casopis **54**(1), 23–43 (2020)
53. S. Moradi, M.G. Oghli, A. Alizadehasl, I. Shiri, N. Oveisi, M. Oveisi, M. Maleki, J. Dhooge, MFP-Unet: A novel deep learning based approach for left ventricle segmentation in echocardiography. Physica Med. **76**, 58–69 (2019). https://doi.org/10.1016/j.ejmp.2019.10.001
54. M.Y. Ng, E.Y. Lee, J. Yang, F. Yang, X. Li, H. Wang, M.M. Lui, C.S.Y. Lo, B. Leung, P.L. Khong, C.K.M. Hui, M.D. Kuo, Imaging profile of the COVID-19 infection: radiologic findings and literature review. Radiology: Cardiothoracic Imaging 2(1):e200034 (2020). https://doi.org/10.1148/ryct.2020200034
55. V.K. Nguyen, S.C. Binder, A. Boianelli, M. Meyer-Hermann, E.A. Hernandez-Vargas, Ebola virus infection modeling and identifiability problems. Front. Microbiol. **6**, 1–11 (2015)
56. V.K. Nguyen, E.A. Hernandez-Vargas, Windows of opportunity for Ebola virus infection treatment and vaccination. Sci. Rep. **7**(1), 8975 (2017)
57. M. Panteghini, Lactate dehydrogenase: an old enzyme reborn as a COVID-19 marker (and not only). Clinical Chemistry and Laboratory Medicine (CCLM), 1(ahead-of-print) (2020)
58. K.A. Pawelek, D. Dor, C. Salmeron, A. Handel, Within-host models of high and low pathogenic influenza virus infections: The role of macrophages. PLoS ONE **11**(2), 1–16 (2016). https://doi.org/10.1371/journal.pone.0215700
59. A.S. Perelson, Modelling viral and immune System dynamics. Nat. Rev. Immunol. **2**(1), 28–36 (2002)
60. A.S. Perelson, R.M. Ribeiro, Modeling the within-host dynamics of HIV infection. BMC Biol. **11**(1), 96 (2013). https://doi.org/10.1186/1741-7007-11-96
61. M. Pinkevych, S.J. Kent, M. Tolstrup, S.R. Lewin, D.A. Cooper, O.S. Sgaard et al., Modeling of experimental data Supports HIV reactivation from latency after treatment interruption on average once Every 5–8 Days. PLoS Pathog. **12**(8), 8–11 (2016). https://doi.org/10.1371/journal.ppat.1005745
62. T.C. Reluga, H. Dahari, A.S. Perelson, Analysis if Hepatitis C Virus infection models with Hepatocyte Homeostasis. SIAM J. Appl. Math. **69**(4), 999–1023 (2009)
63. L. Rong, A.S. Perelson, Modeling HIV persistence, the latent reservoir, and viral blips. J. Theor. Biol. **260**(2), 308–331 (2009). https://doi.org/10.1016/j.jtbi.2009.06.011
64. O.P. Ronneberger, Fischer, T. Brox, U-net: Convolutional networks for biomedical image segmentation, in N. Navab, J. Hornegger, W. Wells, A. Frangi (eds.) *Medical Image Computing and Computer-Assisted Intervention – MICCAI 2015*. Lecture Notes in Computer Science, vol 9351. Springer, Cham (2015). https://doi.org/10.1007/978-3-319-24574-4_28
65. Q. Ruan et al., Clinical predictors of mortality due to COVID-19 based on an analysis of data of 150 patients from Wuhan China. Intensive Care Med. **46**(5), 846–848 (2020)
66. P. Sarzi-Puttini et al., COVID-19, cytokines and immunosuppression: what can we learn from severe acute respiratory syndrome? Clin. Exp. Rheumatol. **38**(2), 337–342 (2020)
67. Y. Shi, Y. Wang, C. Shao, J. Huang, J. Gan, X. Huang, E. Bucci, M. Piacentini, G. Ippolito, G. Melino, COVID-19 infection: the perspectives on immune responses. Cell Death Differ. **27**(5), 1451–1454 (2020). https://doi.org/10.1038/s41418-020-0530-3

68. I. Shiri et al., Direct attenuation correction of brain PET images using only emission data via a deep convolutional encoder-decoder (Deep-DAC). Eur. Radiol. **29**(12), 6867–6879 (2019). https://doi.org/10.1007/s00330-019-06229-1

69. I. Shiri et al., PSFNET: ultrafast generation of PSF-modelledlike PET images using deep convolutional neural network. J. Nucl. Med. **60**(supplement 1), 1369 (2019)

70. I. Shiri et al., Simultaneous Attenuation correction and reconstruction of PET images using deep convolutional encoder decoder networks from emission data. J. Nucl. Med. **60**(supplement 1), 1370 (2019)

71. E. Smistad, A. Ostvik, B. Haugen, L. Lovstakken, 2D left ventricle segmentation using deep learning. 2017 IEEE International Ultrasonics Symposium (IUS). Washington, DC, USA **2017**, 1–4 (2017). https://doi.org/10.1109/ULTSYM.2017.8092573

72. R. Storn, Price K (1997) Differential evolution—a simple and efficient heuristic for global optimization over continuous spaces. J. Global Optim. **11**, 341–359 (1997). https://doi.org/10.1023/A:1008202821328

73. L. Tan et al., Lymphopenia predicts disease severity of COVID-19: a descriptive and predictive study. Signal Transduct. Target. Ther. **5**(1), 1–3 (2020)

74. Tensorflow, (Online). Available: https://www.tensorflow.org/. Accessed 5 Oct 2019

75. J.R. Tisoncik, M.J. Korth, P. Cameron, F.J. SimmonsnCP, T.R. Martin, M.G. Katzea, Into the Eye of the Cytokine Storm. Microbiol Mol Biol Rev. **76**(1), 16–32 (2012)

76. A. Tsuda, N. Filipovic, D. Haberthür, R. Dickie, Y. Matsui, M. Stampanoni, J.C. Schittny, Finite element 3D reconstruction of the pulmonary acinus imaged by synchrotron X-ray tomography. J. Appl. Physiol. **105**, 964–976 (2008)

77. A. Tsuda, F.S. Henry, S. Haber, D. Haberthür, N. Filipovic, D. Milasinovic, J. Schittny, The simultaneous role of an alveolus as flow mixer and flow feeder for the deposition of inhaled submicron particles. J. Biomech. Eng. **134**(12)121001 (2012). https://doi.org/10.1115/1.4007949

78. D.A.J. Tyrrell, S.H. Myint, Coronaviruses. University of Texas Medical Branch at Galveston (1996). Available from: http://www.ncbi.nlm.nih.gov/pubmed/21413266.

79. C. Vannier, Modélisation mathématique du poumon humain. Ph.D. thesis Hal Id: tel-00739462 (2012)

80. D. Wang, B. Hu, C. Hu et al., Clinical characteristics of 138 hospitalized patients With 2019 novel coronavirus-infected Pneumonia in Wuhan, China. Jama **323**(11):1061–1069 (2020). https://doi.org/10.1001/jama.2020.1585

81. Webpage: "Latest informations about COVID-19 in the Republic of Serbia," 8 December 2020. Available: https://covid19.rs/. Accessed 8 December 2020.

82. A.R. Weibel, *Morphometry of the Human Lung.* Springer (1963)

83. R. Woelfel, V.M. Corman, W. Guggemos, M. Seilmaier, S. Zange, M.A. Mueller et al. Clinical presentation and virological assessment of hospitalized cases of coronavirus disease 2019 in a travel-associated transmission cluster (2020). medRxiv. https://doi.org/10.1101/2020.03.05.20030502

84. World Health Organization. (2020) Weekly Operational Update on COVID-19. Accessed 7 December 2020

85. World Health Organization. (2020). Coronavirus disease 2019 (COVID-19): situation report, 46. World Health Organization. https://apps.who.int/iris/handle/10665/331443

86. World Health Organization. (2020). Coronavirus disease 2019 (COVID-19): situation report, 68. World Health Organization. https://apps.who.int/iris/handle/10665/331614

87. D. Wormanns, O.W. Hamer, Glossary of terms for thoracic imaging–German version of the Fleischner Society recommendations. Rofo **187**(8), 638–661 (2015)

88. G. Wu et al., Development of a clinical decision support system for severity risk prediction and triage of COVID-19 patients at hospital admission: an international multicenter study. Europ. Respiratory J. **56**(2) (2020). https://doi.org/10.1183/13993003.01104-2020

89. L. Wynants et al., Prediction models for diagnosis and prognosis of covid-19: systematic review and critical appraisal. BMJ 369 (2020), m1328. https://doi.org/10.1136/bmj.m1328

90. Z. Xu, L. Shi, Y. Wang, J. Zhang, L. Huang, C. Zhang et al., Pathological findings of COVID-19 associated with acute respiratory distress syndrome. Lancet Respiratory Med. **8**(4), 420:422 (2020). https://doi.org/10.1016/S2213-2600(20)30076-X

91. L. Yan et al., An interpretable mortality prediction model for COVID-19 patients. Nature Mach. Intell. **2**, 283–288 (2020). https://doi.org/10.1038/s42256-020-0180-7

92. L. Yan et al., A machine learning-based model for survival prediction in patients with severe COVID-19 infection. MedRxiv (2020). https://doi.org/10.1101/2020.02.27.20028027

93. X. Yang, Y. Yu, J. Xu et al., Clinical course and outcomes of critically ill patients with SARS-CoV-2 pneumonia in Wuhan, China: a single-centered, retrospective, observational study. Lancet Respir. Med. **8**(5), 475–481 (2020). https://doi.org/10.1016/S2213-2600(20)30079-5

94. Z.H. Zhai, *Cell Biology (in Chinese)* (Higher Education Press, Beijing, 1997)

95. F. Zhou, T. Yu, R. Du et al., Clinical course and risk factors for mortality of adult inpatients with COVID-19 in Wuhan, China: a retrospective cohort study. Lancet **395**(10229), 1054–1062 (2020). https://doi.org/10.1016/S0140-6736(20)30566-3

96. P. Zhou, X.L. Yang, X.G. Wang et al., A pneumonia outbreak associated with a new coronavirus of probable bat origin. Nature (2020). https://doi.org/10.1038/s41586-020-2012-7

97. S. Zhou, Y. Wang, T. Zhu, L. Xia, CT Features of Coronavirus Disease 2019 (COVID-19) Pneumonia in 62 Patients in Wuhan. China. Am. J. Radiol. **214**(6), 1287–1294 (2020). https://doi.org/10.2214/ajr.20.22975

98. L. Zou, F. Ruan, M. Huang, L. Liang, H. Huang, Z. Hong et al., SARS-CoV-2 viral load in upper respiratory specimens of infected patients. N. Engl. J. Med. **382**, 1177–1179 (2020). https://doi.org/10.1056/NEJMc2001737

Part II
Advances in Assistive Technologies in Medical Diagnosis

Chapter 5
Towards Personalized Nutrition Applications with Nutritional Biomarkers and Machine Learning

Dimitrios P. Panagoulias, Dionisios N. Sotiropoulos, and George A. Tsihrintzis

Abstract The doctrine of the "one-size-fits-all" approach has been overcome in the field of disease diagnosis and patient management and replaced by a more per patient approach known as *personalized medicine*. Biomarkers are the key variables (*features*) in the research and development of new methods of training prognostic models and neural networks in the scientific field of machine learning and artificial intelligence. Important biomarkers related to metabolism are the *metabolites*. On the other hand, *metabolomics* is a term that refers to the systematic study of unique chemical fingerprints left behind by specific cellular processes. The metabolic profile can provide a snapshot of cell physiology, and, by extension, metabolomics provide a direct "functional reading of the physiological state" of an organism. The goal of this chapter is to formulate a general evaluation chart of nutritional biomarkers, to investigate how to best predict body mass index and to discover dietary patterns with the deployment of neural networks.

5.1 Introduction

5.1.1 Summary

The one-size-fits-all approach has been overcome in the field of patient management and disease classification and has been replaced by a more per patient approach better known as *personalized medicine*.

The most precise way to evaluate patients and personalize patient management, disease classification and medication strategy is through biomarker identification. Biomarkers are medical signs that in a broad sense "provide objective indications of the medical state observed from outside the patient and can be measured accurately

D. P. Panagoulias · D. N. Sotiropoulos · G. A. Tsihrintzis (✉)
Department of Informatics, University of Piraeus, Piraeus 185 34, Greece
e-mail: geoatsi@unipi.gr

D. N. Sotiropoulos
e-mail: dsotirop@unipi.gr

and reproducibly" [76]. Thus, biomarkers include molecules, proteins, antibodies, as well as the biochemical changes discovered in tissue, blood or other bodily fluids, and they indicate the presence or the future expression of a specific disease, condition or/and mutation [27, 28]. In other words, biomarkers can act as a metric for today or as a prognosticator for tomorrow. Biomarkers can be produced by a sole metric from an examination or observation or by a combination of metrics. For example, the well-known body mass index (BMI) is a combined biomarker derived from the mass and height of an individual.

Biomarkers, however, do not present a new reality as they are identified and have been used in medicine for more than 20 years. But why are they important and how can they contribute to improving medical outcomes? At least in theory, the discovery of novel biomarkers will lead to the mapping of more accurate and more effective medical pathways and, thus, reduce costs and optimize health.

At the same time, important social, ethical and economic issues that surround this field may arise and should be thoroughly investigated, discussed and eventually resolved. First, the complexity of the disease biology and of that of the human body could minimize or/and restrict the proper use of pathways discovered through the study of biomarkers. Secondly, the high and, in cases, prohibitive cost of some treatments of cancer, combined with limited available resources, create objective problems on how to best distribute those resources, which groups to prioritize, with what criteria to make selections, and which the corresponding repercussions and outcomes will be. Since the research and the data are dynamic a lot of the issues discussed should be continuously examined and re-evaluated.

Some important biomarkers that are related with metabolism are the *metabolites*. *Metabolomics* refers to the systematic study of the unique chemical fingerprints (metabolites) produced by specific chemical procedures. The metabolic profile is a snapshot of the physiology of the cell and, by extension, metabolomics offers a direct "functional read of the normal state" of an organism [77].

On the other hand, obesity represents a major health problem, not only in the western world, but increasingly in low- and middle-income countries. In order to develop successful weight loss strategies, it is important to understand the molecular pathogenesis of weight change. Some pathways, such as oxidative stress and the basic regulatory framework of insulin, have been implicated in weight gain and energy expenditure regulation. In addition, the role of metabolites produced by the intestinal microbe, particularly short-chain fatty acids, has been highlighted [14].

5.1.2   Chapter Synopsis

A summary of each section of this chapter is outlined for the convenience of the reader.

In Sect. 5.2, some basic concepts regarding biomarkers, personalized medicine and related terms are analyzed for the better understanding of the examined subject.

- **Personalized medicine**: Personalization in medicine aims to provide key *per case* solutions instead of a single solution for all, as such a single solution, when applied, could lead to inadequate or excessive medical treatment.
- **Biomarkers**: The optimal pathway to personalization is through the use of biomarkers. Biomarkers are molecules (proteins or antibodies) or biochemical changes (gene expressions and mutations) found in patients' tissues, blood, or other bodily fluids and are indicators of a specific disease, condition, or mutation in the body.
- **Obesity**: A higher energy intake compared to total energy expended over a long period of time leads to obesity. Although many people, especially in western countries, are overweight or obese, a significant proportion of people with normal weight never become overweight or obese, partly reflecting the large cross-sectional variation in excess caloric intake.
- **Metabolomics** is the systematic study of the unique chemical fingerprints that specific cellular processes leave behind. The metabolic profile can give a snapshot of the physiology of the cell and, by extension, metabolomics provide a direct "functional reading of the physiological state" of an organism.
- **Metabolites** are the substrates, intermediates and products of the metabolic process. A metabolite is usually defined as any molecule less than 1.5 kDa in size. However, there are exceptions to this, depending on the sample and the detection method. For example, macromolecules such as lipoproteins and albumin are reliably detected in plasma NMR-based metabolic studies.
- The goal of **nutritional genomics** is to identify genetic variants that may be important in understanding genetic responses to nutrition.
- **Standard biochemistry profile**: The biochemical profile results from targeted blood tests determine the functional capacity of important organs such as the liver and kidneys. It also determines the difference in the metabolism of proteins and fats, the metabolism of carbohydrates, the production of enzymes, the ability to use compounds, etc.
- **Neural networks** solve classification problems in nonlinearly separable sets using hidden layers in which the neurons are not directly connected to the output. The additional hidden layers can be interpreted geometrically as additional super-levels, which enhance the separation capacity of the network.

In Sect. 5.3, the technical tools and datasets are analyzed that will be used to train neural networks, as well as all utilized related methods, libraries and programming frameworks.

In Sect. 5.4, the results of the executable program and the settings of the neural network are discussed.

Finally, Sect. 5.5 contains some perspectives on the findings of this work and a preview of future works and implementations planned based on this report.

5.1.3   Goals and Perspective

The goal of this work is to capture a general evaluation diagram of nutritional biomarkers, to investigate the predictability of body mass index and to identify dietary patterns using neural networks.

The analysis is conducted through the theoretical description of the problem and the modelling of the solution through the automated application of machine learning and pattern recognition algorithms. Extensive literature has been used to build a basic familiarity with the subject. Python and the appropriate libraries have been deployed.

5.2   Basic Concepts

5.2.1   Personalized Medicine

The efficiency of a medical system is measured by its ability to anticipate and meet the health care needs of the people, ensuring high quality with low cost and utilizing all available financial, technological and human resources [8]. Medical science as well as economics is based on empirical and factual data to make decisions and to form models so that socially just and beneficial solutions can be provided.

In the last 20 years, the medical industry has undergone a significant change. The doctrine of one-treatment-for-all has been overcome by a per-patient approach. Personalized medicine aims to provide key solutions, replacing practices that often led to inadequate or excessive treatment of patients. The goal of medicine today is to adapt treatments to subgroups of patients who share similar genetic characteristics [9, 10].

Comparative effectiveness approaches to evidence-building, including measurements of health outcomes, quality of life, financial analysis, decision modeling, and actual clinical trials, can be used to provide stake holders with a range of information about treatment, guidelines and coverage and assist compensation decisions.

The European Commission under the Luxembourg Presidency has made personalized medicine one of its health priorities and adopted it on 7 December 2015 [19]. The corresponding conclusions are summarized as follows: "personalized medicine refers to a medical modelling that uses the characterization of phenotypes and genotypes of individuals (e.g. molecular profile, medical imaging, lifestyle data) to adapt the correct treatment strategy to the appropriate person at the right time and/or to determine the predisposition of a disease and/or to provide timely and targeted prevention of a disease. Under this premise, the Council calls upon the Member States and the Commission, through STAMP, to analyze issues related to the implementation of European Union legislation in the field of medicine in order to identify ways of maximizing the effective use of existing European Union regulatory instruments. To further improve safe and timely access to medicine, including innovative medicines, and continue to monitor the progress of this adaptive pilot undertaking by the European Medicines Agency and its potential to enable the timely approval of a medicinal product for use in well-defined and targeted population with high medical needs" [9].

As a development of this initiative, the organization ERA perMed was founded and consists of 32 research and investment organizations from 23 countries under the co-financing of the European Commission. Its purpose is to align and fund research strategies, strengthen the competitiveness of European participants in the effort towards personalizing medicine and strengthening the related results [42].

For optimization of execution and for the recognition of the relevant needs, an institutional framework has been developed which consists of 2 main pillars:

A.   Clinical research.

The individualization of the treatment dictates the unification of all health data (medical, imaging, lifestyle) so that essential and effective interventions can be carried out.

In the area of clinical trials there are several challenges that need to be assessed for the enhancement of innovation. Several new, more flexible alternative clinical trials and data generation approaches, statistical methods and analysis tools need to be considered.

B.   Approval.

The joint development of medicines and of medical devices is a complex matter mainly due to the different regulatory systems of the member states of the European Union.

It is worth noting that these drugs are potentially eligible for a number of regulatory pathways/incentives, namely the use of a conditional authorization, and approval in exceptional cases and where standard development is not possible, through rapid evaluation or ten-year exclusivity for orphan drugs for rare diseases. International harmonization of regulatory requirements could also facilitate the acceptance of medical products across borders and should be promoted.

Equally important is the education of health professionals in the field of personalized medicine. In this context, the precise description of the drug characteristics, the call for genetic testing and the exact formulation of the related mutations for the effective use of the drugs are considered particularly important. For example, the molecular characterization of pathogens may be important before prescribing the right antibiotic to patients and, thus, avoiding subsequent resistance to multiple drugs.

Throughout this process, it is necessary to synchronize the pharmaceutical, medical and supervisory bodies, place quality control and thresholds for drug effectiveness and create a framework for rapid interventions.

Patients may respond differently to the same treatment. Indeed, while one treatment brings the desired result in one group of patients, it may not change the condition of other groups at all or lead to adverse consequences as illustrated in Fig. 5.1. The reason is that the genetic composition and metabolic profile of each patient is crucial

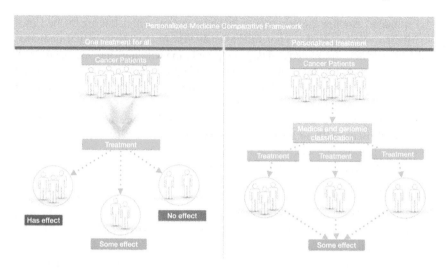

Fig. 5.1 Comparative framework

to the effect of a drug. In personalized medicine, these individual patterns of cellular and metabolic factors are taken into account. The diagnosis of biological markers divides patients into groups with similar characteristics and provides information for optimum individual treatment.

Positive effects in a wider scale may arise through personalized medicine and give way for more effective interventions in local populations based on particular social and lifestyle characteristics. A better understanding of the impact of age, gender, nationality and lifestyle will enable smarter and more efficient allocation of the limited available resources. This presupposes the creation of synergies between different scientific disciplines and the active participation of patients. The rapid development of technology provides and will continue to provide to a greater extent the appropriate environment for the improvement of health provisions. Genomic technologies are already an important part of modern medicine and will hold a more prominent role in the years to come.

Using genetic profiles to identify the best possible drug and treatment and reducing side effects are among the key goals that will benefit the patient. Preventive, personalized and participatory medicine will also help identify the right drug for the right patient at the right time, avoiding the prescription of costly and ineffective drugs and preventing potentially harmful side effects.

At this point it is important for the following assumptions to be made:

A. Medicine is an information-based science.
B. A systematic approach is inextricably linked to the study of the enormous complexity of diseases and its consequent analysis,
C. Mathematical models and the ever-increasing computing power will allow the analysis of very extensive data at speeds that will produce actionable insights.

The widespread application of personalized medical approaches faces two major challenges, namely, the **complex biology** of the human body and the **complex (macro- and micro-) economics** [7] of the medical field and medicine in general [3]. Access to basic health care is a universal and fundamental right. However, the limitations in resources, the high production and distribution cost of drugs and medical applications and the soaring financial and time cost of research activities are often obstacles to making such modern practices more widely available and to establishing a new medical doctrine through the creation of a new social contract [5]. For that reason, it is of great importance that the patients are made to understand the positive effects of such approaches and their inclusion in the **related decisions should be well defined and encouraged** [72].

5.2.1.1 Biomarkers

The most precise way to evaluate patients and personalize patient management, disease classification and medication strategy is through biomarker identification. Biomarkers are medical signs that in a broad sense "provide objective indications of the medical state observed from outside the patient and can be measured accurately and reproducibly" [66].

Biomarkers are molecules (proteins or antibodies) or biochemical changes (gene expressions and mutations) found in patients' tissues, blood, or other bodily fluids that indicate the presence of a specific disease, condition, or mutation in the body.

Biomarkers can be predictive and/or prognostic. They are used to monitor and predict the condition or development of patients' health. They are also utilized to evaluate treatment and a patient's response to a specific medication [27].

In theory, novel biomarker discovery will help to predict and manage the various manifestations of cancer more efficiently and to respond to disease variations proactively. Towards this, the European Union and the US Government have increased funding for ongoing cancer biomarker research as only a small number of biomarkers have clinical application as of today.

In 1998, the working group of the American National Institutes of Health outlined the following biomarker definition: "the characteristic that is objectively measured and evaluated as an indicator of normal biological processes, pathogenic processes or pharmacological responses to therapeutic intervention" [40]. Cancer biomarkers support decisions related to diagnosis and treatment with three main purposes, determining: (1) whom to treat (*predictive* biomarkers), (2) how to treat (*prognostic* biomarkers), and (3) how much treatment will be needed (*pharmacodynamic* biomarkers) [20].

Indeed, biomarkers are used by all medical and paramedical disciplines to answer three key questions: **how much, what and for whom** [46] (Fig. 5.2). Many of the biomarkers are easy to measure and are part of routine medical examinations. Biomarkers in combination or individually can provide an early warning system for the occurrence of diseases in population groups with specific characteristics. Blood pressure, heart rate and pulse are indicators of cardiovascular function. Cholesterol

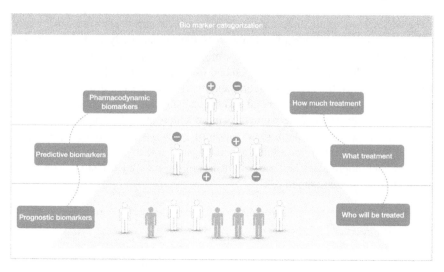

Fig. 5.2 Biomarker categories

and triglycerides provide information about metabolic processes and are used to assess the risk of coronary heart disease. Body measurements such as weight, body mass index (BMI) and waist to hip ratio serve as indicators of obesity, chronic metabolic disorders and fat deposits. T-cell measurements provide information about the state of the immune system. Cortisol, a steroid hormone, is often produced in response to chronic stress [41].

An individual biomarker which exceeds a certain threshold constitutes an indication of risk of a disease to develop in the future [54]. By taking into account other indicators and parameters that are collected as part of a patients record, a new biomarker could be designed and be associated with specific health information. For example, studies show that older men are at higher risk of serious health problems due to a combination of immune system and neuroendocrine disorders. At the same time, in older women, high health risks are usually due to **high systolic blood** pressure in combination with other signals of health issues related to specific biomarkers and metrics (Table 5.1).

Biomarkers are utilized in all medical specialties. In oncology, however, they are in systematic clinical use and considered an important tool for patient management.

As shown in Fig. 5.3, over 50% of the biomarkers in use have practical application in the field of oncology, while over 1/3 of clinical studies in cancer include biomarkers [35].

Genetic testing is crucial in predicting a person's susceptibility to a particular disease and in the detection of specific genetic mutations. Genetic testing combined with virtual disease modelling can produce predictions with regard to a patient's response to certain drugs, lead to treatment enhancements and increase problem management efficiency [6, 21].

Table 5.1 Biomarkers in routine examination (check-up's)

Biomarkers	Description	Health outcome
Systolic blood pressure (SBP)	Cardiovascular activity index: maximum pressure in an artery when the heart supplies the body with blood	Cardiovascular death (CVD), stroke, coronary heart disease (CHD)
Diastolic arterial	Diastolic blood pressure	Cardiovascular death (CVD), stroke, coronary heart disease (CHD)
Resting heart rate	Indicator of heart function and measurement of overall physical fitness	Coronary heart disease
Total cholesterol	Contributes to the synthesis of bile acids and steroid hormones	At middle age: Coronary heart disease (CHD) and mortality of all causes at an older age: u-shaped relation to death
Low density lipoprotein (LDL)	Low density lipoprotein (LDL)	CHD, atherosclerosis, stroke, peripheral vascular disease
High density Lipoprotein (HDL)	Protective cholesterol	Lower atherosclerotic cvd
Triglycerides	Fats stored for energy use	Heart attack, CHD, coronary artery disease, pancreatitis
Fasting glucose	Measures the amount of sugar in the diabetes index	Diabetes, CHD, mortality, poor cognitive function
Body Mass Index (BMI)	Balance ratio between energy intake and energy expenditure	CVD, diabetes, stroke, mortality, some cancers, osteoarthritis
Waist to hip ratio	Abdominal obesity index	Hypertension, CHD, insulin-dependent diabetes and stroke
T-Cells	White blood cells that protect against pathogens and tumors	Cancer, death, atherosclerosis, Alzheimer's
Cortisol	Steroid hormone that reflects the body's response to normal stress	CVD, cognitive impairment, fractures, functional disability, death
Electrocardiogram (EKG)	Measuring electrical pulses in the heart	Cardiovascular risk, stroke, mortality

Genomic medicine will greatly enhance the quality of life of cancer patients. Examples of such cancer treatments include biomarkers for detecting breast cancer that are in treatment under a specific drug (trastuzumab antibody (HerceptinTM)) [27].

Age, race, sex and other exogenous factors such as lifestyle and standard of living are key variables that add valuable information to biomarker applicability [68]. In order to create general-purpose applications and to formulate a more systematic approach, all parameters should be evaluated and investigated [22].

Fig. 5.3 Biomarker
adoption

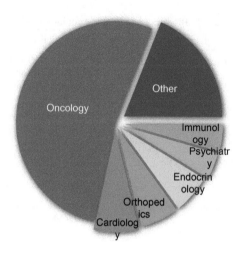

In the USA, for example, people of color are more likely to face stressful situations such as racial discrimination in the labor market or elsewhere. Bad psychology and stress have a biological effect with whatever that entails. Age also increases the risk of developing a disability or chronic illness. There is evidence that, in the long run, the accumulation of environmental and psychological burdens can adversely affect a biological system.

5.2.1.2   Biomarkers

An ideal biomarker should be characterized by accuracy, sensitivity, specificity, patient safety and simplicity. A biomarker test must demonstrate clinical validity, ability to identify defined endpoints of interest, for example patients at risk of recurrent disease, in independent patient groups [12–16]. **Biomarker sensitivity** indicates the ability to distinguish patients with a specific disease [23], while **biomarker specificity** indicates its ability to recognize the patients without a specific disease. An increase in sensitivity implies a decrease in specificity. The biomarker should be able to distinguish cancer patients into two groups based on outcomes, such as responders versus non-responders to a particular treatment or healthy survivors versus patients at risk of premature death from a specific disease. Determining the relative impairment limits of the biomarker test is extremely important, as it has significant implications for patients who may (or may not) be given specific treatment according to the test results [29].

Factors that may affect the sensitivity and specificity of a biomarker include.

- Sample type (e.g. hematological or histological)
- Stability of the sample and processing time
- Proper use of negative controls
- Profile-patient history

The biomarker test should demonstrate **clinical utility**; that is, it should show improvement in the clinical health results of patients for whom the test has been utilized as a management tool, compared to the clinical results of patients for whom no biomarker test has been used [30]. To evaluate clinical utility, the final phase of biomarker development generally consists of a randomized clinical trial, where the biomarker is included as part of an algorithm that allocates patients to treatment A or treatment B, depending on test results.

According to the US Food and Drug Administration (FDA) [44], the ideal biomarker should be.

- Specific. It should relate to a specific disease and be able to distinguish between different physiological conditions
- Safe and measurable
- Fast to deploy, use and analyze, so that faster diagnosis becomes possible
- Cost effective
- Able to provide accurate results
- Consistent so that it can distinguish between different ethnic groups and gender.

The FDA has published a study entitled "Innovation or Stagnation: Challenges and Opportunities for the Critical Path to New Medical Products," which refers to the discrepancy caused by the use of twentieth century tools to evaluate the progress of the twenty-first century and to a need to apply a new approach towards tools used to evaluate new medical products [69]. This is especially true for biomarkers. Janet Woodcock (FDA) noted that despite the hundreds of candidate biomarkers published each year, few reach a high enough level of clinical relevance to allow them to be used in the decision process of product development and patient management [77], as states "obtaining this clinical correlation information we need is very difficult and expensive and simply not possible," adding that "the process of developing biological indicators for various uses is really broken." The same study also calls for more involvement of bioinformatics, with which the exchange of data and databases would be encouraged and enhanced so real knowledge can be gained and information retrieved, instead of being in development limbo. "Terminology should be standardized to allow data collection and the construction of computational quantitative disease models that can incorporate biomarker performance data for modeling and simulation testing" [77].

Evaluating a biomarker and creating a functional and actionable framework is not an easy task [17]. At the same time, the need for new medical incisions that will contribute socially and financially is high and requires redefinitions. The doctor and patient dilemma that often arises is related to the balancing of cost and benefit. For example, the lucky patient who detects a deadly prostate cancer at an early stage with the PSA test (Fig. 5.4) could benefit immensely from the early detection and undesirable side effects. However, most patients who get tested will not enjoy the benefit of early detection. Studies suggest that to prevent a prostate cancer death within 8 years, approximately 1,000 men should undergo a PSA test [73]. These men will be subject to such potential harms as false positive tests, treatment-related stress and erectile dysfunction, incontinence and bowel dysfunction.

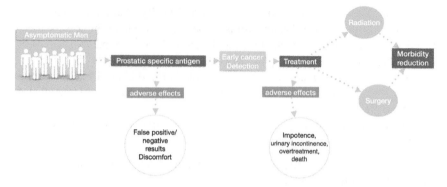

Fig. 5.4 PSA diagnosis and treatment mapping

The ideal biomarker, from an economic policy [63] point of view, must have the following characteristics:

- Contribution to medical decisions
- Patients' health improvement and quality of life (QUALY)
- Contribution to health care systems viability
- National and global accessibility.

The ACCE model (Fig. 5.5) framework [39], originally developed by the CDC for the evaluation of genetic testing and biomarkers, provides a theoretical framework for developing a robust evaluation process that can be applied to any diagnostic test and, consequently, in the evaluation of biomarkers and their integration into the medical practice. One advantage is that it covers issues of both clinical validity, which is a particularly important parameter during the research and development process, and clinical utility prior to the use in a clinical or public setting. The key elements are summarized below [34]:

Analytical validity: The accuracy with which the biomarker test effectiveness is measured.

Clinical validity: The ability of the test to distinguish those who have or will develop a disorder from those who are and will remain healthy. Clinical validity is divided into scientific validity and test performance.

Clinical utility: Evaluation of the risks and benefits of using the test, with reference to the purpose and expediency of delivery.

"ELSI": Ethical, legal and social implications surrounding the tests, including consideration of safeguards and barriers for proper application.

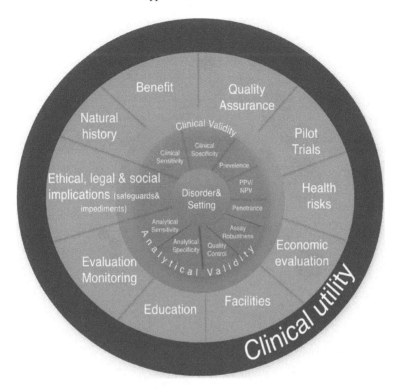

Fig. 5.5 ACCE framework

5.2.1.3 Validation and Evaluation of Novel Bio Markers

The discovery and validation of a biomarker is a long and complex procedure resulting in few biomarkers having found their way into clinical application [25]. But many are now considered successful, such as the HER2 marker, a genetic and protein biomarker currently used in clinical practice for breast cancer patients. The complex development of HER2 started in 1984, when Weinberg et al. [28] identified the human epidermal growth factor 2 gene (HER2 / neu gene) and in 1998 the first version of the HER2 / biomarker treatment package was completed [4]. However, there is a side story to the success of HER2 in breast cancer. Primary resistance to trastuzumab had been observed in patients with elevated HER2 tumors, suggesting that tumor biology is not simple enough to eliminate the possibility of recurrence simply by inhibiting a tumor protein [50]. Therefore, combination therapies including trastuzumab and HER2 have been investigated by the FDA and the European Medicines Agency (EMA) [2] either as treatment for metastatic breast cancer or as adjuvant therapy (treatment given in addition to the initial surgery to prevent recurrent disease) in breast cancer patients.

The image above (Fig. 5.6) depicts a tumor consisting of different elements and their expression (differently colored and graded dots, respectively) in different areas

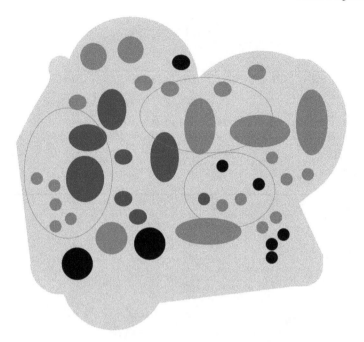

Fig. 5.6 Complexity of tumor sampling

of the tumor. Black elliptical circles represent different tissue sampling areas for biomarker testing. Biomarkers of a molecule will focus on the presence of a specific element in the sampling area (even if there is a lot of data in this area), while complex biomarkers (calculating the sum of the expression of different molecules) are able to collect different elements simultaneously giving a stronger test result that is more representative of the heterogeneity within the tumor [31].

With the advent of new scientific approaches, it is possible to surpass the levels of individual genes and proteins and meet large-scale data at different biological levels (e.g. DNA, epigenetic molecules, metabolites). Based on these new approaches, collaborative initiatives such as the Cancer Genome Atlas Network aim to collect multilevel data from large populations and for many types of cancer.

There are some challenges associated with evaluating biomarkers. Some of the key issues include:

- The increasing complexity of biomarkers, which require knowledge beyond standard doctor training for valid interpretation of results.
- Frequent lack of data on clinical performance (power and utility) due to the cost, size and difficulty of the required studies.
- Lack of agreement on who is responsible for funding studies, data generation and evaluation of evidence for the performance of clinical biomarkers.
- Confusion by biomarker manufacturers and test manufacturers as to the level of clinical data required, leading to different ad hoc standards.

Table 5.2 Cancer bio markers

Marker	Condition	Limit	Sensitivity	Specificity
CEA	Malignant pleural effusion	NA1	58%	79%
CEA	Peritoneal cancer	0.5 ng/ ml	76%	91%
Her-2/neu	Stage IV breast cancer	15 ng/ mL	40%	98%2
Bladder Tumor Antigen	Urothelial cell carcinoma	NA	53%	70%
Thyroglobulin	Thyroid cancer metastases	2.3 ng/ ml3	75%	95%
Alpha- fetoprotein	Hepatocellular carcinoma	20 ng/ ml	50%	70%
PSA	Prostate cancer	4.0 ng/ mL	46%	91%
CA 125	Non-small cell lung cancer	95 IU/ mL	84%	80%
CA19.9	Pancreatic cancer	NA	75%	80%
leptin, prolactin, osteopontin, and IGF-II	Ovarian cancer	NA	95%	95%
CD98, fascin, sPIgR4, and 14-3-3 eta	Lung cancer	NA	96%	77%
Troponin I	Myocardial infarction	0.1 microg /L	93%	81%
B-type natriuretic peptide	Congestive heart failure	8 pg/ mL	98%	92%

- Lack of consensus standards or quality assurance framework for biomarker evaluation [75].

  Examples of FDA-approved biomarkers (Table 5.2).

5.2.2   Next Generation Sequencing

Today, hundreds of millions of DNA fragments per cycle can be analyzed through genome analysis via next generation sequencing (NGS) [61, 74]. The sequencing of the human genome can produce up to 100 times more data than previous methods of analysis, such as the Sanger method [61]. This allowed researchers to complement individual genomes and identify combinations of mutations with specific diseases. As cancer is inherently a disease of the genome, it is not surprising that NGS technology is already being applied to cancer research with promises of greater understanding of carcinogenesis in general. Diagnostics based on genome analysis have tremendous potential in improving cancer management [61].

NGS is based on the "cutting" of DNA into small fragments and the massive parallel sequencing of the fragments. Multiple overlapping sections, called "readings", are assembled in a continuous sequence. To reduce sequencing errors, each genome region must be sequenced several dozen times. This sequencing approach is based on the assumption that genomic DNA fragments are random and sequence independent.

NGS technology can be divided into three phases: model preparation, sequence analysis, and sequencing detection. To prepare the template, it is necessary to produce a representative, non-objective set of fragmented DNA. To facilitate the parallel processing of a large number of sequencing reactions, the fragments are immobilized or confined to spatially separated sites on a solid substrate or in a nanoparticle chamber. The spatially separated patterns are then either clonally amplified or remain as single molecules. Commercially available platforms use a variety of principles for the actual sequencing reaction [55]. These include cyclic reversible termination, the addition of a nucleotide, real-time sequencing, and linkage sequencing.

The genome of a living organism looks like a library full of books, in which the chromosomes that contain information are defined with letters and are called *nucleotides* [64]. The first methods of decoding biological sequences were based on the precise excision of a specific DNA fragment and its accurate reading. An alternative and initially considered inconsistent sequencing method was proposed in 1979. In this method, multiple copies of an entire genome DNA would be cleaved into small fragments, which were then sequenced, and the sequences (called "readings") were assembled into text, based on their overlapping edges.

Nevertheless, the explosive development of automated sequences and advances in computer science have defined the current dominance of this method, better known as *shotgun sequencing*. Medical equipment for modern sequencing are capable of carrying out hundreds of millions of readings per day, where each reading consists of tens or hundreds of nucleotides (i.e. non-random DNA fragmentation).

There are two main types of laboratory approaches, as described above. These are the whole DNA genome sequence and the targeted sequence. The sequence of the whole genome, as the name implies, provides a complete characterization of the whole genome. This allows for greater accuracy in identifying lesions in both coding and non-coding regions. Shotgun sequencing is a lower cost approach based on the sequence of randomly generated fragments. The latter is commonly preferred in research because of the simpler library preparation, lower cost and reduced amount of DNA input required compared to the traditional whole genome sequence. As expected, targeted sequencing sacrifices some information in order to gain speed and availability [58].

Technological developments in NGS are also leading to the development of new diagnostic and therapeutic pathways for cancer treatment [6], as researchers reconstitute their approach by comparing tumors and normal genomes with specific types of cancer [62]. In cancer genome [48], there is great potential for rapid identification of changes, in solid tumors for specific cancer patients [4]. Another interesting product of this technology is the discovery of biomarkers recently developed to detect the presence of tumor-specific gene rearrangements in plasma samples from patients. Genomic rearrangements were specifically located on tumors and did not extent to the normal body tissue. Digital PCR assays were then designed at the rearrangement breakpoints to provide a sensitive cancer biomarker that is successfully used to monitor residual disease after treatment completion [49]. Furthermore, HT-NGS technology also allows the identification of the causative mutations responsible for cancer metastasis and thus generate expectations for improving oncological outcomes [61].

RNA and DNA are nucleic acids and, together with lipids, proteins and carbohydrates, constitute the four major macromolecules necessary for all known life forms. Ribonucleic acid (RNA) is a polymeric molecule necessary in various biological processes and more specifically for the encoding, expression and regulation of genes. Like DNA, RNA is assembled as a nucleotide chain, but unlike DNA, it is more commonly met in nature as single stranded than double stranded. Cellular organisms use messenger RNA (mRNA) to transmit genetic information (using the nitrogenous bases of guanine, uracil, adenine, and cytosine, denoted by the letters G, U, A, and C) that direct the synthesis of specific proteins. Many viruses transfer their genetic information using an RNA genome.

MicroRNA could produce further application in the discovery of new biomarkers. MicroRNAs are involved in the control of protein translation and are present in the blood plasma. NGS could be used to detect tissue or blood plasma to generate miRNA profiles of the entire gene, which could then be mined for specific biomarker signatures.

Figure 5.7 summarizes the process of analyzing the genome. The analysis is performed as described above. The sequences are examined for correlation with existing data to identify relevant findings and mutations [13]. The findings are validated and in the experimental phase the investigation of the probability of clinical usability of the findings takes place [15].

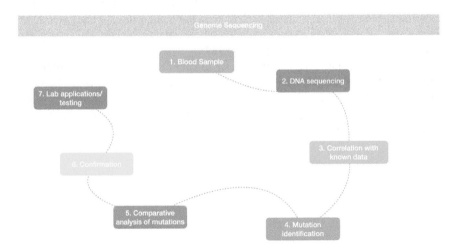

Fig. 5.7 Mapping of the DNA sequencing analysis

5.2.3 Obesity

Obesity represents a major health problem, not only in western countries, but increasingly in low- and middle-income countries as well. In order to develop successful weight loss strategies, it is important to understand the molecular pathogenesis of weight change. Some pathways, such as oxidative stress and the basic regulatory framework of insulin, have been implicated in weight gain and energy expenditure regulation. In addition, the role of metabolites produced by the intestinal microbe, especially short-chain fatty acids, has been highlighted [65].

Obesity or a tendency for obesity is often considered a determining factor of bad health and a preset for a variety of severe health problems. This idea is not new as Hippocrates had soberly stated that "sudden death is more common in those who are naturally fat than in the lean". Type 2 diabetes association with obesity and overweightedness in both male and female populations and all ethnic groups has been the focus of many studies and scientific research. The key finding is that insulin resistance is identified as a fundamental defect discovered in population with type 2 diabetes that is also more commonly obese. Moreover, it is pointed out [56] that over 90% of diabetics are obese or overweight [8, 67]. Studies, also suggest that a number of genes may be linked to the development of obesity. By using sibling-pair linkage analysis, tumor necrosis factor-alpha was a determinant for body fat percentage. Other studies have also established a link between fat cell production of tumor necrosis factor-alpha and obesity [14].

Obesity is a biproduct of long term higher calorie intake and lower relative calorie expenditure. Many different factors contribute to the homeostasis of body weight in humans and, as obesity develops, certain metabolic changes occur, which may not be completely reversed when losing weight [32].

Metabolic syndrome [26] accounts for a significant number of deaths and illnesses in western countries and, increasingly, in lower-income countries as well. The latter countries, like India, face a double burden, with large numbers of malnourished people in rural areas and a growing number of people suffering from obesity and obesity-related diseases.

Almost 40% of adults have a body mass index that characterizes them as overweight, while an additional 13% of adults are characterized as obese [43]. Given the increased risk of diabetes, osteoarthritis and cardiovascular disease caused by obesity, there is a need to understand the molecular determinants of weight change.

5.2.4 Nutritional Biomarkers

5.2.4.1 Metabolomics

The **metabolic profile** refers to the systematic study of the unique chemical fingerprints known as *metabolites* that are left behind by specific cellular processes. The metabolic profile can give a snapshot of the physiology of the cell and, consequently, a direct "functional reading of the physiological state" of an organism.

An overlap has been identified between type 2 diabetes and obesity through the study of the metabolites. Branched chain amino acids (BCAAs), glutamine, proline, cysteine, tyrosine, threonine, phenylalanine, tryptophan, pantothenic acid and choline, have higher values in the obese and diabetics, while asparagine, citrulline and methionine average lower values. In a cross-sectional study of 947 participants, 37 metabolites were significantly correlated with body mass index, including nineteen lipids, twelve amino acids, and six others. Eighteen of these correlations were not previously reported, including histidine and butyrylcarnitine [52].

DNA and genetic related studies are referred to as Genomics [11] (Fig. 5.8). Transcriptomics studies the m.rna, while metabolomics explores the products of the metabolic processes in relation to genetic and enviromental actors and measures the metabolites. The metabolite concentration describes the condition of cells and of tissues and by that metabolomics gives a clearer picture of the molecular phenotype [36].

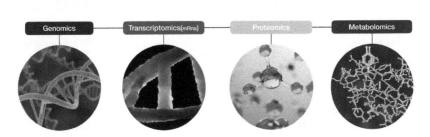

Fig. 5.8 The omics space

Metabolomics uniquely comprises analytical technologies [12] that can provide diagnostic patterns via fingerprinting, absolute quantitation of targeted metabolites via pool analysis, relative quantitation of large portions of the metabolome using metabolite profiling, and tracing of the biochemical fate of individual metabolites through a metabolic system via flux analysis. Each of these technologies is supported by the two most commonly used and powerful techniques currently available for metabolomics: mass spectrometry and NMR(Nuclear magnetic resonance) [45].

5.2.4.2 Metabolites

A *metabolite* is an organic compound of low molecular weight, usually involved in a biological process as a substrate or product. Metabolomics typically studies small molecules on a mass scale of 50–1500 daltons (Da).

Metabolite examples can be seen in Fig. 5.9 and include:

- sugars
- lipids
- amino acids
- phenolic compounds
- alkaloids

There is great variation in metabolites [37] between species. it is estimated that there are about 200,000 metabolites throughout the plant kingdom and somewhere between 7,000 and 15,000 in a single plant species. In humans, on the other hand, it is estimated that there are approximately 3,000 endogenous or common metabolites. The previous are likely under- estimations due to the difficulty in detecting low-abundance molecules. However, it can be concluded that, biochemically, plants are particularly rich compared to many other species. They also usually contain more genes than other eukaryotes.

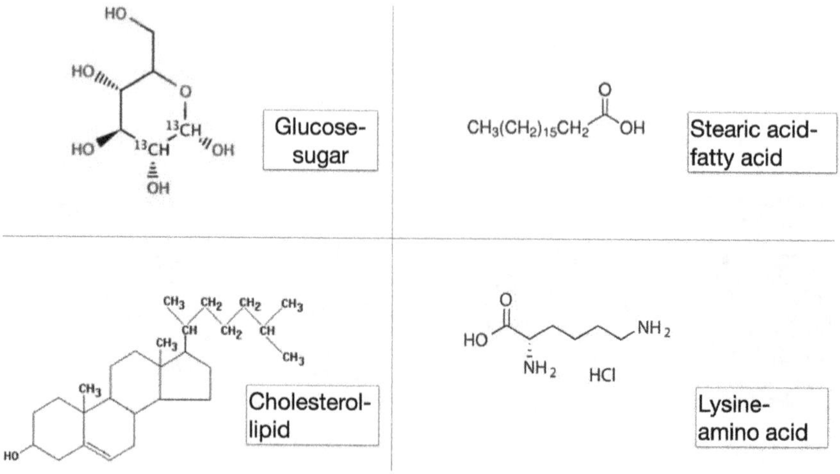

Fig. 5.9 Metabolites

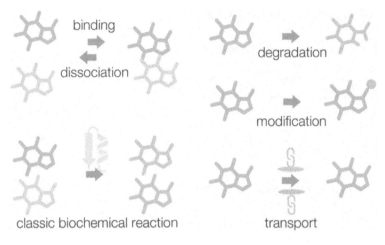

Fig. 5.10 Metabolic reactions (reaktome database)

Metabolism is the complete set of metabolites within a cell, tissue, or biological sample at any given point in time. Metabolism is inherently very dynamic: small molecules are constantly absorbed, synthesized and degraded and interact with other molecules, both within and between biological systems and the environment [60]. The main metabolic reactions are shown in the Fig. 5.10, which has been adapted from the Reactome database.

5.2.4.3 Nutritional Genomics

Nutritional genomics is the application of high-performance functional genomic technologies to nutrition research. These technologies can be integrated into databases of genomic sequences, allowing the process of gene expression to be studied in parallel for many thousands of different genes. Such techniques can facilitate the definition of optimal nutrition at the level of population, specific groups and/or individuals. Diet has a significant impact on chronic disease and health and the study of nutritional indicators could help determine the bio-active components of food [38]. The definition of these activities will lead to the improvement of health through modification and enhancement of diet, creation of new foods, etc. Challenges lie in the optimal design of nutrition studies and the efficient handling of the huge data sets that are created. It is now possible to identify genetic polymorphisms that predispose individuals to disease and thus modify their dietary requirements [51]. The characterization of such gene polymorphisms will allow nutritional counselling and treatment to be targeted on "at risk" groups [1].

5.2.4.4 Standard Biochemistry Profile

For the purposes of this study, databases were used related to the biochemical profile that results from targeted blood tests. Standard Biochemistry Profile (SBP) determines the functional capacity of important organs such as the liver and kidneys. It also determines the difference in the metabolism of proteins and fats, the metabolism of carbohydrates, the production of enzymes, the ability to use compounds, etc. (Table 5.3).

Important variables of the biochemical profile are listed below.

5.2.4.5 Semantic Grouping

Standard biochemistry profile (SBP) includes three targeted examinations, namely comprehensive metabolic panel, liver panel and lipid panel.

Metabolic panel is a test that measures the blood levels of albumin, blood urea nitrogen, calcium, carbon dioxide, chloride, creatinine, glucose, potassium, sodium, total bilirubin and protein, and liver enzymes. It's more commonly subscribed to diagnose diabetes or other conditions such as high blood pressure.

A **liver (hepatic) panel** is a blood test that determines liver functionality. The blood levels of total protein, albumin, bilirubin, and **liver enzymes (aspartame amino transferase, alanine amino transferase, alkaline phosphatase)** are measured.

Finally, **lipid panel** is a blood test that measures fats and fatty substances (lipids) and, in essence, total cholesterol level.

Based on the relation between each measurement and that of the organs, a semantic grouping has been defined as depicted in Figs. 5.11 and 5.12. Semantic similarity and relatedness measures how two concepts are related. It is a critical approach that can improve clustering of biomedical and clinical measurements and variables and can be a defining factor for the definition of biomedical ontologies and the development of related technologies [45].

5.3 Neural Networks, Pattern Recognition and Datasets

5.3.1 Neural Network

Neural networks solve the classification problem for nonlinearly separable sets using hidden layers, the neurons of which are not directly connected to the output. The additional hidden layers can be interpreted geometrically as additional super-levels, which enhance the separation capacity of the network. By connecting artificial neurons to this network through nonlinear activation functions, we can create complex, nonlinear decision boundaries that allow us to deal with problems where different categories are not linearly separated.

Table 5.3 Standard biochemistry profile

Element	Description
Alanine aminotransferase	Alanine aminotransferase measurements are used in the diagnosis and treatment of certain liver diseases (eg viral hepatitis and cirrhosis) and heart disease. Elevated transaminase levels may indicate myocardial infarction, liver disease, muscular dystrophy, or organ damage
Albumin	Albumin measurements are used in the diagnosis and treatment of diseases involving the liver and / or kidneys, and are often used to assess nutritional status because plasma albumin levels are dependent on protein intake
Alkaline phosphatase	Alkaline phosphatase measurements are used in the diagnosis and treatment of liver, bone and parathyroid disease
Aspartate aminotransferase	AST measurements are used to diagnose and treat certain types of liver and heart disease. Elevated transaminase levels can signal myocardial infarction, liver disease, muscular dystrophy or organ damage
Bicarbonate	Along with pH determination, bicarbonate measurements are used in the diagnosis and treatment of many potentially serious disorders associated with acid–base imbalance in the respiratory and metabolic systems
Blood Urea Nitrogen (BUN)	BUN measurements are used to diagnose certain kidney and metabolic diseases. Determination of serum urea nitrogen is the most widely used test to assess renal function. In combination with serum creatinine levels it is used for the differential diagnosis of prenatal, renal and post-renal uremia. High BUN levels are associated with decreased renal function, increased protein catabolism, nephritis, intestinal obstruction, urinary obstruction, metal poisoning, heart failure, peritonitis, dehydration, malignancy, pneumonia, surgical shock and Addiction disease. Low BUN levels are associated with amyloidosis, acute liver disease, pregnancy and nephrosis. Physiological fluctuations are observed depending on the age and sex of a person, at the time of the intake diet day - especially in relation to protein intake
Cholesterol	Elevated cholesterol levels are associated with diabetes, nephrosis, hypothyroidism, biliary obstruction and these rare cases of idiopathic hypercholesterolemia and hyperlipidemia. Low levels are associated with hyperthyroidism, hepatitis and sometimes severe anemia or infection

(continued)

Table 5.3 (continued)

Element	Description
Creatinine	Creatinine measurements are useful in the diagnosis and treatment of kidney disease. Creatine measurements are also used in the diagnosis and treatment of myocardial infarction, skeletal muscle diseases and diseases of the central nervous system
Gammaglutamyl Transaminase (GGT)	GGT measurements are mainly used to diagnose and monitor liver disease. It is currently the most sensitive enzymatic indicator of liver disease. It is also used as a sensitive screening test for occult alcoholism. Elevated levels are found in patients taking chronic medications such as phenobarbital and phenytoin
Globulin	Globulins are a different group of proteins that carry different substances in the blood. They are also involved in various defense mechanisms within the body. Measurements of globin (Total Protein - Albumin) are calculated and used to determine the concentration of globin in the serum
Glucose	Glucose measurements are used in the diagnosis and treatment of pancreatic islet carcinoma and carbohydrate metabolism disorders, including diabetes mellitus, neonatal hypoglycemia and idiopathic hypoglycemia
Iron	Iron measurements are used in the diagnosis and treatment of diseases such as iron deficiency anemia, chronic kidney disease and hemochromatosis (a disease associated with extensive tissue deposition of two pigments containing iron, hemosiderin and hemofuscin, and is characterized by pigmentation of the skin)
Lactate Dehydrogenase (LDH)	LDH measurements are used in the diagnosis and treatment of liver diseases such as acute viral hepatitis, cirrhosis and metastatic liver cancer. heart disease, such as myocardial infarction. and tumors of the lungs or kidneys
Osmolality	Serum osmolality is a measure of the number of particles dissolved in a solution. Osmolality is a calculated value in the chemistry analyzer: $[(1.86 * Na) + (GLUC / 18) + (BUN / 2.8) + 9]$
Phosphorus	There is a reciprocal relationship between serum calcium and inorganic phosphorus. Any increase in the level of inorganic phosphorus causes a decrease in the level of calcium by a mechanism that is not clearly understood. Hyperphosphataemia is associated with vitamin D hypervitaminosis, hypoparathyroidism and renal insufficiency. Hypophosphataemia is associated with rickets, hyperparathyroidism and Fanconi syndrome

(continued)

Table 5.3 (continued)

Element	Description
Potassium, Chloride, and Sodium	Low serum chloride levels are associated with salt-losing nephritis. Addisonian seizures, prolonged vomiting and metabolic acidosis caused by overproduction or decreased acid excretion. High serum chloride levels are associated with dehydration and conditions that cause decreased renal blood flow, such as congestive heart failure Sodium measurements are used in the diagnosis and treatment of diseases involving electrolyte imbalance
Total Bilirubin	Low bilirubin levels are associated with aplastic anemia and certain types of secondary anemia resulting from toxic treatment for cancer and chronic nephritis
Total Calcium	Calcium measurements are used in the diagnosis and treatment of parathyroid disease, bone disease, chronic kidney disease and tetanus. Urine calcium measurement is used in the differential diagnosis of hypercalciuria
Total Protein	Total protein measurements are used in the diagnosis and treatment of a variety of diseases including liver, kidney or bone marrow, as well as other metabolic or eating disorders
Triglycerides	Triglyceride measurements are used in the diagnosis of diabetes, nephrosis, liver obstruction and other diseases including lipid metabolism and various endocrine disorders and in the treatment of patients with these diseases
Uric Acid	Uric acid measurements are used in the diagnosis and treatment of many renal and metabolic disorders, such as renal failure, gout, leukemia, psoriasis, starvation or other wasteful conditions, and in the treatment of patients receiving cytotoxic drugs

The training of neural networks is conducted in a supervised manner. The basic idea is to present the input vector to the network and calculate forward the output of each level and the final output of the network. The desired values are declared and, therefore, the weights can be adjusted for a single layer network in the case of the backpropagation (BP) algorithm following a gradient descent approach [71].

To calculate the weight changes in the hidden layers, the error of the output level is returned according to their respective connection status. This process is repeated for each sample in the training set. A cycle through the training set is called an epoch. The number of epochs required for network training depends on various parameters, especially at the error calculated on the output level. In Fig. 5.13, a representation of the described network is shown. The settings used are the ones that will also be deployed for the training purposes of our dataset.

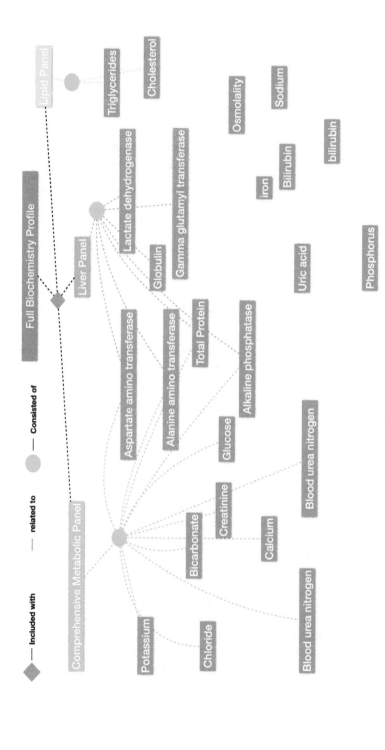

Fig. 5.11 Grouping by tests

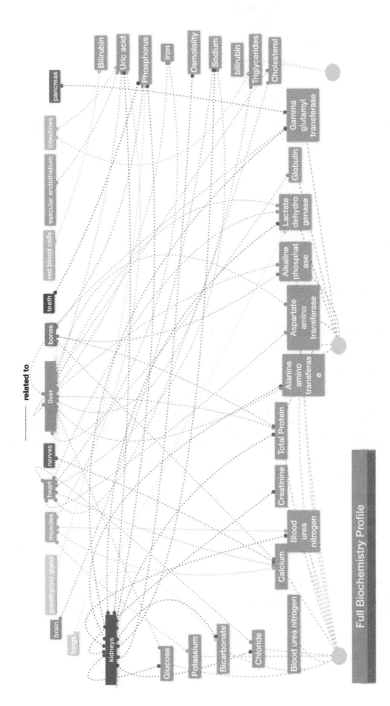

Fig. 5.12 Grouping by relation to organ

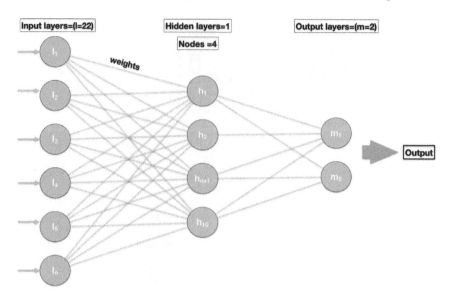

Fig. 5.13 Back propagation neural network

5.3.2 Implementation Environment

Scikit-learn python

Scikit-learn [59] (formerly scikits.learn and also known as sklearn) is a free software learning library for the Python programming language. It includes modules and methods for various sorting procedures, regression models, and grouping algorithms, including support vectors, random forests, gradient amplification, k-means clustering techniques and DBSCAN, and is designed to work with Python NumPy and SciPy numerical and scientific libraries.

Glossary of programming concepts and libraries

- StandardScaler

Transposes the average of the distribution to 0. About 68% of the values will be between -1 and 1.

- sklearn.model_selection.train_test_split
  Separation into tables into random training and test subsets
  Usage Example

  X_train, X_test, y_train, y_test = train_test_split(predict_dataset, target_dataset, test_size = 0.1, random_state = 42)
  # Trains scaler which standarizes all the features in a binary mode
  sc = StandardScaler()
  sc.fit(X_train)

 # Apply the scaler to the X training data
X_train = sc.transform(X_train)

- Cross validation

  In a prediction problem, a model typically receives a set of known data in which the training is being conducted (*training* data set) and a set of unknown data (or data first seen) against which the model is being tested (called a *validation* data set or test set).

  The purpose of cross-validation is to test the model ability (i.e., the estimation of weights that will be deployed) to predict new data that were not used for its training, to identify problems such as over-fitting or selection bias, and to give an idea of how the model will be generalized to an independent data set (i.e., an unknown data set).

  A cross-validation cycle involves splitting a data sample into complementary subsets, performing the analysis on one subset (called a training set), and validating the analysis on the other subset (called a validation set or test set). To reduce alterability, most methods run multiple validation rounds using different partitions taken from sample set and the validation results are combined (averaged) to give an estimate of the predictive performance of the model (Fig. 5.14).

- **Random state**

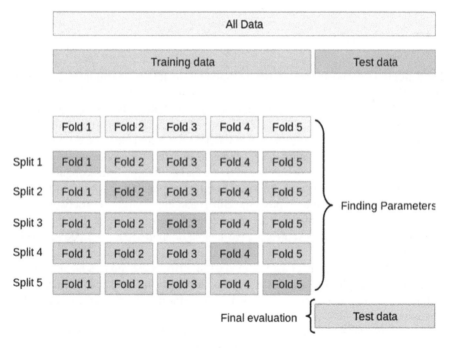

Fig. 5.14 Example of using cross validation with five-fold data separation and model evaluation. TestScores = cross_val_score(ppn, X_test, np.ravel(y_test),cv = 5)

Controls the rearrangement applied to the data before the separation is applied.
- **Solver**
  Optimisation algorithm.
- **Hidden layer size tuple, length = n_layers −2, default = (100)**
  The n element represents the number of neurones in the hidden layer.
- **Activation**
  Activation function for hidden layers.
  "Identity", activation without function, useful for implementing linear congestion, returns $f(x) = x$
  "Logistic", the logistic sigmoid function, returns $f(x) = 1 / (1 + \exp(-x))$.
  "Tanh", the function of excess black, returns $f(x) = \tanh(x)$.
  "Relu", the operation of a corrected linear unit, returns $f(x) = \max(0, x)$
- **Alpha**
  Penalty parameter L2 (normalisation condition).
- **Max_iter**
  Maximum number of repetitions. The solver function is looped until convergence.
- **RadViz visualizer**
  RadViz is a multifactorial data visualization algorithm that depicts each feature dimension around the circumference of a circle and then plots points inside the circle so that the point normalizes its values to the axes from the center to each arc.

5.3.3 Proposed System

The datasets used were retrieved from the National Health and Nutrition Examination Survey (NHANES). NHANES is a study program that aims to assess the health and nutritional status of adults and children in the United States. The research is unique, as it combines interviews and medical exams. NHANES is an important program of the National Center for Health Statistics (NCHS). NCHS is part of the Centers for Disease Control and Prevention (CDC) and is responsible for producing vital health statistics. The NHANES program began in the early 1960s and was conducted in a series of surveys focusing on health issues of different population groups. In 1999, it was reformed into a continuous program with a changing focus on a variety of health and nutrition measurements in order to meet new emerging needs. For the needs of the program a nationally representative sample of approximately 5,000 people is selected each year [47].

All available datasets that range from 2001 to 2016 were acquired. Demographic data, tests (blood, physiology, etc.), diet, laboratory tests, medical and pharmacological patient history and more are included. More information can be reviewed in the appendix.

Our proposed system aims to use the capabilities and strengths of neural networks and machine learning to create a functioning pathway for usable applications in the fields of nutritional bio markers and metabolomics. The design of our model is based

Table 5.4 BMI classes

if	bmi	Greater	Than	30	Obese
else if	bmi	Between	18.5	25	Normal
else if	bmi	Between	25	30	Overweight

on the assumption that observational data (medical history, dietary habits, lifestyle, etc.) and/or laboratory data contain information about a person's condition. As weight is an important indicator of health, BMI (i.e., height/mass index) will be considered as a target for our deployed deep neural network. The standard biochemistry profile results will be used as features/inputs for the training of the neural network [53]. The BMI data will be transposed in a x*2 table and separated into groups, depending on the commonly used ranges to classify weight categories based on the BMI as seen on Table 5.4.

Through a conceptual analysis, a design basis for a machine learning pipeline was depicted and proposed. Implemented improvements and an expanded data set will better test the given hypothesis and outcomes should be improved. The study was conducted in three circles. In the first circle, the productivity of the network was tested between the normal and overweight classes, in the second circle the productivity of the network was tested between the obese and the normal class and in the third circle the productivity of the network was tested between the obese and the overweight. The findings are discussed, and further decisions are made based on the given results. It must be pointed out that through this investigation, even though the medical literature has been used as a guide in the data collection, the main goal was to create a system that can find its own pathways and automatically evaluate outcomes.

Obesity is frequently recognized as a collateral characteristic of major health problems or even the causality of one. Underweightedness is also an obvious outcome of malnutrition that maybe connected to illness, neurological or psychological causes or other factors. In this study, we focus on the search of the ideal weight by grouping our sample in sets of two categories. This classification will help determine the line that separates the population by recognizing patterns in the standard biochemistry profile of each class.

Through a correlation algorithm some values are eliminated. Each feature/input of the system is evaluated based on the contribution to the given classification accuracy. The goal is to slim down the system without compromising on accuracy, while making it faster, more precise and easily scalable. Specifically, our aim is to deploy an artificial neural network-based system that will be capable of classifying an individual based on the given characteristics as they appear in blood exams. A link between blood exam ranges and the dietary profile will formulate a system that recognizes the patterns in those two different populations/weight classes.

The design process of the neural network includes two phases and an optimization intermediary as depicted in Fig. 5.15.

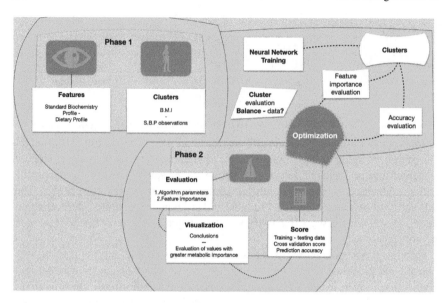

Fig. 5.15 Conceptual analysis

5.4 Implementation and Evaluation of the Proposed System

5.4.1 Deep Back Propagation Neural Network

A deep back propagation neural network is deployed. As already mentioned, python(3.7) and sk-learn libraries are being used for scalability, reusability, flexibility and connectivity with other frameworks. The architecture of the network can be seen in Fig. 5.13. For the deployment of the system and the evaluation of its predictive ability, the BMI data is transposed in a x*2 table and separated in two weight classes, namely overweight and normal (first circle), obese and normal (second circle) and obese and overweight (third circle). The data is balanced to ensure a less biased result. The data corresponds to USA population of men and women of different ages and ethnicity groups [10].

The properties and options of the most accurate model are listed below. On each training optimization phase, the features were examined, and their importance was exported. Based on the importance, those features with zero or negative contribution were drawn out and finally a system was formulated with acceptable accuracy and predictive capabilities with only the necessary elements used as inputs. The features that hold the most information, were then grouped and used as inputs to determine the patterns, if any, in the dietary profile of our sample.

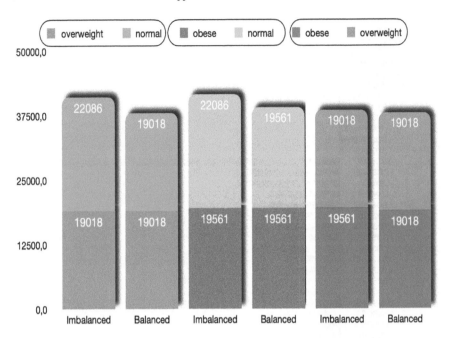

Fig. 5.16 Data analysis

Tests were performed which included biochemistry profile, combination of biochemistry profile and smoking, combination of biochemistry profile and choles-terol, and combination of biochemistry profile, cholesterol and glycohemoglobin. It was observed that the predictive capability did not improve. For this reason, the use of biochemical tests only was preferred as it provides better results at a lower cost.

5.4.2 Standard Biochemistry Profile Neural Network (SBPNN)

1. Data Analysis:

   The data were examined in groups of two. The groups are depicted in Fig. 5.16 along with the related population they occupy. Indexes with null values have been removed. As seen in the same table, the data set is then balanced to the minority group for each case. After running a correlation algorithm to measure the relationship of the features to the BMI, some duplicates are removed. Dupli-cates are data points of the same metric, but using a different measuring method, and have equal or almost equal relationship to the target (BMXBMI). The correlation table is useful for re-evaluating features as per their importance on the predictive capabilities of our proposed **dietary profile neural network** (DPNN). The final features that were used can be seen in Fig. 5.17.

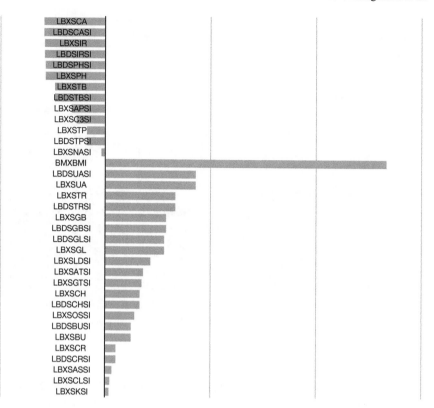

Fig. 5.17 Correlation to BMI

2. Classification Accuracy:

    (a) Circle 1 (Overweight vs normal weight classes): In the first circle, a deep
         neural network with 10 hidden layers was deployed. Our system ran in a
         loop where the accuracies were calculated, and each feature importance
         was computed. The least important features were removed. The accu-
         racies are shown in Fig. 5.18 along with the classification performance
         shown in the corresponding ROC curve Diagram (Fig. 5.19). The training
         average accuracy was 0,69 and the testing average 0,68. This accuracy was
         achieved in the third run of the loop and with 4 features being removed
         that had a negative or close to 0 contribution to the predictive capability
         of the network.

    (b) Circle 2 (Obese vs normal weight classes): In the second circle, we ran a
         deep neural network with 10 hidden layers. Our system ran in a loop where
         the accuracy was calculated, and each feature importance was computed.
         The least important features were removed. The training average accuracy
         (Fig. 5.20) was 0,81 and the testing average 0,80 and the true positive rate
         can be seen in Fig. 5.21. This accuracy was achieved in the second run of

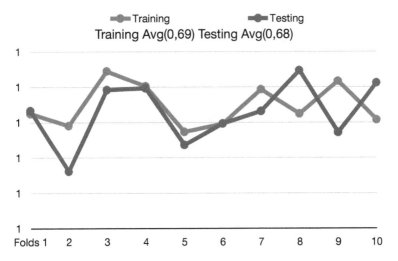

Fig. 5.18 Tenfold cross validation circle 1

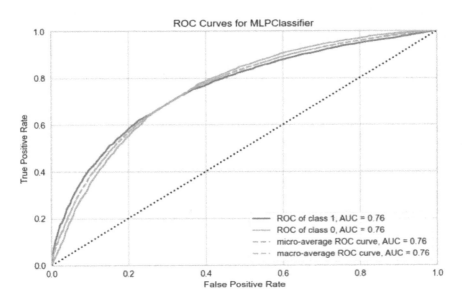

Fig. 5.19 Roc curve circle

the loop and with 4 features being removed which had a negative or close to 0 contribution to the predictive capability of the network.

(c)   Circle 3 (Obese vs overweight weight classes): We repeated the same process for the third circle. The corresponding results are shown in Figs. 5.22 and 5.23.

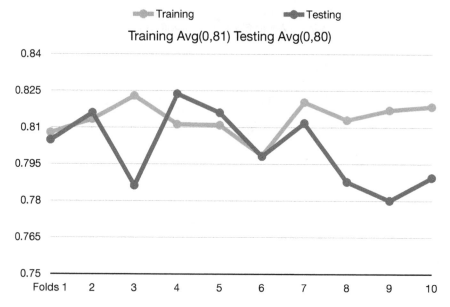

Fig. 5.20 Tenfold cross validation circle 2

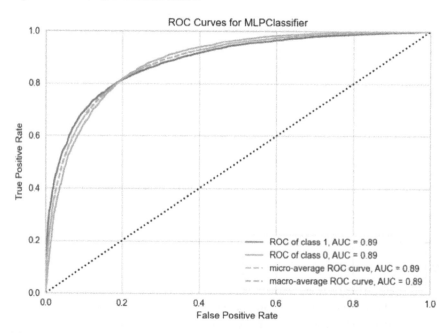

Fig. 5.21 Roc curve circle 2

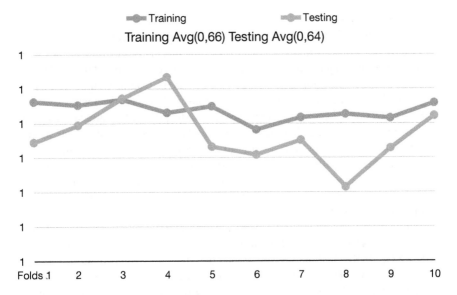

Fig. 5.22 Tenfold cross validation circle 3

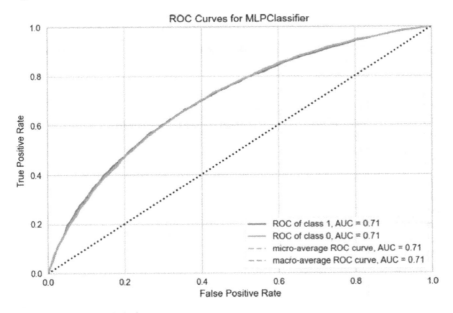

Fig. 5.23 Roc curve circle 3

3. Feature Importance:

    "Permutation feature importance is a model inspection technique that can be used for any fitted estimator when the data is tabular. This is especially useful for non-linear or opaque estimators. The permutation feature importance is defined to be the decrease in a model score when a single feature value is randomly shuffled. This procedure breaks the relationship between the feature and the target; thus, the drop in the model score is indicative of how much the model depends on the feature. This technique benefits from being model-agnostic and can be calculated many times with different permutations of the feature" [59]. The elements identified in this process will be used for the formation of 2 categories related to the underline range of values. Therefore, normal range and out of normal range categories are produced. By comparing the feature importance (Fig. 5.24), we found that 3 features were present in all the circles and occupied one of the top 5 places. These were: Alanine aminotransferase ALT (LBXSATSI), Aspartate aminotransferase AST(LBXSASSI) and Uric acid (LBXDSUASI). It should be mentioned that they are also positively correlated with the BMI. Because uric acid exists in different normal ranges for each gender, we used ALT and AST as a target for the dietary profile neural network

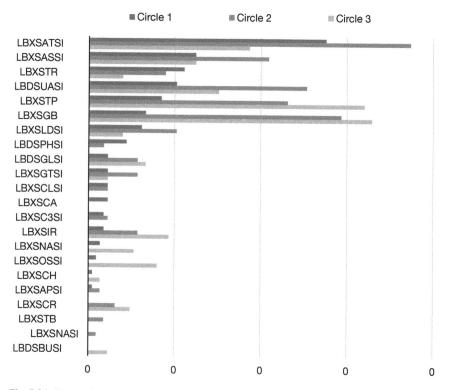

Fig. 5.24 Feature importance

(DPNN). When these two features were examined individually, almost 95% of the values were within the normal range. However, according to the medical literature, if the ratio of ALT:AST is below 1, it is considered normal, but if it is over 1, this is an indication of liver disease. For example, in patients with alcoholic liver disease, the AST:ALT ratio is greater than 1 in 92% of patients and greater than 2 in 70% of the patients [24].

5.4.3   Neural Network Dietary Profile

Data Analysis: In Table 5.5, the features are shown, which were selected for training purposes related to the dietary profile and all null values have been excluded. "For each participant, daily total energy and nutrient intakes from foods and beverages, and whether the amount of food consumed was usual, much more than usual, or much less than usual, are included in the Total Nutrient Intakes files. Daily aggregates of food energy and 64 nutrients/food components from all foods/beverages as calculated using USDA's Food and Nutrient Database for Dietary Studies" [47]. The data were split in two groups, that would become the targets of the NN, based on the ratio of ALT:AST, if below 1, a datapoint was added to the normal group (Fig. 5.25).

(1) *Classification Accuracy*: There is a substantial increase in accuracy from our previous study and first take on this hypothesis [57, 18], that also consisted of a much smaller data sample and a more generic approach in defining the targets of the neural network. The tenfold accuracy shows a testing capability of almost 0.7 that is equal to about 16% (Fig. 5.26) increase compared to our previous study, 0.6 being the testing average of the neural network [57]. Out of the 27 data points that were initially used for the training of the system, in the final loop only 9 were eventually activated, eliminating the rest in a similar method to the one deployed in the S.B.P.N.N (Table 5.5).

Table 5.5 Dietary profile features

Feature	Description
DR1TKCAL	Energy (kcal)
DR1TTFAT	Total fat (gm)
DR1TSUGR	Total sugars (gm)
DR1TPROT	Protein (gm)
DR1TCAFF	Caffeine (mg)
DR1TCARB	Carbohydrate (gm)
DR1TATOC	Vitamin E as alpha-tocopherol (mg)
DR1TACAR	Alpha-carotene (mcg)
DR1TRET	Retinol (mcg)

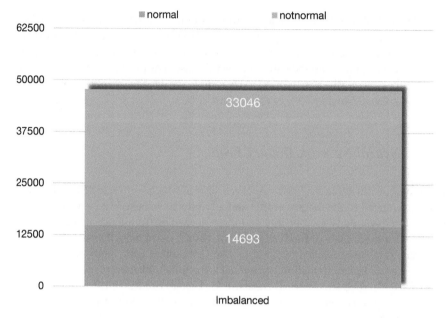

Fig. 5.25 Dietary profile ALT:AST variance

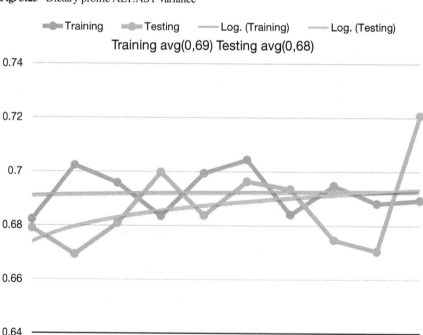

Fig. 5.26 Dietary profile tenfold cross validation

5.5 Conclusions and Future Research

5.5.1 Prevention

Minimizing the occurrence of diseases is the best preventive and cost-effective method for health management. In combination with the best treatment and care, prevention can be and, in many cases, already is a determining factor.

Prevention can be achieved at different levels:

- **Primary** prevention involves reducing exposure to risk factors through legislation, regulation, education and behavior change.
- **Secondary** prevention includes screening, detection, and medical prevention. Avoiding unnecessary treatment and improving the cost-effectiveness of diagnostics can result from identifying high-risk subgroups in the population, including the use of biological markers.
- **Tertiary** prevention aims to prevent recurrence.

The ideal weight is an indicator of health. Controlling and managing it is an important part of preventive health management. Dietary interventions are part of patient and disease management, both qualitatively and quantitatively.

5.5.2 Modelling

Modelling the process of evaluating an individual at the biochemical level not only provides speed and accuracy, but it can also become a useful tool for dealing with problems at the population (macro) level. Cost and accessibility are important efficiency factors, but the most important factor is accessibility to a wealth of data for the design and training of highly predictive A.I systems.

5.5.3 Automation

The usability of the models can be dampened by interfacing with a large portion of the population through interactive web tools and mobile devices. These systems can be enriched with new data from users and primarily create categorizations and dietary suggestions by building "awareness" of the importance of ideal weight and the importance of nutrition.

5.5.4 Perspective

Interpreting and extracting content from complex metabolic datasets is extremely demanding and represents an important area of research. Still, great strides are being made in integrating metabolic data with genomics and proteomics. Metabolomics promises that, in the future, biochemical analysis from genotype to phenotype will be measured and explored for new insights into human body biology and medicine.

The identification of patterns in complex data sets can lead to the discovery of new biomarkers and add precision in medical classification and patient management [70].

5.5.5 Discussion on Feature Research

The complexity of human biology, the medical information and chemical data make the precise process mapping and system design a hard task, where the outline of the problem and the related literature must be carefully considered. At this point of our research, we begin by defining how to best implement a smart system as a basis for more advanced tasks using the available data.

5.6 Appendix

Biochemistry profile—ranges of values

Lab tests and values		
Test	Normal values	
Albumin, serum	3.5–5.5 g/dL	
ALT	45 U/L or less	
AST	40 U/L or less	
Bilirubin, total	1.2–1.3 mg/dL or less	
Blood group (A,B,O)		
BUN, serum	7–25 mg/dL	
Calcium, serum	2.2–2.6 mmol/L	
CBC with differential	Values given with report	
Chloride, serum	96–109 mmol/L	
CPK	Male: 17–148 U/L, female: 10–70 U/L	
Creatinine, serum	0.6–6 mg/dL	
CRP	0.8 mg/dL or less	
Glucose, plasma	70–110 mg/dL	Less than 140 mg/dL

(continued)

(continued)

Lab tests and values	
LDH	Less than 240 U/L
Potassium, serum	3.5−5.3 mmol/L
Prothrombin time	9—12 s
Rapid plasma reagin	Nonreactive
Sedimentation rate	male: 0−15 mm/hr, female: 0−20 mm/hr
Sodium, serum	135−147 mmol/L
Triiodothyronine (T3)	85−205 mg/mL
Thyroxine (T4)	4.5−12 mg/dL
Total protein, serum	6−8.5 g/dL
Triglycerides	Less than 150 mg/dL
Uric acid, serum	Male: 3.9−9 mg/dL, female: 2.2−7.7 mg/dL
Urinalysis, routine	
Bicarbonate	
Clinical use:	
Reference ranges:	22−29 mmol/l
Associated diseases:	Higher than normal levels: Hyperaldosteronism Cushing syndrome Lower than normal levels: Ketoacidosis Lactic acidosis Kidney disease Addison disease
Aspartate Aminotransferase	
	• Males: 10–40 units/L
	• Females: 9–32 units/L
• Serum globulin:	• 2.0–3.5 g per deciliter (g/dL) or 20–35 g per liter (g/L)

Dietary profile variables

Variable name	Variable description
DR1STY	Did {you/SP} add any salt to {your/her/his} food at the table yesterday? Salt includes ordinary or seasoned salt, lite salt, or a salt substitute
DR1TACAR	Alpha-carotene (mcg)
DR1TALCO	Alcohol (gm)
DR1TATOA	Added alpha-tocopherol (Vitamin E) (mg)
DR1TATOC	Vitamin E as alpha-tocopherol (mg)
DR1TB12A	Added vitamin B12 (mcg)
DR1TBCAR	Beta-carotene (mcg)
DR1TCAFF	Caffeine (mg)

(continued)

(continued)

Variable name	Variable description
DR1TCALC	Calcium (mg)
DR1TCARB	Carbohydrate (gm)
DR1TCHL	Total choline (mg)
DR1TCHOL	Cholesterol (mg)
DR1TCOPP	Copper (mg)
DR1TCRYP	Beta-cryptoxanthin (mcg)
DR1TFA	Folic acid (mcg)
DR1TFDFE	Folate as dietary folate equivalents (mcg)
DR1TFF	Food folate (mcg)
DR1TFIBE	Dietary fiber (gm)
DR1TFOLA	Total folate (mcg)
DR1TIRON	Iron (mg)
DR1TKCAL	Energy (kcal)
DR1TLYCO	Lycopene (mcg)
DR1TLZ	Lutein + zeaxanthin (mcg)
DR1TM161	MFA 16:1 (Hexadecenoic) (gm)
DR1TM181	MFA 18:1 (Octadecenoic) (gm)
DR1TM201	MFA 20:1 (Eicosenoic) (gm)
DR1TM221	MFA 22:1 (Docosenoic) (gm)
DR1TMAGN	Magnesium (mg)
DR1TMFAT	Total monounsaturated fatty acids (gm)
DR1TMOIS	Moisture (gm)
DR1TNIAC	Niacin (mg)
DR1TNUMF	Total number of foods/beverages reported in the individual foods file
DR1TP182	PFA 18:2 (Octadecadienoic) (gm)
DR1TP183	PFA 18:3 (Octadecatrienoic) (gm)
DR1TP184	PFA 18:4 (Octadecatetraenoic) (gm)
DR1TP204	PFA 20:4 (Eicosatetraenoic) (gm)
DR1TP205	PFA 20:5 (Eicosapentaenoic) (gm)
DR1TP225	PFA 22:5 (Docosapentaenoic) (gm)
DR1TP226	PFA 22:6 (Docosahexaenoic) (gm)
DR1TPFAT	Total polyunsaturated fatty acids (gm)
DR1TPHOS	Phosphorus (mg)
DR1TPOTA	Potassium (mg)
DR1TPROT	Protein (gm)

(continued)

(continued)

Variable name	Variable description
DR1TRET	Retinol (mcg)
DR1TS040	SFA 4:0 (Butanoic) (gm)
DR1TS060	SFA 6:0 (Hexanoic) (gm)
DR1TS080	SFA 8:0 (Octanoic) (gm)
DR1TS100	SFA 10:0 (Decanoic) (gm)
DR1TS120	SFA 12:0 (Dodecanoic) (gm)
DR1TS140	SFA 14:0 (Tetradecanoic) (gm)
DR1TS160	SFA 16:0 (Hexadecanoic) (gm)
DR1TS180	SFA 18:0 (Octadecanoic) (gm)
DR1TSELE	Selenium (mcg)
DR1TSFAT	Total saturated fatty acids (gm)
DR1TSODI	Sodium (mg)
DR1TSUGR	Total sugars (gm)
DR1TTFAT	Total fat (gm)
DR1TTHEO	Theobromine (mg)
DR1TVARA	Vitamin A as retinol activity equivalents (mcg)
DR1TVB1	Thiamin (Vitamin B1) (mg)
DR1TVB12	Vitamin B12 (mcg)
DR1TVB2	Riboflavin (Vitamin B2) (mg)
DR1TVB6	Vitamin B6 (mg)
DR1TVC	Vitamin C (mg)
DR1TVD	Vitamin D (D2 + D3) (mcg)
DR1TVK	Vitamin K (mcg)
DR1TWS	When you drink tap water, what is the main source of the tap water? Is the city water supply (community water supply); a well or rain cistern; a spring; or something else?
DR1TZINC	Zinc (mg)
DRQSDIET	Are you currently on any kind of diet, either to lose weight or for some other health-related reason?
DRQSPREP	How often is ordinary salt or seasoned salt added in cooking or preparing foods in your household? Is it never, rarely, occasionally, or very often?
DR1_300	Was the amount of food that {you/NAME} ate yesterday much more than usual, usual, or much less than usual?
DR1_320Z	Total plain water drank yesterday - including plain tap water, water from a drinking fountain, water from a water cooler, bottled water, and spring water
DR1_330Z	Total tap water drank yesterday - including filtered tap water and water from a drinking fountain
DR1BWATZ	Total bottled water drank yesterday (gm)

Standard biochemistry profile

Variable name	Variable description
LBDSALSI	Albumin (g/L)
LBDSBUSI	Blood urea nitrogen (mmol/L)
LBDSCASI	Total calcium (mmol/L)
LBDSCHSI	Cholesterol (mmol/L)
LBDSCRSI	Creatinine (umol/L)
LBDSGBSI	Globulin (g/L)
LBDSGLSI	Glucose, refrigerated serum (mmol/L)
LBDSIRSI	Iron, refrigerated serum (umol/L)
LBDSPHSI	Phosphorus (mmol/L)
LBDSTBSI	Total bilirubin (umol/L)
LBDSTPSI	Total protein (g/L)
LBDSTRSI	Triglycerides, refrigerated (mmol/L)
LBDSUASI	Uric acid (umol/L)
LBXSAL	Albumin (g/dL)
LBXSAPSI	Alkaline phosphatase (IU/L)
LBXSASSI	Aspartate aminotransferase AST (IU/L)
LBXSATSI	Alanine aminotransferase ALT (IU/L)
LBXSBU	Blood urea nitrogen (mg/dL)
LBXSC3SI	Bicarbonate (mmol/L)
LBXSCA	Total calcium (mg/dL)
LBXSCH	Cholesterol (mg/dL)
LBXSCK	Creatine Phosphokinase(CPK) (IU/L)
LBXSCLSI	Chloride (mmol/L)
LBXSCR	Creatinine (mg/dL)
LBXSGB	Globulin (g/dL)
LBXSGL	Glucose, refrigerated serum (mg/dL)
LBXSGTSI	Gamma glutamyl transferase (U/L)
LBXSIR	Iron, refrigerated serum (ug/dL)
LBXSKSI	Potassium (mmol/L)
LBXSLDSI	Lactate dehydrogenase (U/L)
LBXSNASI	Sodium (mmol/L)
LBXSOSSI	Osmolality (mmol/Kg)
LBXSPH	Phosphorus (mg/dL)
LBXSTB	Total bilirubin (mg/dL)
LBXSTP	Total protein (g/dL)

(continued)

(continued)

Variable name	Variable description
LBXSTR	Triglycerides, refrigerated (mg/dL)
LBXSUA	Uric acid (mg/dL)
SEQN	Respondent sequence number

References

1. P. Alves Maranhão, G. Marísio Bacelar-Silva, D. Nuno Gonçalves Ferreira, C. Calhau, P. Vieira-Marques, R. João Cruz-Correia, Nutrigenomic Information in the openEHR Data Set
2. Angiogenesis Inhibitors. National Cancer Institute website. http://www.cancer.gov/about-can cer/treatment/types/immunotherapy/angiogenesis-inhibitors-fact-sheet. Reviewed October 7, 2011. Accessed November 30, 2017.
3. N. Bosanquet, K. Sikora, The economics of cancer care in the UK. Lancet Oncol. 5(9), 568–574 (2004 Sep). https://doi.org/10.1016/S1470-2045(04)01569-4 (PMID: 15337487)
4. P.C. Boutros, The path to routine use of genomic biomarkers in the cancer clinic. Genome Res. 25(10), 1508–1513 (2015 Oct). https://doi.org/10.1101/gr.191114.115.PMID:26430161; PMCID:PMC4579336
5. A. Blanchard, Mapping ethical and social aspects of cancer biomarkers (2016)
6. A.J. Brown, MD, Housestaff, C.L. Trimble, MD, Associate Professor, New technologies for cervical cancer screening (2012)
7. D. Callahan, What Price Better Health? Hazards of the Research Imperative, Vol. 9. University of California Press, Berkeley and Los Angeles (2003)
8. E.T. Cirulli, L. Guo, C.L. Swisher, N. Shah, L. Huang, L.A. Napier, E.F. Kirkness, T.D. Spector, C.T. Caskey, B. Thorens et al. Profound perturbation of the metabolome in obesity is associated with health risk. Cell Metab. 29(2), 488–500 (2019)
9. E.T. Cirulli,1,8,* L. Guo, 2, C.L. Swisher,1, N. Shah,1, L. Huang,1, L.A. Napier,1, E.F. Kirkness,1, T.D. Spector, 3, C. Thomas Caskey, 4, B. Thorens, 5, J. Craig Venter, 6, A. Telenti 7, Profound perturbation of the metabolome in obesity is associated with health risk
10. Clinicaltrials.gov. US National Institutes of Health website. http://clinicaltrials.gov. Accessed November 30, 2017
11. Commission staff working document on the use of 'omics' technologies in the development of personalised medicine. European Commission. EC, Brussels (2013)
12. T. Cordier, A. Lanzén, L. Apothéloz-Perret-Gentil, T. Stoeck, Embracing Environmental Genomics and Machine Learning for Routine Biomonitoring
13. J. Cuzick, M. Dowsett, S. Pineda et al., Prognostic value of a combined estrogen receptor, progesterone receptor, Ki-67, and human epidermal growth factor receptor 2 immunohisto-chemical score and comparison with the Genomic Health recurrence score in early breast cancer. J. Clin. Oncol. 29(32), 4273–4278 (2011)
14. M. Daniela Hurtado A1, A. Acosta 2, Precision Medicine and Obesity
15. C. Davies, J. Godwin, R. Gray et al., Early breast cancer trialists' collaborative, relevance of breast cancer hormone receptors and other factors to the efficacy of adjuvant tamoxifen: patient-level meta-analysis of randomised trials. Lancet 378(9793), 771–784 (2011)
16. M.G.M. de Lecea, M. Rossbach, Translational genomics in personalized medicine—scientific challenges en route to clinical practice. HUGO J 6, 2 (2012). https://doi.org/10.1186/1877-656 6-6-2
17. E.P. Diamandis, The failure of protein cancer biomarkers to reach the clinic: why, and what can be done to address the problem? BMC Med. 10, 87 (2012)

18. P. P. Dimitrios, The value of biomarker and genetic testing for patient management and optimization of resources (Under the supervision of Sotirios Bersimis, postgraduate thesis for the University of Pireaus EMBA program).
19. EC 10/3/2016. Regulatory framework applicable in the field of personalised medicine
20. L.M. Fleck, Pharmacogenomics and personalised medicine: wicked problems, ragged edges and ethical precipices. New Biotechnol. **29**(6). European Council conclusions on personalised medicine for patients. Publications Office of the European Union. European Council EC, Luxembourg (2012)
21. A.C. Gelijns, E.A. Halm (Eds.) The Changing Economics of Medical Technology (1991)
22. A. Gentry-Maharaj, PhD, Senior Research Associate, U. Menon, MD FRCOG, Professor in Gynaecological Oncology, Screening for ovarian cancer in the general population (2012)
23. B.J. Geronimus, D. Keene, M. Hicken, Black-white differences in age trajectories of hypertension prevalence among adult women and men, 1999–2002 (2007)
24. E. Giannini, F. Botta, A. Fasoli, P. Ceppa, D. Risso, P.B. Lantieri, G. Celle, R. Testa, Progressive liver functional impairment is associated with an increase in ast/alt ratio. Dig. Dis. Sci. **44**(6), 1249–1253 (1999)
25. D.A. Goldstein, J. Clark, Y. Tu, J. Zhang, F. Fang, R. Goldstein, S.M. Stemmer, E. Rosenbaum, A global comparison of the cost of patented cancer drugs in relation to global differences in wealth (2017)
26. P. González-Muniesa 1,2,3,4, J. Alfredo Martínez 1,2,3,4,5, Precision nutrition and metabolic syndrome management
27. D. Hammerl, M. Smid, A.M. Timmermans, S. Sleijfer, J.W.M. Martens, R. Debets. Breast cancer genomics and immuno-oncological markers to guide immune therapies
28. D. Hanahan, R.A. Weinberg, Hallmarks of cancer: the next generation (2011)
29. M.D. Helen Blumen, MBA, R.N. Kathryn Fitch, MEd, V. Polkus, MSEM, MBA, Comparison of Treatment Costs for Breast Cancer, by Tumor Stage and Type of Service (2015)
30. Hendrik Tobias Arkenau Sarah Cannon Research Institute. PD-L1 in Cancer: ESMO Biomarker Factsheet.
31. F.R. Hirsch, K. Suda, J. Wiens, P.A. Jr. Bunn, New and emerging targeted treatments in advanced non-small-cell-lung-cancer. Lancet **388**(10048), 1012–1024 (2016)
32. A. Hruby, F.B. Hu, The epidemiology of obesity: a big picture. Pharmacoeconomics **33**(7), 673–89 (2015). https://doi.org/10.1007/s40273-014-0243-x. PMID: 25471927; PMCID: PMC4859313.
33. https://academic.oup.com/annonc/article/28/9/2256/3868409
34. https://www.cdc.gov/genomics/gtesting/acce/
35. http://www.clinicaltrials.gov
36. https://gnomad.broadinstitute.org
37. https://hmdb.ca/bmi_metabolomics
38. https://towardsdatascience.com/predicting-micronutrients-using-neural-networks-and-random-forest-part-1-83a1469766d7
39. https://www.cdc.gov/obesity/data/adult.html
40. https://www.ncbi.nlm.nih.gov/medgen/18127
41. https://www.nutrigenetics.net/StartYourResearch.aspx
42. http://www.erapermed.eu
43. https://www.prb.org/wp-content/uploads/2020/11/TRA13-2008-obesity-economics-health-aging.pdf
44. https://www.fda.gov/drugs/biomarker-qualification-program/about-biomarkers-and-qualification
45. https://www.sciencedirect.com/topics/medicine-and-dentistry/metabolomics
46. H. Janes, M.D. Brown, M.S. Pepe, Y. Huang, An Approach to Evaluating and Comparing Biomarkers for Patient Treatment Selection
47. C. L. Johnson, S. M. Dohrmann, V. L. Burt, L. K. Mohadjer, National health and nutrition examination survey: sample design, 2011–2014. US Department of Health and Human Services, Centers for Disease Control and . . . (2014)

48. C. Kandoth, N. Schultz, A.D. Cherniack et al., Cancer Genome Atlas Research, Integrated genomic characterization of endometrial carcinoma. Nature, 497(7447), 67–73 (2013)
49. D. Levenson, Genomic testing update whole genome sequencing may be Worth the money (2012)
50. Lung Adenocarcinoma: The Evolving Role of Chemotherapy. OncLive website. http://www.onclive.com/insights/nondriver-lung-adenocarcinoma/lung- adenocarcinoma-the-evolving-role-of-chemotherapy. Posted July 5, 2017. Accessed December 13, 2017.
51. R. Martín-Hernández, G. Reglero, J.M. Ordovás, A. Dávalos. NutriGenomeDB: a nutrigenomics exploratory and analytical platform
52. S.C. Moore, C.E. Matthews, J.N. Sampson, R.Z. Stolzenberg-Solomon, W. Zheng, Q. Cai, Y.T. Tan, W.H. Chow, B.T. Ji, D.K. Liu, Q. Xiao, S.M. Boca, M.F. Leitzmann, G. Yang, Y.B. Xiang, R. Sinha, X.O. Shu, A.J. Cross, Human metabolic correlates of body mass index. Metabolomics 10(2), 259–269 (2014 Apr 1). https://doi.org/10.1007/s11306-013-0574-1.PMID:25254000; PMCID:PMC4169991
53. C.B. Newgard, J. An, J.R. Bain, M.J. Muehlbauer, R.D. Stevens, L.F. Lien, A.M. Haqq, S.H. Shah, M. Arlotto, C.A. Slentz, J. Rochon, D. Gallup, O. Ilkayeva, B.R. Wenner, W.S. Yancy Jr., H. Eisenson, G. Musante, R.S. Surwit, D.S. Millington, M.D. Butler, L.P. Svetkey, A branched-chain amino acid-related metabolic signature that differentiates obese and lean humans and contributes to insulin resistance. Cell Metab. 9(4), 311–326 (2009 Apr). https://doi.org/10.1016/j.cmet.2009.02.002.Erratum.In:CellMetab.2009Jun;9(6):565-6. PMID:19356713;PMCID:PMC3640280
54. A. Nicolini, P. Ferrari, M.J. Duffy, Prognostic and predictive biomarkers in breast cancer: Past, present and future (2017)
55. M. Olivier, M. Hollstein, P. Hainaut, TP53 Mutations in Human Cancers: Origins, Consequences, and Clinical Use (2010)
56. R. Pallares-Me´ndez, C. A. Aguilar-Salinas, I. Cruz-Bautista, and L. del Bosque-Plata, "Metabolomics in diabetes, a review," Annals of medicine, vol. 48, no. 1–2, pp. 89–102, 2016.
57. D. Panagoulias, D. Sotiropoulos, G. Thihrintzis, Nutritional biomarkers and machine learning for personalized nutrition applications and health optimization
58. C.S. Pareek, Sequencing technologies and genome sequencing (2011)
59. F. Pedregosa, G. Varoquaux, A. Gramfort, V. Michel, B. Thirion, O. Grisel, M. Blondel, P. Prettenhofer, R. Weiss, V. Dubourg, J. Vanderplas, A. Passos, D. Cournapeau, M. Brucher, M. Perrot, É. Duchesnay, Scikit-learn: machine learning in Python 12(85), 2825−2830
60. B. Peng, H. Li, X.-X. Peng, Functional metabolomics: from biomarker discovery to metabolome reprogramming. Protein Cell 6(9), 628–637 (2015)
61. B.A. Perkins et al., Precision medicine screening using whole-genome sequencing and advanced imaging to identify disease risk in adults (2017)
62. G.P. Pfeifer, P. Hainaut, Next-generation sequencing: emerging lessons on the origins of human cancer. Curr Opin. Oncol. 23(1), 62–68 (201). https://doi.org/10.1097/CCO.0b013e328341 4d00 (PMID: 21119514)
63. K.A. Phillips, PhD, J.A. Sakowski, PhD, J. Trosman, PhD, M.P. Douglas, MS, S.-Y. Liang, PhD, P. Neumann, ScD, The economic value of personalized medicine tests: what we know and what we need to know
64. M.S. Poptsova, I.A. Il'icheva, D.Y. Nechipurenko, L.A. Panchenko, M.V. Khodikov, N.Y. Oparina, R.V. Polozov, Y.D. Nechipurenko, S.L. Grokhovsky, Non-random DNA fragmentation in next-generation sequencing. Sci Rep. 31(4), 4532 (2014 Mar). https://doi.org/10.1038/sre p04532.PMID:24681819;PMCID:PMC3970190
65. E. Ravussin, Metabolic differences and the development of obesity. Metabolism 44, 12–14 (1995)
66. M. Roméo, L. Giambérini, History of Biomarkers (2012). https://doi.org/10.1201/b13036-3
67. K.A. Sikaris, The clinical biochemistry of obesity. Clin. Biochem. Rev. 25(3), 165 (2004)
68. K. Strimbu, J.A. Tavel, What are biomarkers? Current Opin. HIV AIDS 5(6), 463 (2010)
69. Stagnation, Innovation OR. "Innovation or Stagnation: Challenge and Opportunity on the Critical Path to New Medical Products." (2004).

70. J. Takahashi, N. Yumoto, AIST Tsukuba, Formulation of the basic grounds for health industry using biomarker database (2008)
71. A. Van Ooyen, B. Nienhuis, Improving the convergence of the back-propagation algorithm. Neural Netw **5**(3), 465–471 (1992)
72. E. Vayena, U. Gasser, Between Openness and Privacy in Genomic
73. M. Verma, P. Patel, M. Verma, Biomarkers in prostate cancer epidemiology. Cancers **3**(4) (2011).https://doi.org/10.3390/cancers3043773
74. A. von Bubnoff, Next-Generation Sequencing: The Race Is On (2008)
75. M. Wiesweg, S. Ting, H. Reis, K. Worm, S. Kasper, Feasibility of preemptive biomarker profiling for personalised early clinical drug development at a comprehensive cancer center (2012)
76. A.N. Winn, MPP, D.U. Ekwueme, PhD, G.P. Guy, Jr., PhD, P.J. Neumann, ScD., Cost-utility analysis of cancer prevention, treatment, and control
77. J. Woodcock, S. Buckman, F. Goodsaid, M.K. Walton, I. Zineh, Qualifying biomarkers for use in drug development: a US Food and Drug Administration overview. Expert Opin Med Diagn. **5**(5), 369–374 (2011 Sep). https://doi.org/10.1517/17530059.2011.588947 (PMID: 23484625)

Chapter 6
Inductive Machine Learning and Feature Selection for Knowledge Extraction from Medical Data: Detection of Breast Lesions in MRI

Evangelos Karampotsis, Evangelia Panourgias, and Georgios Dounias

Abstract This paper presents an approach to the problem of breast cancer diagnosis through the data analysis of magnetic mammography observations (MRi Data), developing corresponding hybrid classification models of patient cases into specific classes (e.g. Benign and Malignant). The aim of this work is the contribution of machine learning to the diagnostic process of breast cancer, offering a supportive intelligent tool that can be used by expert doctors as a medical decision-making aiding tool. Data were collected in collaboration with expert doctors and consist of 77 patient cases. The development of the presented classification models is a combination of inductive decision trees, clustering and feature selection techniques. Specifically, nine (9) different classification models were developed and evaluated by using statistical criteria, medical expert knowledge and where possible, using the Chi-Square statistical test. The performance achieved is considered encouraging for application in real-world practice, while further research is underway for associating MR imaging data with data from invasive examinations (biopsies).

6.1 Introduction

Breast cancer is the fifth leading cause of cancer death, but the most common cancer type for women with an incidence of 2.1 million cases per year and the first cause of cancer death in women, listing 627,000 reported deaths in 2018 [1].

E. Karampotsis · G. Dounias (✉)
Management and Decision Engineering Laboratory (MDE-Lab), School of Engineering, University of the Aegean, 41 Kountouriotou Street, 8210 Chios, Greece
e-mail: g.dounias@aegean.gr

E. Karampotsis
e-mail: ekarampotsis@aegean.gr

E. Panourgias
First Department of Radiology ('Aretaieio' Hospital), Medical School, National and Kapodistrian University of Athens, 76 V. Sofias, 11523 Athens, Greece

© The Author(s), under exclusive license to Springer Nature Switzerland AG 2022
G. A. Tsihrintzis et al. (eds.), *Advances in Assistive Technologies*, Learning and Analytics in Intelligent Systems 28, https://doi.org/10.1007/978-3-030-87132-1_6

The diagnostic process of breast cancer—like any cancer—involves two basic stages: (*i*) the calibration and (*ii*) the staging of histological lesion. During calibration, the degree of similarity between cancer and normal cells is determined (1: high similarity, 4: very low similarity), while during staging, the stage of cancer metastasis is determined (0: abnormal cells that have not spread to some nearby tissue—IV: cancer cells that have spread to distant sites) according to specific staging protocols such as TNM [2–4].

One of the most common and widely accepted techniques used in the diagnostic process and for staging of breast cancer is Magnetic Resonance Imaging (MRI). MRI is an established complementary mammography technique used to evaluate suspected breast histological lesions. The diagnostic act of breast cancer with the MRI method, although in the diagnosis of invasive cancers is very sensitive (89–100%), is characterized by low specificity ($\approx$73%) in the distinction between benign and malignant lesions, leading to further laboratory tests and surgeries, which in some cases are unnecessary [5].

In this paper, the efficient modelling of breast cancer diagnosis from magnetic mammography data through the development of hybrid intelligent techniques is presented. The aim is not only to obtain high classification accuracy but also to verify existing expert knowledge and if possible, to induce new knowledge for medical practice and then validate it through standard statistical tests. The results of the present research also aim to contribute in reducing costly and painful laboratory tests or unnecessary surgeries and ultimately, helping in the reduction of mortality from breast cancer.

6.2 Detailed Literature Review

In the recent years, various machine learning methods have been used in the development of intelligent systems for detecting, classifying, diagnosing and predicting breast cancer. In the respective approaches presented in literature, were used data collections resulting from

(i)   clinical observations (CLi Data—Clinical image Data),
(ii)   mammography observations (DMi Data—Digital Mammography image Data),
(iii)   observations of breast ultrasounds (USi Data—Ultrasound image Data)
(iv)   observations of microscope images collected through biopsy (MCi Data—Microscope image Data),
(v)   breast elastography observations (IRi Data—Infrared Thermography image Data and
(vi)   breast magnetic resonance observations (MRi Data—Magnetic—resonance image Data) [6, 7].

Wolberg and Mangasarian [8], presented one of the first approaches to the problem of classifying benign and malignant breast lesions. Specifically, by using the Multi-surface method (MSM) and with the utilization of MCi Data, they developed a corresponding classification model that managed to achieve a classification accuracy of 95.1%.

Wu et al. [9], in their approach to the problem of classifying benign and malignant breast lesions, presented a corresponding classification model analyzing DMi Data and using error Back Propagation Feed Forward Artificial Neural Networks (BP–FNN).

Furundzic et al. [10], in their attempt to optimize the eBP–FNN method yields in the classification of benign and malignant breast lesions, presented a corresponding classification model, which, apart from the classification process, included a feature selection using empirical knowledge. This model was developed using DMi Data and achieved a classification accuracy of 94.8%, while Pendharkar et al. [11] proposed a similar eBP–FNN classification model that was developed using CLi Data.

However, from the beginning of the decade 2000–2010, similar approaches were presented in the international literature at a faster pace, which were mainly based on both MCi Data and DMi Data. A typical example is the approach to the problem of breast cancer diagnosis, presented by Chou et al. [12], which approaches the problem of classifying benign and malignant breast lesions with a corresponding hybrid model in combination with the Multivariate Adaptive Regression Splines (MARS) and Artificial Neural Networks (ANN) approaches, applied on MCi Data, which reached a classification accuracy of 97.66%.

Nevertheless, according to Yassin et al. [6], the study of Sadaf et al. [13] is considered as a focal point of the development of corresponding intelligent models as well as of integrated Computer Aided Diagnostic (CAD) systems for breast cancer. This study examined the use of the widely used CAD version 7.2 (iCAD Inc.) system in full field mammograms (FFDM) and demonstrated that this CAD system has a 100% sensitivity to detect breast cancer masses, which initially manifested as microcatalysis and 86% for other mammographic breast lesions.

However, in the same year, Horsch et al. [14] reported that, when using intelligent models or integrated CAD systems of breast cancer, the sensitivity of detecting various breast lesions on mammograms increases from 80 to 90%. This fact boosted the respective field of research, focusing the researchers' interest on the development of new corresponding intelligent systems, utilizing data derived from other medical imaging techniques, such as MRI Data.

Dietzel et al. [15] classified the benign and malignant breast lesions using an eBP FNN classifier, developed using MRI Data from 1012 patients and showing an AUC (Area Under the ROC Curve) of 88.8%.

Hassanien and Kim [16], utilizing MRi Data from 120 patients and using the method of SVM (Support Vector Machines), developed a classification model which classifies the findings of magnetic mammograms into normal and abnormal with a 98% classification accuracy.

Milenković et al. [17], applying the LS–SVM (Least-Squares Support Vector Machine) classification method to data from Dynamic Contrast–Enhanced Magnetic-Resonance Imaging (DCE–MRI) observations from 115 patients, presented a classification model with a sensitivity equal to 95% and a specificity equal to 78.19%.

Baltzer et al. [18], utilizing MRi Data from 1084 patients and using the CHAID (Chi-squared Automatic Interaction Detection) method, presented a classification tree of benign and malignant breast lesions with a classification accuracy of 98.4% while Hoffmann et al. [19], using the same type of data from 84 patients, approached the problem of diagnosing breast lesions that do not increase cancerous mass, using the SVM method.

Bhooshan et al. [20], utilizing magnetic mammography data from both High Spatial and Spectral resolution MRI (HiSS–MRI) and Dynamic Contrast-Enhanced MRI (DCE MRI) from 51 patients and using the hybrid method B–ANN (Bayesian Artificial Neural Networks), presented a classification model of benign and malignant breast lesions, which achieves an AUC equal to 92% (HiSS–MRI Data) and AUC = 90% (DCE–MRI Data).

Weiss et al. [21], using magnetic mammography data (HiSS–MRI Data) from 23 patients and using the FCM method (Fuzzy c-Means), presented a model of classification rules corresponding to benign and malignant breast lesions, which achieves an AUC of 88%.

Agner et al. [22], utilizing the MRI Data from 76 patients, developed a hybrid model combining Linear Discriminant Analysis (LDA) as feature selection method and SVM as classification method. The approach manages to correctly classify (AUC = 77%) the cases of patients according to four (4) specific categories (triple-negative, ER-positive, HER2-positive and fibroadenoma) of breast lesions.

Soares et al. [23] presented an assessing model for breast lesion through the use of MRI. This model was developed using MRI Data from 70 patients and classified the detected breast lesions by the method of SVM (Acc = 94%), in two (2) large classes, namely "Probably Malignant" and "Probably Benign".

Huang et al. [24] evaluated the findings of magnetic resonance imaging in benign and malignant lesions using the FCM method and using data from similar observations from 61 patients. This classification model obtained a classification accuracy of 100%. Similar performance was achieved by Yang et al. [25] who presented a corresponding classification model (AUC = 0.919), which was developed utilizing data from magnetic mammography observations from 115 patients and using the SVM method.

Gubern-Mérida et al. [26] presented a classification model of benign and malignant breast lesions, which was developed using MRI Data from 209 patients using the Random Forest (RF) method. This classification model achieved a classification accuracy of 89%.

Waugh et al. [27] presented a model for classifying three (3) subtypes of non-invasive breast lesions. This model was developed using the KNN (K-Nearest Neighbor) method utilizing MRI Data from 200 patients and reached a classification accuracy of 74.7%.

Vidić et al. [28] presented a model for classifying benign and malignant breast lesions developed using MRI Data from 51 patients using the SVM method. This classification model reached a classification accuracy of 96%.

Truhn et al. [29], analyzed MRI Data from 447 patients using Convolutional Neural Networks (CNN), and came up with a model for classifying benign and malignant breast lesions, achieving an AUC of 88%.

Ha et al. [30] presented a model for classifying four (4) subtypes of breast cancer (Luminal A, Luminal B, HER2+ , and Basal Breast Tumors), which was developed using MRI Data from 216 patients and the CNN method. This classification model reached a classification accuracy of 70%.

Ji et al. [31], using MRI Data from 1979 patients and the SVM method, presented a classification model of benign and malignant breast lesions with an AUC of 88%.

D'Amico et al. [32], analyzed MRI Data from 68 patients using the KNN method, and managed to classify benign and malignant breast lesions with a classification accuracy of 94.12%.

Ellmann et al. [33], proposed the hybrid method named Polynomial Kernel Function Support Vector Machine (PFK-SVM) and applied it on MRI Data from 178 patients, building a classification model of benign and malignant breast lesions with an AUC of 90.1%.

Finally, Parekh et al. [34] approached the problem of classifying benign and malignant breast lesions with a corresponding hybrid classification model, which was developed by combining the SVM and Sparse Autoencoder (SA) methods, applied on MRI Data from 195 patients. This model achieved an AUC of 90%.

According to the above literature review, most of the approaches regarding the diagnostic process modelling of breast cancer using available MRI Data, were based on Artificial Neural Networks and Support Vector Machines. Methods that are either considered black–box methods (difficult to be understood by medical experts), or do not offer the necessary medical knowledge that could improve the medical decision-making process. This paper uses intelligent techniques based on inductive machine learning algorithms, aiming at producing handy and reliable results (i.e. sets of rules) but also comprehensible and capable of brining diagnostic practice on step beyond by attempting to discover new useful medical knowledge.

6.3   Presentation of the Data

This section presents the data collected for this paper. Specifically, the following subsections describe the data collection process, the diagnostic attributes and the diagnostic classification (decisions) and the data preprocessing.

6.3.1 Data Collection

The original dataset has been collected by expert doctors of Aretaieio Hospital during the follow-up of 112 patients with suspicious breast MRI lesions (BI-RADS 4 or 5) and correlated with pathology over the last five (5) years (2015–2020).

The American College of Radiology Breast Imaging Reporting & Data System lexicon (ACR BI-RADS) was formulated to rate breast lesions and to predict both benign and malignant disease based on imaging features. BI-RADS is not a clinical decision rule, differentiating a benign lesion from a malignant tumor, but rather a lexicon that provides a common language for lesion description and standardized interpretation criteria.

Breast magnetic resonance imaging (b-MRI) exams were acquired on a 1.5 T and a 3 T scanner (Phillips Healthcare, Best, Netherlands) using a surface coil on the 1.5 T machine and a dedicated double breast coil, with the patient in a prone position.

The MRI protocol included: axial fat-suppressed T2-Weighted images (TR/TE, 10,000/70 ms); Field of View (FOV) 360 × 360 mm; Matrix, 288 × 288; Slice thickness (ST), 2.5 mm, Axial T2-Weighted Turbo Spin Echo (TSE) Images (TR/TE, 5000/120 ms); FOV, 359 × 359 mm; Matrix, 560 × 560; ST, 2.0 mm), Axial T1-Weighted Images (TR/TE), 550/8 ms; FOV, 360 × 360 mm; Matrix, 512 × 512; ST, 2.5 mm and Axial Diffusion-Weighted Images (DWI) (TR/TE, 2400/63 ms); FOV, 391 × 391 mm: Matrix, 288 × 288; ST, 2.5 mm. Diffusion-Weighted Imaging and Apparent Diffusion Coefficient (ADC) values were calculated using b-values of 0 and 800 s/mm^2. Diffusion-Weighted Imaging Sequences were performed before the Dynamic Contrast-Enhanced (DCE) imaging sequences. Axial high spatial resolution T1-weighted with fat suppression DCE images were also obtained, including one native and five (5) acquisitions after gadolinium injection (TR/TE, 5/2 ms); flip angle 10°; FOV, 362 × 362 mm; Matrix, 640 × 640; ST, 1.2 mm; time of acquisition of each dynamic series, 70 s). Post-processing of the dynamic study included the calculation of time–signal intensity curves and subtraction of unenhanced images from the contrast-enhanced dynamic images.

The MRI studies were reviewed by two (2) independent readers (7 and 14 years of experience in breast imaging, respectively), who were blinded to the histology results and clinical data of each patient. In cases of disagreement, consensus was reached.

Diagnostically important lesions show pathological enhancement on MRI studies and should be classified morphologically according to the BI-RADS lexicon into a mass (3D-enhancing structure), area of non-mass enhancement (NME) or a focus (<5 mm area of enhancement).

In every lesion, 13 descriptors published in the BI-RADS lexicon were assessed and documented in a database.

6.3.2 Description of Variables

In the context of the data collection process, various cases of patients are continuously collected and examined in collaboration with field experts, having as ultimate goal to create a database that can be used for future medical data analysis. The dataset that was used in this paper is a sample of the first 77 patient cases collected so far and includes:

– clinical observations (1 variable),
– imaging observations (13 variables) derived from magnetic mammography,
– histological observations (4 variables) derived from biopsy examinations, and
– diagnostic observations (4 variables).

Age (AGE)

The variable Age (AGE) expresses the age of the patient in years. It is a numerical variable with discrete values $[0.1.2, 3..., n]$.

Morphology (MORPH)

The variable Morphology (MORPH) characterizes the morphology of the lesion located in the magnetic mammography images. It is a nominal variable and takes the following values:

– MASS: It is a three-dimensional (3D), space-occupying lesion within the breast parenchyma that infiltrates or displaces it. If it has irregular or spiculated margins it is more often associated with an invasive cancer, usually Invasive Ductal Cancer (IDC),
– MASS & NON-MASS: This includes a 3D lesion and an area on non-space occupying enhancement. It is commonly due to an invasive cancer with an associated Ductal Carcinoma in Situ (DCIS) component,
– NON-MASS: A regional or segmental in nature, non-space occupying and respects the surrounding breast tissue. Several benign or malignant processes are associated with this morphology, usually DCIS.

Borders (BDS)

The variable Borders (BDS) describes the outer shell (Margins or Borders) of a lesion. It is a nominal variable and receives the values:

– IRR: Irregular or spiculated margins, commonly seen in invasive cancers),
– SPIC: Irregular or spiculated margins, commonly seen in invasive cancers (spiculations are due to a desmoplastic process),
– SMTH: Well-defined, smooth margins, indicative of a benign process such as a fibroadenoma or a cyst.

Tumor Size (TUMS)

The variable Tumor Size (TUMS) is a numerical variable that expresses the diameter of the localized tumor in centimeters.

Peritumoral Edema (PRED)

The variable Peritumoral Edema (PRED) characterizes the presence of edema of the localized lesion. It is a nominal variable and receives the values:

- NO: It is not observed peritumoral edema (usually indicates a slower growing tumor)
- YES: It is observed peritumoral edema (an ominous sign in breast cancer and usually indicative of a high-grade process).

T2-Weighted Imaging (T2-Wi)

The variable T2-Wi represents the T2-weighted image. On T2 weighted images, water has a very high signal intensity. Hypercellular lesions such as cancer display hypointensity on T2-weighted images. It is a nominal variable and gets the following values:

- NONE: Usually blends in with the rest of the breast parenchyma and is non-specific,
- LOW & INTER: Low and intermediate intensity on T2 weighted images usually indicates a process that is more cellular and has less extracellular fluid,
- HIGH: High water content usually indicates a more benign process.

Curve Morphology (CRM)

The variable Curve Morphology (CRM) characterizes the type of kinetic curve that describes the behavior of the localized lesion over time after the administration of the shading fluid. It is a nominal variable and takes the following values:

- TYPE_1: Persistent or type 1 curve is more often seen in benign lesions or low-grade tumors,
- TYPE_2: Type 2 curve is seen 60% in malignant tumors and 40% benign, highly proliferative tumors),
- TYPE_3: Type 3 kinetic curve is seen in malignant lesions in over 95% times.

Breast Density (BD)

The variable Breast Density (BD) expresses the quantity of fibroglandular tissue in relation to breast fat. It is categorized as A to D according to the amount of breast tissue. in the ACR BI-RADS lexicon. It is a nominal variable and takes the following values:

- A: The breasts are almost entirely fatty,
- B: Scattered areas of fibroglandular tissue,

- C: Areas of relatively dense fibroglandular tissue,
- D: Extremely dense fibroglandular tissue.

Background Parenchymal Enhancement (BPE)

The variable Background Parenchymal Enhancement (BPE) expresses the contrast enhancement of fibroglandular tissue. The BI-RADS lexicon provides descriptors for the level of enhancement. the level of enhancement is estimated in the first post-contrast acquisition (approximately 90 s). it is a nominal variable and takes the following values:

- MIN: There is minimum background parenchymal enhancement,
- MILD: There is mild background parenchymal enhancement,
- MOD: There is moderate background parenchymal enhancement,
- MARK: There is marked background parenchymal enhancement.

Feeding Vessel (FV)

The variable Feeding Vessel (FV) is a descriptor for differentiating malignant from benign breast lesions on breast MRI. Malignant lesions >2 cm more often present with a FV. It is a nominal variable and has the following values:

- NO: It is not observed feeding vessel (small tumor or benign lesion),
- YES: It is observed feeding vessel (malignant lesion).

Internal Enhancement (INTEN)

The variable Internal Enhancement (INTEN) interprets the enrichment (of shading fluid) of a lesion. It is a nominal variable and has the following values:

- HOMOG: Homogeneous enhancement. It is usually seen in benign lesions,
- INHOMOG/HETTER: Rim, peripheral or inhomogeneous/heterogeneous enhancement. It is most common in cancerous tumors.

Diffusion (DF)

The Diffusion (DF) rate of water molecules is restricted in cases of lesions with a high volume of cells, such as cancerous tumors and assists in differentiating between benign and malignant processes. It is a nominal variable and takes the following values:

- N/A: Not available,
- LOW: Low diffusion (malignant tumor most frequently, less common -hematoma or abscess),
- HIGH: High diffusion (benign lesions).

Apparent Diffusion Coefficient (ADC)

The variable Apparent Diffusion Coefficient (ADC) provides information that depends on the molecular motion of water between tissue compartments. The diffusion rate of water molecules is restricted in cases of lesions with a high volume of cells, such as cancerous tumors and assists in differentiating between benign and malignant processes. It is a nominal variable and takes the following values:

– N/A: Not available,
– LOW: Low apparent diffusion coefficient,
– HIGH: |high apparent diffusion coefficient.

Focality (FC)

The variable Focality (FC) characterizes the focus of the detected fault. It is a nominal variable and has the following values:

– U: Unifocal focality,
– MC: Multicentric focality,
– MF: Multifocal focality.

Malignant Lesions (ML)

The variable Malignant (ML) represents the malignant lesions. The malignant lesions that exist in the used data are:

– DCIS: Ductal carcinoma in situ,
– IDC: Invasive *ductal* carcinoma,
– DCIS & IDC: Ductal carcinoma in situ and invasive *ductal* carcinoma,
– ILC: Invasive lobular carcinoma,
– SOLID PAPILLARY: Carcinoma.

BIRADS

The variable Breast Imaging-Reporting and Data System (BIRADS) represents the classification of the detected faults, according to the BI-RDADS protocol. The BIRADS categories that exist in the used data are:

– CATEGORY_3: Probably benign, requiring short term follow-up in 6 months,
– CATEGORY_4; Suspicious finding requiring biopsy,
– CATEGORY_5: Highly suspicious, biopsy mandatory,
– CATEGORY_6: Known, histologically-verified cancer.

Tumor Grade (TUMG)

Tumor Grade (TUMG) is an indicator of how quickly a tumor is likely to grow and spread. It is a nominal variable and takes the following values:

- GRADE_1: Cancer cells look like normal breast cells and is usually slow-growing,
- GRADE_2: Cancer cells look less like normal cells and is growing faster,
- GRADE_3: Cancer cells looks different to normal breast cells and is usually fast-growing.

Estrogen Receptors and Progrsterone Receptors (ER & PR)

If breast cancer cells have estrogen receptors, the cancer is called ER-positive breast cancer. If breast cancer cells have progesterone receptors, the cancer is called PR-positive breast cancer. If the cells do not have either of these 2 receptors, the cancer is called ER/PR-negative. They are nominal variable and has the values NO and YES.

Protein Cerb-B2

Cerb-B2 is a protein that appears on the surface of some breast cancer cells. Overexpression of Cerb-B2 is detected in up to 20% of patients and is characterized by rapid tumor growth, lower survival rate and increased disease progression. It is a nominal variable and has the values NO and YES.

Cell Proliferation Ki-67

The expression of Ki-67 (%) is strongly associated with cell proliferation and is widely used in routine pathology. Ki-67 is well characterized at the molecular level and extensively used as a prognostic and predictive marker in cancer (Table 6.1).

6.3.3 Dataset Preprocessing

As it has already been mentioned, there are different types of variables (quantitative/numerical and qualitative/nominal). Also, as observed in the frequency charts (Annex 6.2), the quantitative variables (Age and Tumor Size) do not follow a normal distribution and the groups of values from most variables are not balanced. These are two facts that can add extra statistical noise to the data and make the adaptive of machine learning algorithms more difficult.

In an attempt to reduce the statistical noise of the data and at the same time to increase the adjustment of machine learning algorithms to them, a new clustering process of the attribute values was performed in collaboration with field experts (MDs). The table in Annex 6.3 presents the adopted clustering process, from which the final dataset came up.

The final dataset consists of 23 qualitative (nominal) attributes of which the two (2) variables (BOM—Benign or Malignant & BAM—Benign and Malignant) are composite variables derived from the variables Benign and Malignant.

Table 6.1 Summary table describing the variables of the dataset used in this paper

Variables' description

S. no.	Type of observations	Variable	Full name of variable	Observations Missing	Total
1	Clinical	AGE	Age (years)	0	77
2	Imaging	MORPH	Morphology	0	77
3		BDS	Borders	0	77
4		TUMS	Tumor Size (cm)	0	77
5		PRED	Peritumoral Edema	0	77
6		T2-Wi	T2 Weighted image	0	77
7		CRM	Curve Morphology	0	77
8		BD	Breast Density	0	77
9		BPE	Background Parenchymal Enhancement	0	77
10		FV	Feeding Vessel	0	77
11		INTEN	Internal Enhancement	0	77
12		DF	Diffusion	0	77
13		ADC	Apparent Diffusion Coefficient	0	77
14		FC	Focality	0	77
15	Diagnostic	BIRADS	Breast Imaging-Reporting and Data System	0	77
16		BN	Benign	0	19
17		ML	Malignant	2	58
18		TUMG	Tumor Grade	1	58
19	Histological	ER	Estrogen Receptors	1	58
20		PR	Progesterone receptors	1	58
21		Cerb-B2	Cerb-B2 Protein	5	58
22		Ki-67	Cell Replication Rate	5	58

6.4 Methodology

Below the classification methodology adopted in this paper for detecting breast lesions in MRI, is presented. Specifically:

(a) the classification problems arising from the dataset used, are given,
(b) the feature selection process and the classification method used in the development of the respective classification models, and

(c) their validation methodology is provided, which was carried out on the basis of statistical evaluation criteria and based on medical significance by utilizing the knowledge and experience of collaborating expert doctors.

6.4.1 Modeling Methodology

The methodology adopted in this study aims to develop various classification models that will contribute to the diagnostic process of breast cancer by offering relevant new medical knowledge and/or will confirm existing relevant medical knowledge. The classification models presented in this work were developed utilizing the final dataset previously mentioned. Table 6.2 presents the input variables used in all classification models and Table 6.3 presents the output variables from each classification model. The variables included in Table 6.2 have been classified according to their clinical significance (as defined by the collaborating expert doctors) in ascending order (The variable BPE—Background Parenchymal Enhancement has the lower clinical significance and the variable AGE has the higher clinical significance).

Table 6.3 shows that there are nine (9) different classification problems in which exactly the same variables (14 variables) were used as input variables (Table 6.2) and in each classification problem the variables of Table 6.3 (9 variables) were used as target variables.

Table 6.2 Input variables of classification approaches

Input variables

S. no.	Variable type	Variable	Full name of variable	Cases	
				Missing	Total
1	Nominal	BPE	Background Parenchymal Enhancement	0	77
2		BD	Breast Density	0	77
3		TUMS	Tumor Size	0	77
4		FC	Focality	0	77
5		FV	Feeding Vessel	0	77
6		T2-Wi	T2 Weighted image	0	77
7		DF	Diffusion	0	77
8		ADC	Apparent Diffusion Coefficient	0	77
9		CRM	Curve Morphology	0	77
10		INTEN	Internal Enhancement	0	77
11		PRED	Peritumoral Edema	0	77
12		BDS	Borders	0	77
13		MORPH	Morphology	0	77
14		AGE	Age (years)	0	77

Table 6.3 Target variables of each classification approach

Target variables

S. no.	Variable type	Variable	Full name of variable	Cases	
				Missing	Total
1	Nominal	BOM	Benign or Malignant	0	77
2		BAM	Benign and Malignant	0	77
3		ML	Malignant	19	58
4		BIRADS	Breast Imaging-Reporting and Data System	0	77
5		TUMG	Tumor Grade	21	56
6		ER	Estrogen Receptors	20	57
7		PR	Progesterone receptors	20	57
8		Cerb-B2	Cerb-B2 Protein	20	57
9		Ki-67	Cell Replication Rate	24	53

In the *1st Classification Problem*, the aim was to classify all of patients' cases into Benign and Malignant (2–class classification problem), i.e. having findings pointing the existence of benign or malignant tumor. In this classification problem 77 patterns were used to develop and validate the corresponding classification model that has as target variable the composite variable which represents the existence of Benign or Malignant tumor (Target Variable: BOM—Benign or Malignant).

In the *2nd Classification Problem*, a classification model was developed utilizing the data of 77 patient cases and with the aim to classify these patient cases into benign or into specific types of malignancy (6–class classification problem). The target variable of this problem was the composite variable that represents the existence of both Benign and specific types of Malignant tumors (Target Variable: BAM—Benign and Malignant).

In the *3rd Classification Problem*, the aim was to classify the malignant cases of patients into specific classes of malignancy (5–class classification problem). The target variable of this problem was the variable which represents the type of Malignant (Target Variable: ML—Malignant) and the development of this model was made using 58 patient cases with malignant lesions.

In the *4th Classification Problem*, the problem of classifying 77 cases of patients in specific BIRADS classes was approached (2–class classification problem). The target variable of this problem was the variable that includes specific classes of the Breast Imaging-Reporting and Data System (Target Variable: BIRADS—Breast Imaging-Reporting and Data System).

In the *5th Classification Problem*, the aim was the classification of the malignant cases of patients into specific tumor grade classes (3–class classification problem). The target variable of this problem was the variable that includes specific classes of Tumor Grade (Target Variable: TUMG—Tumor Grade) and the classification model was developed using 56 pattern cases,

In the *6th Classification Problem*, the malignant cases of patients (57 patient cases) were classified into ER classes (2–class classification problem. The target variable of this problem was the variable which represents the existence of estrogen receptors (Target Variable: ER—Estrogen Receptors).

In the *7th Classification Problem*, the aim was the classification of the malignant cases of patients (57 patient cases) into PR classes (2–classes classification problem. The target variable of this problem was the variable which represents the existence of progesterone receptors (Target Variable: PR—Progesterone Receptors).

In the *8th Classification Problem*, the malignant cases of patients were classified cording to the presence of Cerb-B2 protein (2–class classification problem. The target variable of this problem was the variable which represents the existence of Cerb-B2 protein (Target Variable: Cerb-B2).

Finally, in the *9th Classification Problem*, using as target variable the variable that expresses the replication rate of cancer cells (Target Variable: Ki-67), a classification model was developed that classifies 53 cases of patients with malignant lesions into 3 specific Ki-67 classes (3-class classification problem).

6.4.2 Feature Selection Process

The feature selection process used in all classification problems aims at identifying the best possible combination of input variables, which will give the highest classification accuracy (Acc) in each classification problem and consists of two (2) parts: (*i*) classification of variables and (*ii*) search of the best input variables combination.

In the first part, the candidate input variables are classified according to their clinical significance (Table 6.2). The candidate input variables are classified in ascending order (from least important to most important), according to the knowledge and experience of the collaborating expert doctors.

In the second part, a search technique of the best input variables combination is applied. In this search technique, classification experiments are performed, using a common classification algorithm (in standard settings for all experiments) and a common validation method.

In the first phase of experiments (Exp_n), the classification vectors, that have been derived from the data set used in this paper (MRI77N), are classified into the classes of each classification problem using all candidate input variables (n variables) and the classification accuracy of the produced model is recorded (Acc_n).

In the second phase of experiments (Exp_{n-1}), all candidate input variables are used, except the variable with the least clinical significance ($n-1$ input variables). The classification accuracy of the model with $n-1$ input variables (Acc_{n-1}) is recorded and compared with the classification accuracy of the model with n variables (Acc_n):

- If the classification accuracy of the model with $n-1$ input variables is higher than the classification accuracy of the model with n input variables ($Acc_{n-1} > Acc_n$), then we move on to the next phase of experiments (Exp_{n-2}), where one more input variable (the one with the second less clinical significance) is subtracted.

- If the classification accuracy of the model with $n-1$ input variables is less than or equal to the classification accuracy of the model with n variables ($Acc_{n-1} \leq Acc_n$), then the removed variable is repositioned as input variable, the second least clinically significant variable is removed and the classification experiment is repeated. This process is repeated until the classification accuracy of the model with $n-1$ input variables becomes higher than the classification accuracy of the model with n input variables ($Acc_{n-1} > Acc_n$) and terminates when all possible input variables are exhausted.

In the third phase of experiments (Expn-2), all candidate input variables are used, except the variable resulting from the previous experiments phase (Expn-1) and the variable which has higher clinical significance than the variable resulting from the previous experiments phase (Expn-1). The classification accuracy of the model with $n-1$ input variables (Acc_{n-1}) is recorded and compared with the classification accuracy of the previous models with n variables (Acc_n and Acc_{n-1}):

- If the classification accuracy of the model with n-2 input variables is higher than the classification accuracy of the previous models ($Acc_{n-2} > Acc_{n-1}$, Acc_n), then we move on to the next phase of experiments (Exp_{n-3}), where one more input variable (the third most less clinical significance) is subtracted.
- If the classification accuracy of the model with n-2 input variables is less than or equal to the classification accuracy of the previous models ($Acc_{n-2} \leq Acc_{n-1}$, Acc_n), then the removed variable is repositioned as input variable, the variable which has more clinical significance than the variable removed in this experiments phase is subtracted and the classification experiment is repeated. This process is repeated until the classification accuracy of the model with n-2 input variables becomes higher than the classification accuracy of the previous models ($Acc_{n-1} > Acc_{n-1}$, Acc_n) and terminates when all possible input variables are exhausted.

The above process terminates when all possible input variables are checked or when the number of input variables reaches a certain limit (e.g., if we have ten possible input variables and we want to use nine of them, the tests will end in the first phase of experiments) (Fig. 6.1).

6.4.3   Classification Method

In all classification problems presented in this work, during the classification process, the method of inductive decision trees (C5.0) was used [35]. The C5.0 algorithm is an updated commercial version of the C4.5 algorithm [36]. Compared to more advanced machine learning methods (such as Artificial Neural Networks and Support Vector Machines), this algorithm performs equally well (especially on nominal data), develops corresponding classification models much more easily and quickly, and is well understood and accepted by field experts [35].

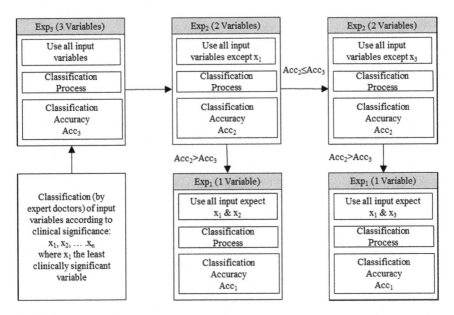

Fig. 6.1 Flowchart of proposed feature selection process (Example with three (3) variables)

Generally C5.0 algorithm, in order to address the basic challenge of induction decision trees, which corresponds in finding the root and the branches of a tree, chooses the most suitable feature (variable) in every development stage of a decision tree, using entropy information criteria, specifically the information gain criterion [36]:

$$\text{Gain}(S, A) = E(S) - \sum_{i=1}^{m} \Pr(A_i) \cdot E(S_{Ai}) \qquad (6.1)$$

where:

S: cases set,
A: variable,
m: number of values of variable A in S,
$\Pr(A_i)|$: frequency of cases that have A_i value in S,
$E(S_{Ai})|$: subset of S with items that have A_i value,
$E(S)$: information entropy of S, and
$\text{Gain}(S, A)$: gain of S after a split on attribute A.

Entropy is calculated before information gain is calculated and used to determine how "informative" an attribute is. The basic formula of entropy:

$$E(S) = - \sum_{i=1}^{n} \Pr(G_i) \cdot \log_2 \Pr(G_i) \qquad (6.2)$$

where:

S: cases set,
G_i: frequency of class C_i in S,
n: number of classes in S, and
E(S): information entropy of S.

A key feature of C5.0 algorithm over other inductive decision trees algorithms is its pruning process of the under-construction decision tree. Specifically, algorithm C5.0 constructs firstly a large tree that fits closely into the input data and then removes parts of the tree that have a large effect on the overall classification error. Using the specific pruning procedure of the C5.0 algorithm significantly reduces the chances of the phenomenon of over-fitting of the tree on its training data [35].

Finally, an important feature of C5.0 algorithm is the ability to transform a classification tree to classification rules called ruleset. The transformation of the tree into a ruleset consists of the process of converting the tree paths into simple "IF/THEN" rules and the process of pruning each rule (having as an evaluation criterion the classification accuracy of each rule) to yield the final ruleset. The main advantage of the ruleset is that they are generally more understandable than trees, as it describes with simple logic sentences a specific context associated with a class. Other advantages of the ruleset are that they contain fewer classification rules than the rules derived from tree paths and in some cases are more accurate predictors than trees [37].

6.4.4 Validation Process

The validation process of the produced classification models was the same in each classification problem and was chosen to be the "Leave–One–Out" approach in order to utilize the maximum possible information available during the training phase. The evaluation statistical measures used in this paper were the well-known classification accuracy, the sensitivity and the specificity of the respective classification models [38] (Table 6.4):

$$\text{Sensitivity}(\%) = \frac{\text{TP}}{\text{TP} + \text{FN}} \cdot 100 \tag{6.3}$$

$$\text{Specificity}(\%) = \frac{\text{TN}}{\text{TN} + \text{FP}} \cdot 100 \tag{6.4}$$

Table 6.4 Example of confusion matrix of two (2) classes

Confusion matrix of class A and B		
	Predicted class A	Predicted class B
Actual class A	True Positive (TP)	False Negative (FN)
Actual class B	False Positive (FP)	True Negative (TN)

$$\text{Accuracy}(\%) = \frac{\text{TP} + \text{TN}}{\text{TP} + \text{TN} + \text{FP} + \text{FN}} \cdot 100 \qquad (6.5)$$

The validation of the ruleset was carried out both with the use of specific statistical criteria, as well as with the use of the knowledge and experience of the collaborating expert doctors.

To use the following statistical evaluation criteria, the rules of the ruleset were transformed into association rules of the $X \rightarrow Y$ format, where:

X: the antecedent (the combination of input variables with the corresponding values, for each rule), and
Y: the consequent (the class of each rule) from the ruleset.

Specifically, the statistical criteria used to evaluate the ruleset presented in this paper were:

(i)   the support for a $X \rightarrow Y$ rule, which is the percentage of cases (to the total of cases) that contain X and Y at the same time (Eq. 6.6) [39],

(ii)   the confidence for a $X \rightarrow Y$ rule, which defines how many of the cases containing X, also contain Y as a percentage of the total number of cases containing X. The C5.0 algorithm uses the following equation (Laplace ratio) to estimate the confidence of each rule (Eq. 6.7) [40].

The lift (or interest), which is a value that represents the number of times X and Y appear at the same time compared to the number of times Y would appear if it were statistically independent of X (Eq. 6.8) [39, 41].

$$\text{supp}(X \rightarrow Y) = P(X \cup Y) = \frac{N(X \cup Y)}{N(\Omega)} \qquad (6.6)$$

Rules with small support are often uninteresting as they do not describe significantly large populations [39].

$$\text{conf}(X \rightarrow Y) = \frac{(X \cup Y) + 1}{X + 2} \qquad (6.7)$$

A rule with high confidence (e.g. close to 1) is quite important as it gives the impression that it can classify new cases with correspondingly high classification accuracy [39].

$$\text{lift}(X \rightarrow Y) = \frac{\text{conf}(X \rightarrow Y)}{\text{supp}(Y)} \qquad (6.8)$$

When:

– lift < 1: X and Y appear less frequently together in the data than expected. X and Y are negatively interdependent.

- lift $= 1$: X and Y appear as frequently together as expected. X and Y are negatively interdependent. X and Y are independent of each other.
- lift > 1: X and Y appear more frequently together in the data than expected. X and Y are positively interdependent.

Also, in this paper, we attempt to confirm statistically the produced classification models using known statistical tests. Specifically, because the data used in this work are nominal, the Chi-Square test was used, where possible. The aim of this approach in the evaluation of classification models was to utilize the above statistical criteria (support, confidence and lift) with ultimate goal the better definition of the statistical significance of the classification rules that emerged from each classification model using known and more understandable statistical evaluation measures, such as the value X^2.

Chi-Square Analysis is a well-known statistical method commonly used for the assessment of the independence of two (2) random variables. In the case of a ruleset this method is used to estimate the independence between the antecedent X and the consequent Y of a rule, using the Eq. (6.9) [42], see next page.

Chi–Square Analysis (X^2), when applied to the classification rules, that have been transformed into association rules, does not assess the correlation strength of the two (2) parts (antecedent X and consequent Y) of a rule, but helps to make the decision related to the independence of the two (2) parts of a rule [41].

A necessary condition for the application of X^2 test is that the 80% of the cells of the expected-frequency table should have a value >5, see [39].

During the X^2 test, the calculated value X^2 is compared with a critical value X^2 (X_c^2). The critical X^2 depends on the degree of freedom (df) of X^2 test relevance table. If the calculated value X^2 is less than the critical value X^2, then the two (2) under checking variables are independent [42].

In the case of using this test in a rule, the degree of freedom is equal to the unit (df $= 1$), the trial probability is equal to 0.05 and the critical X^2 is equal to 3.841 $(X_c^2 = 3.841)$, as the correlation table between the two (2) parts (antecedent X and consequent Y) of each rule is 2×2. If the calculated value X^2 is <3.841, then the two (2) parts (antecedent X and consequent Y) of under check rule are independent [42] (Tables 6.5 and 6.6).

$$X^2(X \rightarrow Y) = N \cdot (\text{lift} - 1)^2 \frac{\text{supp}\cdot\text{conf}}{(\text{conf}-\text{supp})\cdot(\text{lift}-\text{conf})}\# \qquad (6.9)$$

Table 6.5 Observed frequencies for a rule $X \rightarrow Y$

Observed frequencies			
N = size of dataset		Consequent Y	
		Υ	$\neg\Upsilon$
Antecedent X	X	$N \cdot \text{sup}$	$N \cdot \frac{\text{supp}}{\text{conf}} \cdot (1 - \text{conf})$
	$\neg X$	$N \cdot \left(\frac{\text{conf}}{\text{lift}} - \text{supp}\right)$	$N \cdot \left[1 - \frac{\text{supp}}{\text{conf}} \cdot (1 - \text{conf}) - \frac{\text{conf}}{\text{lift}}\right]$

Table 6.6 Expected Frequencies for a rule X → Y, where: supp = supp(X → Y), conf = conf(X → Y) and lift = lift(X → Y)

Expected frequencies			
N = size of dataset		Consequent Y	
		Y	¬Y
Antecedent X	X	$N \cdot \frac{supp}{lift}$	$N \cdot \frac{supp}{conf} \cdot \left(1 - \frac{conf}{lift}\right)$
	¬X	$N \cdot \left(1 - \frac{supp}{conf}\right) \cdot \frac{conf}{lift}$	$N \cdot \left(1 - \frac{supp}{conf}\right) \cdot \left(1 - \frac{conf}{lift}\right)$

where:

$$supp = supp(X \rightarrow Y)$$
$$conf = conf(X \rightarrow Y)$$
$$lift = lift(X \rightarrow Y).$$

Finally, in addition to the statistical evaluation, an evaluation of the overall approach was carried out, utilizing the knowledge and experience of the collaborating doctors, who judged the approach to the new produced medical knowledge, to the verification of the existing medical knowledge and to the usefulness of the overall approach during the diagnostic operation.

6.5 Modeling Approaches

This section presents the experimental process, the results as well as the conclusions that emerged from the proposed approach to the problem of optimizing the diagnostic process of breast cancer using intelligent data analysis methods.

6.5.1 Experimental Process

As mentioned in subsection of Modeling Methodology, nine (9) different classification problems for the breast cancer data under investigation, were defined in this work. In each of these approaches, a hybrid classification method was used, blending an inductive learning algorithm and a feature selection process. In addition, in all classification approaches, the input variables used in the feature selection process were the same (Table 6.2) while the number of training and validation cases varied according to the number of cases that each target variable had, in each classification problem. Finally, in all approaches, the produced classification models were validated using the "Leave–One–Out" validation method.

6.5.2   Experimental Results

The initial objective of the experimental process of this work was to apply the proposed feature selection technique in combination with an inductive machine learning approach (C5.0 algorithm was used) to all classification problems presented in section of Modeling Methodology, in order to obtain accurate, comprehensible and if possible handy but also reliable and useful outcomes for medical/clinical practice.

In the *1st Classification Problem*, as it has already been mentioned, the aim was the classification of all patient cases (77 cases) into two (2) classes (Benign or Malignant). The best combination of input variables that emerged after the proposed feature selection process included nine (9) diagnostic variables (AGE, MORPH, BDS, PRED, NTEN, CRM, FC, BD & BPE) (see Table 6.2).

In the *2nd Classification Problem*, where the task was the classification of all patient cases (77 cases) into six (6) classes (Benign Ductal carcinoma in situ, Invasive ductal carcinoma—DCIS, Ductal carcinoma in situ and invasive ductal carcinoma—DCIS & IDC, Invasive lobular carcinoma—ILC and Solid papillary carcinoma), the best combination of input variables that arose after the proposed feature selection process included eight (8) variables (AGE, MORPH, BDS, PRED, ADC, DF, T2WI & FV) (see Table 6.2).

In the *3rd Classification Problem*, where the aim was to classify only the patient cases with malignant lesions (58 cases) into five (5) specific malignant classes (Ductal carcinoma in situ, Invasive ductal carcinoma—DCIS, Ductal carcinoma in situ and invasive ductal carcinoma—DCIS & IDC, Invasive lobular carcinoma—ILC and Solid papillary carcinoma), the best combination of input variables found after the proposed feature selection process included 10 variables (AGE, MORPH, BDS, PRED, INTEN, CRM, ADC, DF, T2-Wi & BPE) (see Table 6.2).

Then, in the *4th Classification Problem*, where all the cases of patients (77 cases) are classified into four (4) classes (BIRADS_3, BIRADS_4, BIRADS_5 or BIRADS_6), the best combination of input variables that came out after the proposed feature selection process consists of 11 variables (MORPH, BDS, PRED, INTEN, CRM, ADC, DF, T2WI, FV, FC & BD) (see Table 6.2) and in *the 5*th *Classification Problem*, where the aim was to classify the malignant patient cases (56 cases) into specific classes of Tumor grade (Tumor Grade 1, Tumor Grade 2 or Tumor Grade 3), the best combination of input variables includes seven (7) variables (AGE, BDS, INTEN, CRM, ADC, T2-Wi & FC).

For the *6th Classification Problem*, in which the problem of classification of patients with malignant lesions (57 cases) into 2 classes (breast cancer cells have estrogen receptors or breast cancer cells do not have estrogen receptors) was approached, best combination of input variables consists of 12 variables (MORPH, BDS, INTEN, CRM, ADC, DF, T2-Wi, FV, FC, TUMS, BD & BPE) (see Table 6.2), while in the *7th Classification Problem*, where the aim was to classify the same patient cases (57 cases) into two (2) classes (breast cancer cells have progesterone receptors or breast cancer cells do not have progesterone receptors), the best combination of input variables that came out after the proposed feature selection process included eight (8) variables (MORPH, PRED, INTEN, CRM, DF, T2-Wi, FC, & BD) (see Table 6.2).

Finally, in the *8th & 9th classification problems*, where the respective objectives of the approaches of each problem were the classification of patients with malignant lesions into two (2) classes related to the presence of Cerb-B2 protein (whether or not Cerb-B2 protein is present) and in three (3) classes representing the degree of regeneration of cancer cells (Ki-67_1: Cell Replication Rate under 15%, Ki-67_2: Cell Replication Rate between 16 and 26% or Ki-67_3: Cell Replication Rate over 26%), the best combination of input variables for the eight classification problem includes 11 variables (AGE, MORPH, BDS, PRED, INTEN, CRM, ADC, DF, T2-Wi, FV & FC) (see Table 6.2) and the best combination of input variables for the ninth classification problem consists of 12 variables (AGE, MORPH, BDS, PRED, INTEN, CRM, ADC, DF, T2WI, FV, FC & TUMS) (see Table 6.2).

As presented in Table 6.7, in eight (8) out of the nine (9) classification problems presented in this paper the classification accuracy of the respective classification models is relatively low (below 80%) due to the small number of patient cases collected so far, regarding some of the classes. Regarding medical knowledge mining, not only from the classification process, but also from the feature selection process presented in this paper, it is observed that the variables Morphology (MORPH) Borders (BDS), Curve Morphology (CRM), Diffusion (DF), Internal Enhancement (INTEN) & T2-Wi are used as input variables in eight (8) out of nine (9) classification approaches and the Apparent Diffusion Coefficient (ADC) & Peritumoral Edema (PRED0) variables in seven (7) out of (9) classification approaches. This fact confirms the clinical significance of these variables, as defined by the collaborating expert doctors (Table 6.2).

Completing the experimental process and comparing its results, the most efficient classification model is represented by a classification tree (Annex 6.4) of two (2) classes (Benign or Malignant) has nine (9) input variables (AGE, MORPH, BDS, PRED, NTEN, CRM, FC, BD & BPE) (see Table 6.2) and was developed and validated using 77 training/validation vectors (77 patient cases). This classification tree (see Annex 6.4) has a classification accuracy equal to 97.4% in the training set and 87% in the test set using "Leave–One–Out" approach. The ruleset of this classification tree consists of seven (7) different classification rules, which are presented and analyzed below and presents a classification accuracy equal to 92.2% in training set and in its validation process (using "Leave–One–Out" method) achieves a classification accuracy equal to 86.7%.

According to the classification tree shown in the figure in Annex 6.4 anfd the ruleset derived from this tree, it is observed that the variables Curve Morphology (CRM) and Peritumoral Edema (PRED) have a high frequency of occurrence in the ruleset (i.e. they appear in more than 3 rules) and in the classification tree together with the variable root, which in this case is the variable Borders (BDS), are the first three (3) variables used by the algorithm to develop the specific tree. This fact

Table 6.7 Summary table of classification problems and approaches

Classification problems

S. no.	Input variables	Target variable	Tr./Val. set	Acc (%)
1	AGE, MORPH, BDS, PRED, INTEN, CRM, FC, BD & BPE	Benign or Malignant (BOM)	77	87
2	AGE, MORPH, BDS, PRED, ADC, DF, T2-Wi & FV	Benign and Malignant (BAM)	77	<80
3	AGE, MORPH, BDS, PRED, INTEN, CRM, ADC, DF, T2-Wi & BPE	Malignant (ML)	58	<80
4	MORPH, BDS, PRED, INTEN, CRM, ADC, DF, T2-Wi, FV, FC & BD	BIRADS	77	<80
5	AGE, BDS, INTEN, CRM, ADC, T2-Wi & FC	Tumor Grade (TUMG)	56	<80
6	MORPH, BDS, INTEN, CRM, ADC, DF, T2-Wi, FV, FC, TUMS, BD & BPE	Estrogen Receptors (ER)	57	<80
7	MORPH, PRED, INTEN, CRM, DF, T2-Wi, FC, & BD	Progesterone Receptors (PR)	57	<80
8	AGE, MORPH, BDS, PRED, INTEN, CRM, ADC, DF, T2WI, FV & FC	Cerb-B2	57	<80
9	AGE, MORPH, BDS, PRED, INTEN, CRM, ADC, DF, T2WI, FV & FC	Ki-67	53	<80

confirms both the statistics (as confirmed during the feature selection process in the overall approach of all classification problems), as well as the clinical significance of these variables, as defined by medical experts (see Table 6.2).

Finally, for the assessment of malignancies the above classification tree can classify with 93.1% accuracy (sensitivity) cases of malignant lesions in new (unseen) patients with corresponding lesions and at the same time has the ability to estimate the benign lesions in these new patients with corresponding lesions with a certainty of 73.7% accuracy (specificity), while the corresponding sensitivity and specificity of the ruleset from the specific classification tree is 93.1% and 63.1%, respectively.

For example, in a total of 10 patients with malignant lesions, the comparative classification tree and ruleset will correctly classify nine (9) of them as having malignant lesions and one (1) will erroneously assess it as a benign lesion, whereas in the case of this patient group consisting of cases of benign lesions, the specific tree will correctly assess the seven (7) cases as benign lesions and incorrectly the remaining (3) as malignant lesions, while the ruleset of this tree will correctly assess the six (6) cases as benign lesions and incorrectly the remaining (4) as malignant lesions (Tables 6.8 and 6.9).

As mentioned above, the classification tree (see Annex 6.4) of the 1st classification problem (classification problem 2-category: Benign or Malignant) led to the following seven (7) classification rules.

Rule 1

If there is no peritumoral edema, the borders of the finding are smooth and areas with relatively or extremely dense fibroglandular tissue are present (Breast Density category C or D), then the finding is benign (IF BDS = BDS_2 AND PRED = PRED_1 AND BD = BD_2 THEN BOM = BOM_1).

Transforming the above classification rule into association rule ($X_1 \rightarrow Y_1$, where $X_1 = \{BDS\_2, PRED\_1, BD\_2\}$ and $Y_1 = \{BOM\_1\}$), it is observed that X_1 collects six (6) patient cases ($X_1 = 6$), Y_1 collects 19 patient cases ($Y_1 = 19$) and this rule covers six (6) patient cases ($X_1 \cup Y_1 = 6$).

The support of this rule amounts to 7.78% (supp ($X_1 \rightarrow Y_1$) = 0.0778), the confidence amounts to 85.5% (conf ($X_1 \rightarrow Y_1$) = 0.855) and the lift of this association rule amounts to 3.5 (lift ($X_1 \rightarrow Y_1$) = 3.5).

In the statistical validation frame of the ruleset and according to the values of the abovementioned statistical measures (support, confidence and lift), the Rule 1,

Table 6.8 Confusion matrix of the classification tree of the most efficient model that was used in the estimation of specificity and sensitivity presented in cases of malignant lesions

Classification tres's confusion matrix			
N = 77 Leave–One–Out validation process		Malignant (predicted values)	
		YES	NO (Benign)
Malignant (actual values)	YES	True Positive = 53	False Negative = 5
	NO (Benign)	False Positive = 5	1True Negative = 14

Table 6.9 Confusion matrix of the classification ruleset of the most efficient model that was used in the estimation of specificity and sensitivity presented in cases of malignant lesions

Ruleset's confusion matrix			
N = 77 Leave–One–Out validation process		Malignant (predicted values)	
		YES	NO (Benign)
Malignant (actual values)	YES	True Positive = 54	False Negative = 4
	NO (Benign)	False Positive = 7	1True Negative = 12

although appears more frequently in the dataset than expected (lift $= 3.5 > 1$), does not describe significantly large population of the dataset (supp $= 0.0778$) and gives the impression that it can classify new patient cases with a classification accuracy of 85.5% (i.e. equal to its confidence: $Acc_{R1} = conf = 0.855$).

According to the collaborating experts, Rule 1 confirms existing medical knowledge, as the presence of a smooth border or a well-defined margin, without peritumoral edema is 97% specific and 35% sensitive for a benign diagnosis [43].

Rule 2

If there is no peritumoral edema, the signal intensity persists two (2) minutes after the shading fluid is injected (kinetic curve type 1) and there is unifocal focality then the finding is benign (IF PRED $=$ PRED_1 AND CRM $=$ CRM_1 AND FC $=$ FC_3 THEN BOM $=$ BOM_1).

Transforming the above classification rule into association rule ($X_2 \rightarrow Y_2$, where $X_2 = \{PRED\_1, CRM\_1, FC\_3\}$ and $Y_2 = \{BOM\_1\}$), it is observed that X_1 collects three (3) patient cases ($X_2 = 3$), Y_2 collects 19 patient cases ($Y_2 = 19$) and this rule covers three (3) patient cases ($X_2 \cup Y_2 = 3$).

The support of this rule amounts to 3.90% (supp ($X_2 \rightarrow Y_2$) $= 0.039$), the conference amounts to 80% (conf ($X_2 \rightarrow Y_2$) $= 0.800$) and the lift of this association rule amounts to 3.2 (lift ($X_2 \rightarrow Y_2$) $= 3.242$).

In the statistical validation frame of the ruleset and according to the values of above statistical measures (support, confidence and lift), the Rule 2, although appears more frequently in the dataset than expected (lift $= 3.2 > 1$), does not describe significantly large population of the dataset (supp $= 0.0390$) and gives the impression that it can classify new patient cases with a classification accuracy equal to its confidence ($Acc_{R2} = conf = 0.800$).

According to the collaborating expert doctors, this rule verifies existing medical knowledge, as it is known from the literature that the shape of the time-signal intensity curve is an important factor in differentiating benign and malignant enhancing lesions in dynamic breast MR imaging. A type 1 (type I) curve is a strong indicator of a benign lesion and observed in at least 83% of benign lesions [44].

Rule 3

If there is no peritumoral edema, the signal intensity of the lesion continues to increase two (2) minutes after the contrast fluid is injected (kinetic curve type 1) and there is marked background parenchymal enhancement then the finding is benign (IF PRED $=$ PRED_1 AND CRM $=$ CRM_2 AND BPE $=$ BPE_2 THEN BOM $=$ BOM_1).

Transforming the above classification rule into association rule ($X_3 \rightarrow Y_3$, where $X_3 = \{PRED\_1, CRM\_2, BPE\_2\}$ and $Y_3 = \{BOM\_1\}$), it is observed that X_3 collects three (3) patient cases ($X_3 = 3$), Y_3 collects 19 patient cases ($Y_3 = 19$) and this rule covers three (3) patient cases ($X_3 \cup Y_3 = 3$).

The support of this rule amounts to 3.90% (supp ($X_3 \rightarrow Y_3$) $= 0.0390$), the conference amounts to 80% (conf ($X_3 \rightarrow Y_3$) $= 0.800$) and the lift of this association rule amounts to 3.2 (lift ($X_3 \rightarrow Y_3$) $= 3.242$).

In the statistical validation frame of the ruleset and according to the values of above statistical measures (support, confidence and lift), the Rule 3, although appears more frequently in the dataset than expected (lift $= 3.2 > 1$), does not describe significantly large population of the dataset (supp $= 0.0390$) and gives the impression that it can classify new patient cases with a classification accuracy equal to its confidence (Acc_{R3} $= conf = 0.800$).

According to the collaborating expert doctors, this rule confirms the corresponding medical knowledge confirmed by Rule 2.

Rule 4

If internal enhancement is homogeneous, the signal intensity of the lesion continues to increase two (2) minutes after the contrast fluid is injected (kinetic curve type 1) and the breasts are almost entirely fatty or there are scattered areas of fibroglandular tissue (Breast Density category A or B), then the finding is benign (IF INTEN $=$ INTEN_1 AND CRM $=$ CRM_1 AND BD $=$ BD_1 THEN BOM $=$ BOM_1).

Transforming the above classification rule into association rule ($X_4 \rightarrow Y_4$, where $X_3 = \{$INTEN_1, CRM_1, BD_1$\}$ and $Y_4 = \{$BOM_1$\}$), it is observed that X_4 collects two (2) patient cases ($X_4 = 2$), Y_4 collects 19 patient cases ($Y_4 = 19$) and this rule covers two (2) patient cases ($X_4 \cup Y_4 = 2$).

The support of this rule amounts to 2.60% (supp ($X_4 \rightarrow Y_4$) $= 0.0260$), the conference amounts to 75% (conf ($X_4 \rightarrow Y_4$) $= 0.750$) and the lift of this association rule amounts to 3 (lift ($X_4 \rightarrow Y_4$) $= 3.0395$).

In the statistical validation frame of the ruleset and according to the values of the abovementioned statistical measures (support, confidence and lift), Rule 4, although appears more frequently in the dataset than expected (lift $= 3 > 1$), does not describe significantly large population of the dataset (supp $= 0.0260$) and gives the impression that it can classify new patient cases with a classification accuracy equal to its confidence ($Acc_{R4} = conf = 0.750$).

According to the collaborating expert doctors, homogeneous enhancement and kinetic curve type 1 (type I) are imaging signs associated with benignity according to the BI-RADS lexicon. Women with a lower breast density (category A and B) are less likely to suffer from breast cancer as increased breast density (Category C and D) is a risk factor for developing breast cancer. Therefore, according to the above, this rule verifies existing medical knowledge.

Rule 5

If the patient is over 50 years old and contrast washout is observed two (2) minutes after gadolinium is injected ((kinetic curve type 3), then the finding is malignant (IF AGE $=$ AGE_2 AND CRM $=$ CRM_3 THEN BOM $=$ BOM_2).

Transforming the above classification rule into association rule ($X_5 \rightarrow Y_5$, where $X_3 = \{$AGE_2, CRM_3$\}$ and $Y_5 = \{$BOM_2$\}$), it is observed that X_5 collects 21 patient cases ($X_5 = 21$), Y_5 collects 58 patient cases ($Y_5 = 58$) and this rule covers 21 patient cases ($X_5 \cup Y_5 = 21$).

The support of this rule amounts to 27.27% (supp $(X_5 \rightarrow Y_5) = 0.2727$), the conference amounts to 95.7% (conf $(X_5 \rightarrow Y_5) = 0.956$) and the lift of this association rule amounts to 1.3 (lift $(X_5 \rightarrow Y_5) = 1.27$).

In the statistical validation frame of the ruleset and according to the values of above statistical measures (support, confidence and lift), Rule 5 occurs as almost often as expected (lift $= 1.3 > 1$), describes significantly large population of the dataset (supp $= 0.2727$) and gives the impression that it can classify new patient cases with a classification accuracy equal to its confidence (Acc$_{R5}$ = conf $= 0.956$).

According to the collaborating expert doctors, this rule verifies with a high degree of confidence (conf $= 95.7\%$) existing medical knowledge, as it is known from the literature that the risk of breast cancer increases with age and a sharp increase in incidence of breast cancer is observed after the age of 50 years old. Also, a type 3 time-signal intensity curve is a strong indicator of malignancy in dynamic breast MR imaging and is independent of other criteria. The sensitivity, specificity and diagnostic accuracy for signal intensity time curves is respectively 91, 83 and 86% [45].

Rule 6

If the borders of the finding are irregular or spiculated and areas with relatively or extremely dense fibroglandular tissue are presented (Breast Density category C or D) and there is minimum or mild background parenchymal enhancement then the finding is malignant (IF BDS = BDS_1 AND BD = BD_2 AND BPE = BPE_1 THEN BOM = BOM_2).

Transforming the above classification rule into association rule ($X_6 \rightarrow Y_6$, where $X_3 = \{BDS\_1, BD\_2, BPE\_1\}$ and $Y_6 = \{BOM\_2\}$), it is observed that X_6 collects 24 patient cases ($X_6 = 24$), Y_6 collects 58 patient cases ($Y_6 = 58$) and this rule covers 23 patient cases ($X_6 \cup Y_6 = 23$).

The support of this rule amounts to 29.9% (supp $(X_6 \rightarrow Y_6) = 0.2987$), the conference amounts to 92.3% (conf $(X_6 \rightarrow Y_6) = 0.9230$) and the lift of this association rule amounts to 1.2 (lift $(X_6 \rightarrow Y_6) = 1.225$).

In the statistical validation frame of the ruleset and according to the values of above statistical measures (support, confidence and lift), Rule 6 occurs as almost often as expected (lift $= 1.2 > 1$), describes significantly large population of the dataset (supp $= 0.2987$) and gives the impression that it can classify new patient cases with a classification accuracy equal to its confidence (Acc$_{R6}$ = conf $= 0.923$).

According to the collaborating expert doctors, morphologic features are considered to be the most important constituent of breast MRI in terms of assessing the likelihood of malignancy. Spiculated or irregular margins of a breast lesion in dynamic breast MRI examinations is a strong indicator of malignancy, according to the BIRADS lexicon. In addition, increased breast density (Category C and D) is a significant risk factor for developing breast cancer [46]. This rule verifies the specific medical knowledge with a high confidence index (conf $= 92.3\%$).

Rule 7

If the internal enhancement is inhomogeneous then the finding is malignant (IF INTEN = INTEN_2 THEN BOM = BOM_2).

Transforming the above classification rule into association rule ($X_7 \rightarrow Y_7$, where $X_3 = \{INTEN\_2\}$ and $Y_7 = \{BOM\_2\}$), it is observed that X_7 collects 61 patient cases ($X_7 = 61$), Y_7 collects 58 patient cases ($Y_7 = 58$) and this rule covers 49 patient cases ($X_7 \cup Y_7 = 49$).

The support of this rule amounts to 63.6% (supp $(X_7 \rightarrow Y_7) = 0.6363$), the conference amounts to 79.4% (conf $(X_7 \rightarrow Y_7) = 0.7936$) and the lift of this association rule amounts to 1.1 (lift $(X_7 \rightarrow Y_7) = 1.053$).

In the statistical validation frame of the ruleset and according to the values of above statistical measures (support, confidence and lift), Rule 7 occurs as often as expected (lift $= 1.1 \approx 1$), describes a large population of the dataset (supp $= 0.6363$) and gives the impression that it can classify new patient cases with a classification accuracy equal to its confidence ($Acc_{R7} = conf = 0.7936$).

According to medical experts, this rule verifies existing medical knowledge, as inhomogeneous internal Enhancement is an important sign indicative of malignancy on breast MRI [47].

As can be observed in Table 6.7 (statistical evaluation of the classification rules) and according to the evaluation of the collaborating expert doctors, all the produced rules confirm with sufficient certainty existing medical knowledge (Table 6.10).

Comparing specific rules with each other, it turns out that Rules #5 and #6 seem to be the strongest classification rules of this ruleset. According to the Chi-Square (X^2) test of Rules #5 and #6, by checking and confirming the necessary conditions and using Eq. (6.9) to estimate the value of X^2 for each rule, the specific rules are further strengthened (there is a high dependency between the two parts of the rule) as the estimated value of X^2 in each rule is greater than the critical value of X^2 for the distribution of the examined data ($X_5^2 = 3.95.X_6^2 = 5.70.X_5^2, X_6^2 > X_c^2 = 3.84$).

Concluding on the experimental results of the entire approach, we see that due to the small number of patient cases used in this study and also because of the statistical noise of data caused by the unequal value groups of each variable, the models obtained through inductive learning, although not achieving very high accuracy classification (below 90%), yield some important initial information about the statistical behavior of the variables used in this work, which will be used in future similar approaches to breast cancer diagnosis.

Table 6.10 Ruleset of most efficient classification model

Ruleset

Rule	Support (%)	Confidence (%)	Lift (%)
Rule 1	7.8	87.5	3.5
Rule 2	3.9	80.0	3.2
Rule 3	3.9	80.0	3.2
Rule 4	2.6	75.0	3.0
Rule 5	27.3	95.7	1.3
Rule 6	29.9	92.3	1.2
Rule 7	63.6	79.4	1.1

However, comparing the produced classification models with each other (taking into account the classification accuracy that each presents when validating them with the "Leave–One–Out" method), the most efficient classification model is the one that emerged when approaching the classification problem of benign and malignant patient cases (2–class classification problem). This model is a classification tree for benign and malignant breast lesions, which has a classification accuracy of 87%. From this classification tree emerge (7) different classification rules, which have classification accuracy equal to 86.7%. Also, this classification model (classification tree and ruleset), although, because of small number of benign cases ($N_{Benign} = 19$) included in the dataset used in this paper, it presents great ability (average sensitivity $= 96.5\%$) in diagnosing malignancies but is relatively weak (average specificity $= 68.4\%$) in diagnosing benign lesions.

Finally, according to the statistical validation and the evaluation made with the collaborating medical experts about the clinical significance of each rule of the above ruleset, it is concluded that all rules confirm -with relatively high certainty—known existing medical knowledge. The most powerful rules prove to be #5 and #6, which are of great statistical interest as they can confirm existing medical knowledge with a confidence level of 95.7% and 92.3%, respectively.

6.6 Conclusions and Further Search

This paper presents an intelligent methodology combining inductive machine learning and feature selection for modelling the diagnostic process of breast cancer through the analysis of related medical (MRI) data collected from magnetic mammographies.

The aim of the study was not only to acquire a satisfactory diagnostic modelling in terms of accuracy, but also to discover—if possible—new medical knowledge, or to confirm existing medical knowledge in the context of the diagnostic process of breast cancer.

Specifically, utilizing the data collected from observations of breast MRI scans and invasive examinations from 77 patients, nine (9) different classification problems were defined for which the corresponding models were developed. The development of the specific classification models was carried out using a hybrid methodology that combines inductive decision trees and feature selection processes. Attribute values were defined with the aid of medical experts. The "Leave–One–Out" validation method is used for measuring classification accuracy.

Results show that classification accuracy can be improved and it is expected to do so, with the enrichment of the available dataset in the following time.

Nevertheless, comparing the produced models with each other, it turned out that the classification model discriminating malignant from benign lesions (2–class classification model) performs much better than the other classification models. This model corresponds to a simple classification tree which was trained with 77 patient cases (19 cases of benign lesions and 58 cases of malignant lesions) and uses nine (9) imaging input variables for classifying new cases in two classes (Benign and Malignant).

The abovementioned classification tree can diagnose new patient cases with a classification accuracy of 87% and in case of the assessment of malignancies it can achieve an accuracy of 93.1% (sensitivity) for the cases of malignant lesions in patients with corresponding lesions. At the same time it has the ability to estimate the benign lesions in patients with corresponding lesions with a certainty of 73.7% accuracy (specificity).

The ruleset arising from the above classification tree imparts medical knowledge in the form of classification rules, which were evaluated using specific statistical criteria (support, confidence and lift of each rule) as well as the knowledge and experience of the collaborating experts.

According to the statistical evaluation of each rule of above ruleset, two strong rules (confidence > 90% and lift > 1) have been formed, which, according to the medical evaluation, confirm existing medical knowledge that is quite useful for the diagnosis of malignant lesions in MRI scans of the breast. Also, the specific two (2) classification rules, during the Chi–Square test, are of great statistical interest, as the values of X for both rules are close to the X critical of the respective distribution, which indicates complete dependence of the antecedent (IF) and the consequent (THEN) of each rule and a very good adaptation of the specific rules to corresponding values of the variables that included in the parts (antecedent & consequent) of this rule.

In conclusion, this work is the basis for further experimentation in the future, around breast lesions detection in MRI. The proposed approach in this paper gives rather satisfactory and encouraging results, defining the current research report as a trigger for further research in collaboration with specialists.

Future research presupposes the enrichment of the collection of relevant medical data in order to be able to induce new reliable medical knowledge from all nine types of classification problems defined and presented in this paper.

In addition, different attribute value areas might be tested in collaboration to MDs and also with the aid of statistical analysis of the data. And of course, other existing or innovative computational intelligence approaches will be tested, in an attempt to increase classification (diagnostic) accuracy, comprehensibility of results and— if possible—discovery of new medical knowledge (confirmed also with statistical techniques), regarding the diagnostic act of breast cancer.

Annex 6.1—Abbreviations

Abbreviations	Descriptions
ACR BI-RADS	American College of Radiology Breast Imaging Reporting & Data System
ADC	Apparent Diffusion Coefficient
ANN	Artificial Neural Networks
AUC	Area Under the ROC Curve
B–ANN	Bayesian Artificial Neural Networks
CAD	Computer Aided Diagnostic systems
CHAID	Chi-squared Automatic Interaction Detection
CLi Data	Clinical image Data
CNN	Convolutional Neural Network
DCE–MRI	Dynamic Contrast–Enhanced Magnetic-Resonance Imaging
DMi Data	Digital Mammography image Data
eBP–FNN	error Back Propagation Feed Forward Artificial Neural Networks
FCM	Fuzzy c-Means
FFDM	Full Field Mammograms
FOV	Field of View
HiSS–MRI	High Spatial and Spectral resolution MRI
IRi Data	Infrared Thermography image Data
KNN	K Nearest Neighbor
LDA	Linear Discriminant Analysis
LS–SVM	Least-Squares Support Vector Machine
MARS	Multivariate Adaptive Regression Splines
MCi Data	Microscope image Data
MRi Data	Magnetic–resonance image Data
MSM	Multisurface method
PFK-SVM	Polynomial Kernel Function Support Vector Machine
RF	Random Forest
SA	Sparse Autoencoder
ST	Slice thickness
SVM	Support Vector Machines
TNM	T category describes the primary tumor site, N category describes the regional lymph node involvement and M category describes the presence or otherwise of distant metastatic spread (UNICC- Union for International Cancer Control
TSE	Turbo Spin Echo

Annex 6.2—Variables Frequency Charts (Original Dataset)

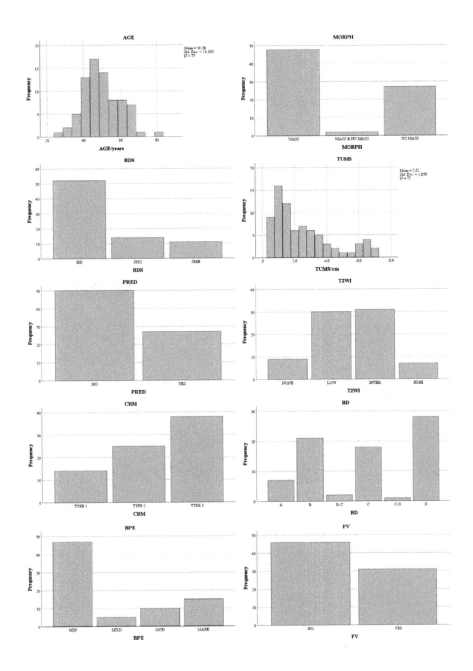

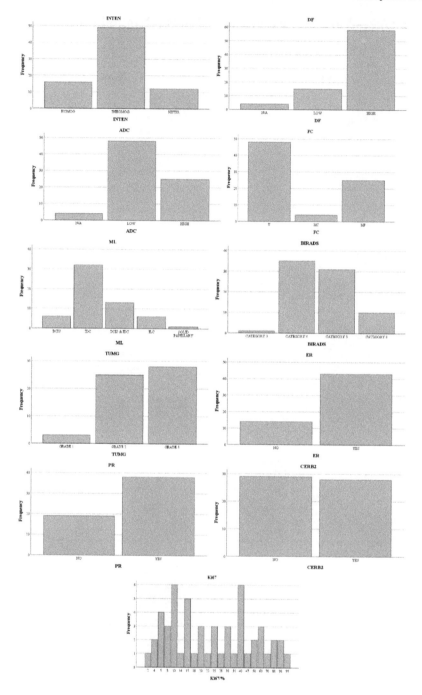

Annex 6.3—Variables' Values Range

Variables' values range

S. no.	Variable	Values	Group	Frequency (N)	Frequency (%)
1	AGE	[0, 49]	AGE_1	39	50.6
		[50, ∞]	AGE_2	38	50.4
		Total		**77**	**100**
2	Morphology (MORPH)	MASS	MORPH_1	48	62.3
		MASS & NON MASS	MORPH_2	2	2.6
		NON MASS	MORPH_3	27	35.1
		Total		77	100
3	Borders (BDS)	IRR	BDS_1	66	85.7
		SPIC			
		SMH	BDS_2	11	14.3
		Total		**77**	**100**
4	Tumor Size (TUMS)	[0, 1.3]	TUMS_1	28	36.4
		[1.4, 2.5]	TUMS_2	19	24.7
		Over 2.6	TUMS_3	30	39.0
		Total		**77**	**100**
5	Peritumoral Edema (PRED)	NO	PRED_1	50	64.9
		YES	PRED_2	27	35.1
		Total		**77**	**100**
6	T2 Weighted image (T2-Wi)	NONE			
		LOW	T2WI_1	70	90.9
		INTER			
		HIGH	T2WI_2	7	9.1
		Total		**77**	**100**
7	Curve Morphology (CRM)	TYPE_1	CRM_1	14	18.2
		TYPE_2	CRM_2	25	32.5
		TYPE_3	CRM_3	38	49.4
		Total		**77**	**100**
8	Breast Density (BD)	A	BD_1	28	36.4
		B			
		B-C	BD _2	49	63.6
		C			
		C-D			
		Total		**77**	**100**
9	Background Parenchymal Enhancement (BPE)	MIN	BPE_1	52	67.5

(continued)

(continued)

Variables' values range

S. no.	Variable	Values	Group	Frequency (N)	Frequency (%)
		MILD			
		MOD	BPE_2	25	32.5
		MARK			
		Total		**77**	**100**
10	Feeding Vessel (FV)	NO	FV_1	46	59.7
		YES	FV_2	31	40.3
		Total		**77**	**100**
11	Internal Enhancement (INTEN)	HOMOG	INTEN_1	16	20.8
		INHOMOG	INTEN_2	61	79.2
		HETER			
		Total		**77**	**100**
12	Diffusion (DF)	N/A	DF_1	4	5.2
		LOW	DF_2	15	19.5
		HIGH	DF_3	58	75.3
		Total		**77**	**100**
13	Apparent Diffusion Coefficient (ADC)	N/A	ADC_1	4	5.2
		LOW	ADC_2	48	62.3
		HIGH	ADC_3	25	32.5
		Total		**77**	**100**
14	Focality (FC)	U	FC_1	48	62.3
		MC	FC_2	4	5.2
		MF	FC_3	25	32.5
		Total		**77**	**100**
15	Benign or Malignant (BOM)	BENIGN	BOM_1	19	24.7
		MALIGNANT	BOM_2	58	75.3
		Total		**77**	**100**
16	Benign and Malignant (BAM)	BENIGN	BAM_1	19	24.7
		DCIS	BAM_2	6	7.8
		IDC	BAM_3	36	46.8
		DCIS & IDC	BAM_4	9	11.7
		ILC	BAM_5	6	7.8
		SOLID PAPILLARY	BAM_6	1	1.3
		Total		**77**	**100**
17	Malignan (ML)	DCIS	ML_1	6	7.8
		IDC	ML_2	32	41.6

(continued)

(continued)

Variables' values range

S. no.	Variable	Values	Group	Frequency (N)	Frequency (%)
		DCIS & IDC	ML_3	13	16.9
		ILC	ML_4	6	7.8
		SOLID PAPILLARY	ML_5	1	1.3
		Total		**58**	**75.3**
		Missing	System	**19**	**24.7**
		Total		**77**	**100**
18	BIRADS	CATEGORY 3	BIRADS_2	36	46.8
		CATEGORY 4			
		CATEGORY 5	BIRADS_3	41	53.2
		CATEGORY 6			
		Total		**77**	**100**
19	Tumor Grade (TUMG)	GRADE 1	TUMG_1	3	3.9
		GRADE 2	TUMG_2	25	32.5
		GRADE 3	TUMG_3	28	36.4
		Total		**56**	**72.7**
		Missing	System	**21**	**27.3**
		Total		**77**	**100**
20	Estrogen Receptors (ER)	NO	ER_1	14	18.2
		YES	ER_2	43	55.8
		Total		**57**	**74.0**
		Missing	System	**20**	**26.0**
		Total		**77**	**100**
21	Progesterone Receptors) (PR)	NO	PR_1	19	24.7
		YES	PR_2	38	49.4
		Total		57	7 4.0
		Missing	System	**20**	**26.0**
		Total		**77**	**100**
22	Cerb-B2	NO	CERB2_1	29	37.7
		YES	CERB2_2	28	36.4
		Total		**57**	**74.0**
		Missing	System	**20**	**26.0**
		Total		**77**	**100**
23	Ki-67	[0, 15]	KI67_1	22	28.6
		[16, 25]	KI67_2	8	10.4
		[26, 100]	KI67_3	23	29.9

(continued)

(continued)

Variables' values range					
S. no.	Variable	Values	Group	Frequency (N)	Frequency (%)
		Total		53	68.8
		Missing	System	24	31.2
		Total		77	100

Annex 6.4—Classification Tree (Benign or Malignant)

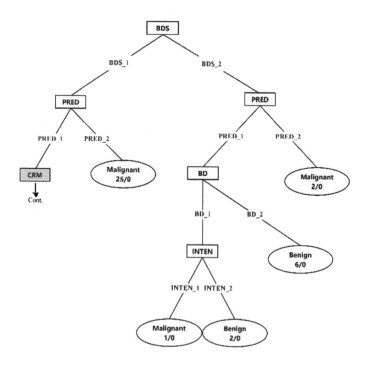

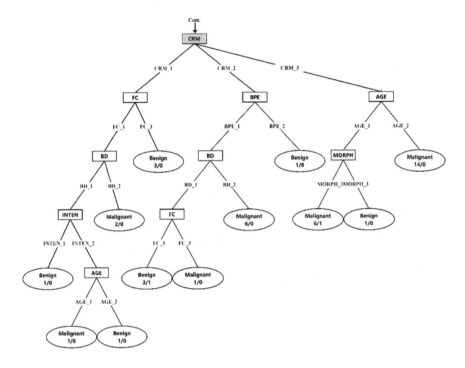

References

1. World Health Organisation, Latest global cancer data. *IARC*, 13–15, 2018.
2. R. Kraus, M. Espy, P. Magnelind, P. Volegov, *Ultra-Low Field Nuclear Magnetic Resonance: A New MRI Regime* (Oxford University Press)
3. C.S. Sureka, C. Armpilia, *Radiation Biology for Medical Physicists*, 1st edn. (CRC Press, Taylor & Francis Group, Florida, USA, 2017)
4. M. B. Amin et al. (eds.), *AJCC Cancer Staging Manual*, 8th edn. (Springer International Publishing, 2017)
5. A. Aydiner, A. İgci, A. Soran (eds.), *Breast Cancer: A Guide to Clinical Practice* (Springer International Publishing, 2019)
6. N.I.R. Yassin, S. Omran, E.M.F. El Houby, H. Allam, Machine learning techniques for breast cancer computer aided diagnosis using different image modalities: a systematic review. Comput. Methods Programs Biomed. **156**, 25–45 (2018). https://doi.org/10.1016/j.cmpb.2017.12.012
7. W. Yue, Z. Wang, H. Chen, A. Payne, X. Liu, Machine Learning with Applications in Breast Cancer Diagnosis and Prognosis. Designs **2**(2), Art. no. 2, June 2018. https://doi.org/10.3390/designs2020013
8. W.H. Wolberg, O.L. Mangasarian, Multisurface method of pattern separation for medical diagnosis applied to breast cytology. Proc. Natl. Acad. Sci. USA **87**(23), 9193–9196 (1990)
9. Y. Wu, M.L. Giger, K. Doi, C.J. Vyborny, R.A. Schmidt, C.E. Metz, Artificial neural networks in mammography: application to decision making in the diagnosis of breast cancer. Radiology **187**(1), 81–87 (1993). https://doi.org/10.1148/radiology.187.1.8451441

10. D. Furundzic, M. Djordjevic, A. Jovicevic Bekic, Neural networks approach to early breast cancer detection. J. Syst. Archit. **44**(8), 617–633, April 1998. https://doi.org/10.1016/S1383-7621(97)00067-2

11. P.C. Pendharkar, J.A. Rodger, G.J. Yaverbaum, N. Herman, M. Benner, Association, statistical, mathematical and neural approaches for mining breast cancer patterns. Expert Syst. Appl. **17**(3), 223–232 (1999). https://doi.org/10.1016/S0957-4174(99)00036-6

12. S.-M. Chou, T.-S. Lee, Y.E. Shao, I.-F. Chen, Mining the breast cancer pattern using artificial neural networks and multivariate adaptive regression splines. Expert Syst. Appl. **27**(1), 133–142 (2004). https://doi.org/10.1016/j.eswa.2003.12.013

13. A. Sadaf, P. Crystal, A. Scaranelo, T. Helbich, Performance of computer-aided detection applied to full-field digital mammography in detection of breast cancers. Eur. J. Radiol. **77**(3), 457–461 (2011). https://doi.org/10.1016/j.ejrad.2009.08.024

14. A. Horsch, A. Hapfelmeier, M. Elter, Needs assessment for next generation computer-aided mammography reference image databases and evaluation studies. Int. J. CARS **6**(6), 749 (2011). https://doi.org/10.1007/s11548-011-0553-9

15. M. Dietzel et al., Artificial neural networks for differential diagnosis of breast lesions in mr-mammography: a systematic approach addressing the influence of network architecture on diagnostic performance using a large clinical database. Eur. J. Radiol. **81**(7), 1508–1513 (2012). https://doi.org/10.1016/j.ejrad.2011.03.024

16. A.E. Hassanien, T. Kim, Breast cancer MRI diagnosis approach using support vector machine and pulse coupled neural networks. J. Appl. Log. **10**(4), 277–284 (2012). https://doi.org/10.1016/j.jal.2012.07.003

17. J. Milenković, K. Hertl, A. Košir, J. Žibert, J.F. Tasič, Characterization of spatiotemporal changes for the classification of dynamic contrast-enhanced magnetic-resonance breast lesions. Artif. Intell. Med. **58**(2), 101–114 (2013). https://doi.org/10.1016/j.artmed.2013.03.002

18. P.A.T. Baltzer, M. Dietzel, W.A. Kaiser, A simple and robust classification tree for differentiation between benign and malignant lesions in MR-mammography. Eur. Radiol. **23**(8), 2051–2060 (2013). https://doi.org/10.1007/s00330-013-2804-3

19. S. Hoffmann, J.D. Shutler, M. Lobbes, B. Burgeth, A. Meyer-Bäse, Automated analysis of non-mass-enhancing lesions in breast MRI based on morphological, kinetic, and spatio-temporal moments and joint segmentation-motion compensation technique. EURASIP J. Adv. Signal Process. **2013**(1), 172 (2013). https://doi.org/10.1186/1687-6180-2013-172

20. N. Bhooshan et al., Potential of Computer-Aided Diagnosis of High Spectral and Spatial Resolution (HiSS) MRI in the Classification of Breast Lesions. J. Magn. Reson. Imaging **39**(1), 59–67 (2014). https://doi.org/10.1002/jmri.24145

21. W.A. Weiss, M. Medved, G.S. Karczmar, M.L. Giger, Residual analysis of the water resonance signal in breast lesions imaged with high spectral and spatial resolution (HiSS) MRI: a pilot study. Med. Phys. **41**(1), 012303 (2014). https://doi.org/10.1118/1.4851615

22. S.C. Agner et al., Computerized image analysis for identifying triple-negative breast cancers and differentiating them from other molecular subtypes of breast cancer on dynamic contrast-enhanced mr images: a feasibility study. Radiology **272**(1), 91–99 (2014). https://doi.org/10.1148/radiol.14121031

23. F. Soares, F. Janela, M. Pereira, J. Seabra, M.M. Freire, Classification of breast masses on contrast-enhanced magnetic resonance images through log detrended fluctuation cumulant-based multifractal analysis. IEEE Syst. J. **8**, 929–938 (2014). https://doi.org/10.1109/JSYST.2013.2284101

24. Y.-H. Huang, Y.-C. Chang, C.-S. Huang, J.-H. Chen, R.-F. Chang, Computerized breast mass detection using multi-scale hessian-based analysis for dynamic contrast-enhanced MRI. J Digit Imaging **27**(5), 649–660 (2014). https://doi.org/10.1007/s10278-014-9681-4

25. Q. Yang, L. Li, J. Zhang, G. Shao, B. Zheng, A new quantitative image analysis method for improving breast cancer diagnosis using DCE-MRI examinations. Med Phys **42**(1), 103–109 (2015). https://doi.org/10.1118/1.4903280

26. A. Gubern-Mérida et al., Automated localization of breast cancer in DCE-MRI. Med. Image Anal. **20**(1), 265–274 (2015). https://doi.org/10.1016/j.media.2014.12.001

27. S.A. Waugh et al., Magnetic resonance imaging texture analysis classification of primary breast cancer. Eur. Radiol. **26**(2), 322–330 (2016). https://doi.org/10.1007/s00330-015-3845-6

28. I. Vidić et al., Support vector machine for breast cancer classification using diffusion-weighted MRI histogram features: preliminary study. J. Magn. Reson. Imaging **47**(5), 1205–1216 (2018). https://doi.org/10.1002/jmri.25873

29. D. Truhn, S. Schrading, C. Haarburger, H. Schneider, D. Merhof, C. Kuhl, Radiomic versus convolutional neural networks analysis for classification of contrast-enhancing lesions at multi-parametric breast MRI. Radiology **290**(2), 290–297 (2018). https://doi.org/10.1148/radiol.201 8181352

30. R. Ha et al., Predicting breast cancer molecular subtype with MRI dataset utilizing convolutional neural network algorithm. J Digit Imaging **32**(2), 276–282 (2019). https://doi.org/10.1007/s10 278-019-00179-2

31. Y. Ji et al., Independent validation of machine learning in diagnosing breast cancer on magnetic resonance imaging within a single institution. Cancer Imaging, **19**, September 2019. https:// doi.org/10.1186/s40644-019-0252-2

32. N.C. D'Amico et al., A machine learning approach for differentiating malignant from benign enhancing foci on breast MRI. European Radiology Experimental **4**(1), 5 (2020). https://doi. org/10.1186/s41747-019-0131-4

33. S. Ellmann et al., Implementation of machine learning into clinical breast MRI: potential for objective and accurate decision-making in suspicious breast masses. PLoS One **15**(1), January 2020. https://doi.org/10.1371/journal.pone.0228446

34. V.S. Parekh et al., Multiparametric deep learning tissue signatures for a radiological biomarker of breast cancer: preliminary results. Med. Phys. **47**(1), 75–88 (2020). https://doi.org/10.1002/ mp.13849

35. P. Pandya, P. Jayati, C5. 0 algorithm to improved decision tree with feature selection and reduced rrror pruning. Int. J. Comput. Appl. **117**(16), 18–21 (2015). https://doi.org/10.5120/ 20639-3318

36. J.R. Quinlan, *C4.5: Programs for Machine Learning* (Morgan Kaufmann, 1993)

37. J.R. Quinlan, Generating production rules from decision trees, in *Proceedings of the 10th International Joint Conference on Artificial intelligence—Volume 1*, San Francisco, CA, USA, August 1987, pp. 304–307. Accessed 23 Aug 2020 [Online]

38. A. Baratloo, M. Hosseini, A. Negida, G. El Ashal, Part 1: Simple definition and calculation of accuracy, sensitivity and specificity. Emerg (Tehran) **3**(2), 48–49 (2015)

39. V. Jaiswal, A. Jitendra, The evolution of the association rules. Int. J. Model. Optim. **2**(6), 726–729 (2012)

40. G.I. Webb, OPUS: an efficient admissible algorithm for unordered search. J. Artif. Intell. Res. **3**, 431–465 (1995). https://doi.org/10.1613/jair.227

41. P.J. Azevedo, A.M. Jorge, Comparing rule measures for predictive association rules, in *Machine Learning: ECML 2007* (Berlin, Heidelberg, 2007), pp. 510–517. https://doi.org/10.1007/978-3-540-74958-5_47

42. S. Alvarez, Chi-squared computation for association rules: preliminary results. Technical Report BC-CS-2003–01, July 2003. Accessed 24 August 2020 [Online]. https://www.aca demia.edu/11560769/Chi_squared_computation_for_association_rules_preliminary_results

43. C.S. Leong et al., Characterization of breast lesion morphology with delayed 3DSSMT: an adjunct to dynamic breast MRI. J. Magn. Reson. Imaging **11**(2), 87–96 (2000). https://doi.org/ 10.1002/(SICI)1522-2586(200002)11:2%3c87::AID-JMRI3%3e3.0.CO;2-E

44. A.G. Sorace et al., Distinguishing benign and malignant breast tumors: preliminary comparison of kinetic modeling approaches using multi-institutional dynamic contrast-enhanced MRI data from the International Breast MR Consortium 6883 trial. JMI **5**(1), 011019 (2018). https://doi. org/10.1117/1.JMI.5.1.011019

45. C.K. Kuhl et al., Dynamic breast MR imaging: are signal intensity time course data useful for differential diagnosis of enhancing lesions? Radiology **211**(1), 101–110 (1999). https://doi. org/10.1148/radiology.211.1.r99ap38101

46. S.D. Edwards, J.A. Lipson, D.M. Ikeda, J.M. Lee, Updates and revisions to the BI-RADS magnetic resonance imaging Lexicon. Magn. Reson. Imaging Clin. **21**(3), 483–493 (2013). https://doi.org/10.1016/j.mric.2013.02.005
47. M. Goto et al., Diagnosis of breast tumors by contrast-enhanced MR imaging: comparison between the diagnostic performance of dynamic enhancement patterns and morphologic features. J. Magn. Reson. Imaging **25**(1), 104–112 (2007). https://doi.org/10.1002/jmri.20812

Chapter 7
Learning Paradigms for Neural Networks for Automated Medical Diagnosis

Smaranda Belciug

Abstract As time passes, we find ourselves facing a new era of automated medical diagnosis, a time when not only accuracy in establishing a diagnosis is important, but also the speed of the entire process. Luckily enough, the extensive development of intelligent systems (IS) algorithms, and especially of neural networks (NNs), have led to the discovery of fast and efficient models for automated medical diagnosis. One direction in building new robust and trustworthy NNs is designing new learning paradigms, with focus on the initialization of the network's synaptic weights. In this study, our aim is to provide an overview of some learning paradigms for NNs that can be applied in automated medical diagnosis, highlighting their benefits in this crucial domain.

Keywords Neural networks · Learning paradigms · Intelligent systems · Automated medical diagnosis

7.1 Introduction

Medicine is the final frontier. Fortunately, the days when bloodletting was the cure for all diseases have passed. Nowadays, we are collecting more and more data from the healthcare domain, whether we are talking about the pharmaceutical, biotechnological, or healthcare equipment, distribution, facilities, and management industries. To process all this data is not easy, hence scientists all over the world developed 'Intelligent healthcare'. This name might sound pompous, even sci-fi. Truth to be told, the healthcare system cannot exist without Artificial Intelligence (AI). This statement does not imply the idea that there will be doctorless hospitals, because the human sixth sense shall never be replaced.

S. Belciug (✉)
Faculty of Sciences, Department of Computer Science, University of Craiova, 200585 Craiova, Romania
e-mail: sbelciug@inf.ucv.ro

© The Author(s), under exclusive license to Springer Nature Switzerland AG 2022
G. A. Tsihrintzis et al. (eds.), *Advances in Assistive Technologies*, Learning and Analytics in Intelligent Systems 28, https://doi.org/10.1007/978-3-030-87132-1_7

We can measure AI's success in the healthcare system by the amount of money that is invested in it. The AI market is supposed to reach 6 billion dollars in 2021, according to Frost and Sullivan (https://ww2.frost.com/news/press-releases/600-m-6-billion-artificial-intelligence-systems-poised-dramatic-market-expansion-health care—accessed May, 14, 2020).

A differential medical diagnosis is established by highlighting the connections between a patient's record or history, physical examination, results of different tests, and/or other clinical, radiological, laboratory data, etc., and a certain disease.

Throughout history, scientists have created 'artificial' models to solve humanity's problems. As time passed by, their focus turned towards the supreme model architect, Mother Nature. By mimicking the way different biological, physical or chemical processes work, they took problem solving to another level. One of the most used tools in computational intelligence are the artificial neural networks (NNs) [1, 2]. The aim of this study is to present different learning paradigms for NNs that enhance their performance in assisting physicians in dealing with the huge amount of medical data. This chapter is addressed to all the professionals that work in the healthcare domain, as well as for the data scientists who develop such new paradigms.

7.2   Classical Artificial Neural Networks

For someone to understand the concept behind an artificial NN, one must have a little bit of know-how regarding the way the human brain works. In a human brain we have nerve cells that propagate electrochemical signals. These cells are called *neurons*, and they are interconnected with other neurons through *synapses*. Each one of these synapses converts the activity that is passed by a neuron through the axon into different electrical effect that excite or inhibit the other neurons. Each neuron has an inhibitory input, which, if surpassed, it sends an electrical spike, or in other words it fires. This process is called *learning*.

In 1943, the first artificial neuron was developed by the neuropsychiatrist McCulloch and by the mathematician Walter Pitts. In the last years, NNs have started developing at a fast pace, with a particular emphasis in the medical care domain.

Technically, we can consider a NN as a massive parallel distributed computing structure. NNs learn from examples, exactly as the human brain does. The learning process happens when during the *training* phase, the *synaptic weights* are tuned in order to store gained knowledge from the presented examples (Fig. 7.1).

Let us presume that we have a two-class decision problem, Ω_1 and Ω_2. We can denote as input:

- $Train(n) = \{\mathbf{x}_1, \mathbf{x}_2, \ldots, \mathbf{x}_n\}$—training dataset, where $\mathbf{x}_i = \left(1, x_i^1, x_i^2, \ldots, x_i^p\right)$ is an input training vector;
- $\mathbf{w}_i = \left(1, w_i^1, w_i^2, \ldots, w_i^p\right)$—the synaptic weight vector;
- $o(n)$—the network's output;
- $y(n)$—the ground truth.

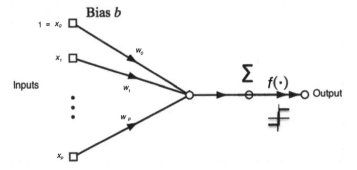

Fig. 7.1 The architecture of the perceptron

The steps that must be undertaken for the training to happen are described in the following algorithm.

1. *Initialization.* Assigning random values to the synaptic weight vector.
    Run for a certain number of epochs or until there is no more improvement achieved.
2. *Activation.* Each input $\mathbf{x}(i), i = 1, 2, \ldots, n$ is weighted using the synaptic weight vector, w_i, summed up and passed through an activation function f.

$$s_j = \sum_{i=1}^{p} w_{ji} \cdot x_i$$

3. *Updating the synaptic weights.* Taking into account the network's response, $o(i), i = 1, 2, \ldots, n$, and the ground truth $y(i), i = 1, 2, \ldots, n$, we update the weight vector as it follows:

$$\mathbf{w}(i + 1) = \mathbf{w}(i) + \eta \cdot [y(i) - o(i)] \cdot \mathbf{x}_i,$$

where η is the learning rate parameter.

The learning process begins with the activation function. If the function's output surpasses a threshold, the neuron fires, if not, the neuron is inhibited. Besides the scalar product of the weighted sum—the linear combiner, another parameter is used during the activation: the *bias*—b_i. We use the bias as a possibility of increasing or decreasing the input on the activation. Technically speaking, we add an extra input 1 and also an extra weight w_i^0.

There are numerous activation functions that can be used, but in practice the most encounter ones are:

* The *sigmoid* activation function:

$$f(s) = \frac{1}{1 + \exp(-ass \cdot s)}.$$

Fig. 7.2 Sigmoid function plot

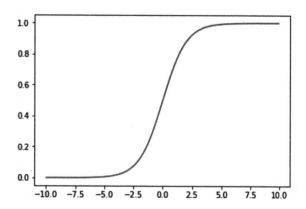

It can be seen from Fig. 7.2. that the sigmoid takes a real number and squashes it into a value between 0 and 1.

- The *hyperbolic tangent* activation function:

$$f(s) = \tanh(s) = \frac{e^s - e^{-s}}{e^s + e^{-s}}.$$

For several years now, the hyperbolic function has started to be preferred to the sigmoid, due to the fact that as Fig. 7.3. shows it takes a real number and squashes it into a number between -1 and 1 (Fig. 7.4).

- The *Rectified Linear Unit (ReLU)* activation function:

$$f(s) = \max(0, s).$$

The ReLU plot seems quite simplistic, but when dealing with image processing it is very powerful and should be definitely used.

Fig. 7.3 The hyperbolic tangent function plot

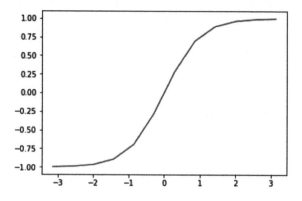

Fig. 7.4 ReLU function plot

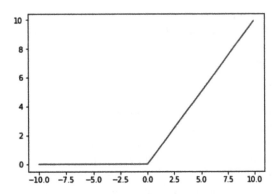

We have covered so far the basic architecture of NNs and also some types of activation functions. Now it is high time to discuss different state-of-the-art learning paradigms. The learning paradigm refers to the way the weights are being updated. Depending on the problem at hand there are three type of learning that can be utilized:

- *Supervised learning*: the training model learns from historical data, previously labeled data.
- *Unsupervised learning*: the training model does not have a previously labeled training set. It learns to discriminate data, searching for different patterns.
- *Reinforcement learning*: the training model wants to achieve a specific objective. It has rules set up for it, and it is trying to find the best strategy in order to reach its goal.

In this chapter, we are going to cover only supervised learning methods. As stated above, the network learns from previously labeled examples. Taking into account the ground truth (label), the NN learns how to distinguish a certain class from raw unlabeled data. Using the training set, the algorithm tunes its parameters in order to obtain the correct answer. An important part of the learning process is to create a network that can generalize well, so it can be used on new data with high performance.

One of the most used training methods for NNs is the backpropagation algorithm. It is used for training types of NN such as the Multi-layered Perceptron (MLP) and the Convolutional Neural Networks (CNN). The backpropagation algorithm computes the error between the ground truth and the network's output, and propagates this error backwards through the network, changing the weights. These errors represent the *loss function, L*, of the network, function that needs to be minimized. L's formula is:

$$L = \frac{1}{n} \sum_i |y_i - d_i|^2.$$

To minimize L we need to find the optimum parameter weights. In order to obtain this the backpropagation algorithm shows us the correct direction where we should start the search, and what direction to follow, through the *gradient descent*, a mathematical method. In this process the learning rate plays an important part, since it determines at what pace the search will be done. The best real-life comparison of the gradient descend is the famous example regarding hiking: imagine that you are hiking in the mountains, and unfortunately you find yourself lost in the woods at the top of the mountain. The plot thickens, as you find out that you have not reception on your smartphone, no compass, no map, and the day turned into night. You can only remember that on your way up, you passed by a village at the bottom of the mountain. Hence, the only viable solution remains to try to descend carefully, without dying of hunger or thirst (yes, you also have no more food or water left in your backpack), and also without falling into the abyss. Since you have no provisions, you must get there quickly, so you follow the sharpest path. The size of the step (too small you will never get to the village; too large you might miss the village and end up on the other side of the mountain) is the learning rate.

Choosing the learning rate is an optimization process also. In general, its value ranges between 0.0 and 1.0. A good idea is to start with a high valued learning rate and slowly decrease it as epochs go by Smith [3] proposed that at every batch to record the learning rate and also the loss. By plotting them in the same graph, we can choose the value to the learning rate that gives us the smallest loss.

Mathematically, the gradient is a vector of derivatives. We compute the network's loss through a continuous, differentiable function of its weights. The standard gradient, or vanilla gradient, is computed as such:

$$\nabla L = \left(\frac{\partial L}{\partial w_1}, ..., \frac{\partial L}{\partial w_p} \right).$$

The weights will be updated taking into account the learning rate, increment and error function:

$$\Delta w_i = -\eta \frac{\partial L}{\partial w_i},$$

where $\Delta w_i = w_{i+1} - w_i$.

More details regarding the backpropagation learning paradigm and its applications in medicine can be found in Belciug [4].

7.3 Learning Paradigms

7.3.1 Evolutionary Computation Learning Paradigm

Another way for setting the right parameters is using optimization techniques such as evolutionary computation [5]. In this subsection, we are going to discuss a special class of evolutionary computation—genetic algorithms (GAs). GA can find optimal solutions using the following operators: selection, crossover or recombination, and mutation. In what regards NNs, GAs can be used for setting the values of the weights, learn the network's topology, interpret its behavior, etc.

Being bio-inspired, GAs mimic the natural selection process. Hence, we start with a *population* made out of randomly generated individuals. Each one of these individuals, called a *chromosome*, can be considered as a solution of the optimization problem. The chromosome consists of *genes*. Depending on the genes' value domain, we can have different kinds of chromosome representations: binary, integer or real-coded.

For the evaluation of the chromosomes, we use a function called the *fitness function*, $f(x_i)$, $i = 1, 2, \ldots, n$, where n is the number of chromosomes in the population. The fitness function measures how well a chromosome adapts to the environment, creating a competition between them. The better a chromosome adapts to the environment, the higher chances it has to be selected for reproduction or to be part of next generation.

To compute the overall performance of a certain population we use the following formula:

$$F = \sum_{i=1}^{n} f(x_i).$$

Selection can be used either for parent selection or for the replenishing of the population. There are several types of selection that can be used:

1. *The tournament selection*: this type of selection implies a contest between k chromosomes. The fitness score of each chromosome is computed and compared with the other's scores. The best one is chosen. Due to the fact that each chromosome has an equal chance of competing, the diversity can be preserved. Razali and Geraghty [6] signaled the fact that the tournament selection leads to a decrease in the convergence speed. The probability of selecting a certain chromosome is:

$$P(i) = \begin{cases} \frac{C_{n-1}^{k-1}}{C_n^k}, & \text{if } i \in [1, n - k - 1], \\ 0, & \text{if } i \in [n - k, n]. \end{cases}$$

2. *The Fitness proportional selection*: is also known under the name of the *Roulette Wheel*. The probability of a chromosome to be selected is proportionate to its fitness. We can compute this probability according to the following equation:

$$P(i) = \frac{f(i)}{\sum_{j=1}^{n} f(j)}.$$

3. *The Ranking selection*: takes into account the rank of a chromosome. The rank is assigned after the fitness score has been computed. Thus, the chromosome with the highest fitness score has the highest rank, followed by the second, etc. The selection probability is given by:

$$P(i) = \frac{rank(i)}{n \cdot (n-1)}.$$

4. *The Exponential rank selection*: similar to the rank selection, in this type of selection the probabilities of the ranked chromosomes are weighted exponentially. We use a parameter c for the base of the exponent, which takes values between $0 < c < 1$. The probability is given by:

$$P(i) = \frac{c^{N-i}}{\sum_{j=1}^{N} c^{N-j}}.$$

For more selection operators and explanations, please see [4, 7, 8].

The survival selection for the new population can be performed in two manners: either we use the *fitness based survival selection* or the *age-based selection*. The first is similar to the reproduction selection presented above. In what regards the latter, the individuals do not survive based upon their fitness, they survive based upon their age.

The crossover or recombination operator is used for obtaining new offspring from two parents. In this type of operator, we use a parameter called the *recombination probability*, $p_c \in [0, 1]$. For each pair of selected parents, a random number is generated, taking values between 0 and 1. If the generated number's value is lower than the recombination probability, then the crossover starts, otherwise we are dealing with an asexually crossover, implying a cloning process of the parents. The general crossover scheme is $p = q = 2$. Taking into account the population's type of representation, we have the following types of crossover:

- For the *binary representation*: one-point crossover, n-point crossover, uniform crossover;
- For the *integer representation*: one-point crossover, n-point crossover, uniform crossover;
- For the *real-coded representation*: one-point crossover, n-point crossover, uniform crossover, simple arithmetic crossover, single arithmetic crossover, total arithmetic crossover, blend crossover, linear BGA crossover, Wright's heuristic crossover. For more details please see Eiben and Smith [9, 10].

The last GA operator is the mutation. In order to perform mutation, one must use a new parameter, the *mutation probability*. By using this parameter an original chromosome can be modified by changing one or multiple genes. In some cases, the entire original chromosome can suffer mutations. Just like in the case of recombination, mutation depends on the chromosome's representation. Accordingly, we have the following:

- For the *binary representation:* bitwise mutation;
- For the *integer representation*: random setting, creep mutation;
- For the *real-coded representation*: uniform mutation, normally distributed mutation.

The general scheme of GA consists of the next steps:

1. Population initialization
2. Obtain fitness score by applying the fitness function to all the chromosomes
3. Until the stopping criterion is met do:
   3.1. Parent selection
   3.2. Crossover
   3.3. Mutation
   3.4. Compute fitness scores of the new individuals
   3.5. Select next generation.

The evolutionary learning paradigm was introduced by Belciug and Gorunescu in 2013. Initially the method was designed for a two class decision problem, but can be extended to multiple classes [11]. The authors chose the following MLP's architecture: the input layer of the network contains n neuron units, where n is the number of features from the dataset; one hidden layer that has an equal number of hidden units as the number of classes; whereas the output layer has only one neuron, which computes the network's output through the *winner-takes-all* rule. The sigmoid plays the role of the activation function. The author replaced the classical backpropagation with a GA. Technically, the synaptic weights represent the genes of a chromosome. The GA finds the optimal weights.

The learning paradigm has been compared with the classical backpropagation using four breast cancer publicly available datasets. The evolutionary learning paradigm made it possible for the NN to juggle with different types of attributes (numerical, categorical, or mixed). In terms of comparison, this new approach proved to be superior to the classical algorithm.

For further reading on mixing NNs with GAs see Shaffer et al. [12–14].

7.3.2 Bayesian Learning Paradigm

Another interesting alternative to the backpropagation algorithm is the Bayesian learning paradigm. This approach was first designed for a MLP [15], and secondly for a single-hidden layer feedforward NN [16]. The paradigm uses the Bayes' theorem for classification problems. Mathematically speaking, if we have N items that have the following features $\{A_1, A_2, ..., A_p\}$, we need to determine which class they belong to. Let us suppose we have Ω_1, Ω_2, ..., Ω_q classes, then according to Bayes' formula, the correct class will be the one that maximizes the following probability $P\{A_1, A_2, ..., A_p|\Omega_i\}$, $i = 1, 2, ..., q$. Due to *Idiot's Bayes*, that presumes that all the attributes are independent from each other, we can transform this formula into:

$$P\{A_1, A_2, \ldots, A_p|\Omega_i\} = P\{A_1|\Omega_i\} \cdot P\{A_2|\Omega_i\} \cdot \cdots \cdot P\{A_p|\Omega_i\}.$$

The proposed network has the following architecture: the input layer contains examples from the training dataset, which have been previously normalized and shuffled; the batch mode is used in the training phase [17, 18], the activation function is a modified version of the hyperbolic function, which converges faster [17], $f(u) = 1.7159 \cdot tanh\left(\frac{2}{3} \cdot u\right)$; the number of hidden units from the hidden layer equals the number of classes; the output is computed using the softmax function; the weights are initialized by computing the rank correlation Goodman–Kruskal Gamma between the input and the decision class, $P\{A_{ij}\} = \Gamma(X_i, Y_j)$, $i = 1, 2, \ldots, p$, $j = 1, 2, \ldots, q$.

The synaptic weight between two neurons can be interpreted as belief measure [19, 20]. The subjective Bayesianism implies the fact that the domain value of the weights ranges from 0 to 1, and each weight practically encodes the connection strength between neurons [21]. If we consider the weights as being posterior probabilities, we can update them according to the Bayesian rule:

$$w_{ij}(n+1) = P\{A_{ij}|E(n)\} = \frac{P\{E(n)|A_{ij}\} \cdot P\{A_{ij}\}}{\sum_{ij} P\{E(n)|A_{ij}\} \cdot P\{A_{ij}\}},$$

where $P\{E(n)\}$ is the *evidence*, $P\{A_{ij}\}$ is the *prior probability*, and $P\{E(n)|A_{ij}\}$ is the *likelihood*, for $i = 1, 2, ..., p$ and $j = 1, 2, ..., q$.

The likelihood formula can be rewritten as the Goodman–Kruskal correlation rank between each attribute and the error at step n:

$$P\{E(n)|A_{ij}\} = \Gamma(X_i, E(n)), i = 1, 2, \ldots, p, j = 1, 2, \ldots, q.$$

In what follows, we shall present the steps that need to be undertaken in order to use the Bayesian learning paradigm:

1. The mean attribute value m_j^i is computed per attribute A_i, $i = 1, 2, ..., p$, and per class Ω_j, $j = 1, 2, ..., q$.
2. The weights are initialized using the formula:

$$w_{ij} = \Gamma\left(\left(x_i^k - m_i^j\right), y_k\right), \quad i = 1, 2, \ldots, p, \quad j = 1, 2, \ldots q, \quad k = 1, \ldots, N.$$

3. The discriminant linear function u_j, $j = 1, 2, .., q$, for each unit in the hidden layer, is computed as follows:

$$u_j = \sum_{i=1}^{p} \left(x_i^k \cdot w_{ij} \cdot \frac{1}{(x_i^k - m_i^j)^2} \right),$$

where $i = 1, 2, ..., p$, $j = 1, 2, ..., q$ and $k = 1, 2, ..., N$.

4.   Apply the non-linear activation function for each hidden unit:

$$f(u_j) = 1.7159 \cdot tanh\left(\frac{2}{3} \cdot u_j\right), \, j = 1, 2, ..., q.$$

5.   The hidden layer is interpreted as a discrete random variable that is being governed by the probability mass function $g_j = g(f(u_j))$:

$$g_j = \frac{exp(f(u_j)) - max\{f(u_i)\}}{\sum_{j=1}^{p}(exp(f(u_j)) - max\{f(u_i)\})}$$

$j = 1, 2, ..., q$.

6.   The error is computed for each object $\mathbf{x}_k$ in the training set:

$$error_k = \sum_{j=1}^{q} \left(y_j^k - g_j^k\right), \quad k = 1, 2, ..., N.$$

7.   The weights are being update according to the Bayesian learning paradigm:

$$w_{ij}(n+1) = w_{ij}^* = \frac{\Gamma(X_i, E(n)) \cdot \Gamma(X_i, y)}{\sum_i \Gamma(X_i, E(n)) \cdot \Gamma(X_i, y)},$$

where $E = (error_1, error_2, ..., error_N)$, $i = 1, 2, ..., p$ and $j = 1, 2, ..., q$.

The above process is repeated until the stopping criterion is met.

The Bayesian learning paradigm has been applied on breast and lung cancer, diabetes, and heart attack real medical datasets. The reported results showed that this paradigm equals and even sometimes exceeds other results reported in literature on the same datasets.

7.3.3 Markovian Stimulus-Sampling Learning Paradigm

In Gorunescu and Belciug [22] is presented an enhancement of the backpropagation algorithm by applying a Markovian stimulus-sampling process. The stimulus-sampling being Markovian, implies the fact that is memoryless, that is at each step the choice of a stimulus depends only on the network's output at the previous step. This learning paradigm makes the learning process gradual and cumulative, and prone to suffer modifications due to environmental changes. The NN uses the observed behavior that is variable throughout the training process and applies a reward or a penalty accordingly. Mathematically speaking, each output neuron has a counter c_j that is initially set to 0. This counter computes all the samples that have been correctly classified. After all the training data has been presented, each one of the output neurons receives a stimulus s_j according to its performance. The performance is measured as the percentage of correctly classified cases.

In what follows we shall present the steps that need to be undertaken in order to perform the Markovian stimulus-sampling learning paradigm:

1. In the forward pass, each input sample is propagated through the network. The output values are normalized and using the *winner-takes-all* rules determines the networks answer.
2. Update the counters c_j, $j = 1, 2, \ldots q$, according to the performance of neuron j.
3. Use the backpropagation algorithm to update the synaptic weights.
4. After all the training samples have been presented to the network, each output neuron receives a stimulus s_j. Multiply the values of the output neurons for the next epoch with s_j.

This learning paradigm was applied on six publicly available datasets regarding breast, lung, and colon cancer, heart attack and diabetes, and proved its robustness and competence.

7.3.4 Logistic Regression Paradigm

The logistic regression paradigm for a two-class decision problem was proposed in Belciug [4, 23]. The learning paradigm was developed for a single-hidden layer NN, and it basically embeds the logistic regression so that the hidden nodes of the network are problem dependent. Besides this, the paradigm has a filtering module that is based on the significance of each feature to the output, so that the features that are not statistically significant enough are removed from the network. This approach is important due to the fact that it can easily avoid the *'curse of dimensionality'*.

The input weights are computed by estimating the regression coefficients using the maximum likelihood. The logistic regression can be written mathematically using the following equation:

$$logit(p) = a + X \cdot b,$$

where:

$$X = \begin{pmatrix} 1 & x_1^1 & \cdots & x_m^1 \\ \vdots & & \ddots & \vdots \\ 1 & x_1^N & \cdots & x_m^N \end{pmatrix}.$$

We denote p the proportion of subjects that have a certain feature, $logit(p)$ transformation formula is:

$$logit(p) = \ln\left(\frac{p}{1 - p}\right).$$

The filtering module uses the statistical significance level p. By computing p, we find out what features are not significant, and we remove them from the network. In general, for the p-level threshold it is used 0.05 as cut-off value.

The steps for applying the logistic regression learning paradigm are outlined below:

1. Initialize the synaptic weights of the network by assigning them the computed values of the regression coefficients b_i, $i = 1, ..., N$, and intercept b_0.
2. Compute the corresponding p-level for each feature.
3. Remove from the network all the connections that did not surpass the cut-off value 0.05.
4. Compute the hidden layer output matrix, M, using as activation function the hyperbolic tangent.
5. Compute the hidden-output weights, $\lambda = M^+ \cdot y$, using the Moore–Penrose pseudo-inverse matrix M^+ and the ground truth y.

The logistic regression paradigm was applied successfully on gene expression datasets regarding breast, colon and lung cancer. The method proved that indeed it can filter features, increase diagnostic performance, and decrease computational cost and time. The method has been extended to multiple classes in Belciug [24].

7.3.5 Ant Colony Optimization Learning Paradigm

Another learning paradigm adds a filtering module based on the ant colony optimization (ACO) method to an Extreme Learning Machine (ELM) NN. ELM is a new type of neural network, that caused a huge debate in the scientific community. The architecture of an ELM contains only one hidden layer. The weights between the input and hidden layer are randomly initialized, and the weights between the hidden and the output layer are computed in one step using the Moore–Penrose generalized pseudo-inverse [25–27] and the ground truth.

Ant colony optimization approach has been added as feature selection to the learning paradigm of an ELM in Berglund and Belciug [28]. ACO is another bio-inspired metaheuristic that uses artificial ants to optimize different problems [29]. The idea behind ACO is to simulate ants' behavior when they are gathering food. Real ants communicate within their community through an odorous substance, pheromone. When gathering food, ants place a certain amount of pheromone each time they find a source of food. This quantity is dependent on the distance between the nest and the food. While randomly moving through the search space, another ant might sense the pheromone laid by other ants, making it decide to follow that path. Once it arrives at the food source, it will lay itself more pheromone, rising the pheromone quantity, therefore making that path more appealing to other ants in the colony.

The steps that one needs to undertake in order to perform the ACO learning paradigm are:

1. Presume that we have n ants in the colony. Randomly initialize a subset of characteristics from the training data set, having the restriction that each one of the ants in the colony has a unique characteristic set.
2. Set the values of the pheromone matrix T to 0.
3. For each ant compute pheromone level. Transform pheromone values into probabilities. The probability of a feature to belong to a certain ant is:

$$p_i = \frac{T_i}{\sum_{i=0}^{n} T_i}.$$

4. Denote by m the number of selected features for each ant. At each step, m decreases by 1. If m decreases below 2, another value is generated for it using the Monte Carlo roulette procedure.
5. Evaluate each feature set using the ELM algorithm. The best feature set will be used during the testing phase.
6. Update the pheromone using the mean square error (MSE):

$$\Delta T_i = \begin{cases} max(MSE) - MSE_j, & if \quad f_i \in S_j \\ 0, & otherwise \end{cases}$$

where T_i is the pheromone of feature f_i, S_j is the features' subset.

The method was applied successfully on high dimensional datasets regarding breast and lung cancer, and thyroid disease.

7.4   Conclusions and Future Outlook

Medical databases store large amount of information regarding patients and their medical conditions. This amount of data causes the well-known *'curse of dimensionality'* [30]. The advances made in medical equipment made possible the collection of features that can lead to the creation of datasets with a relatively small number of attributes, or to high dimensional datasets that contain medical data obtained through DNA microarray analysis (e.g. approximately 25,000 features). Classical machine learning algorithms sometimes are 'suffocated' by high dimensional datasets, and that is why it is imperative for data scientists to develop new strategies, such as new learning paradigms, in order to process faster the data.

In this context, the state-of-the-art paradigms presented in this chapter are designed to deal with this 'big data' in order to discover patterns that are mandatory in diagnosing, treating and managing diseases. Future research's goals should be to cross-fertilize even further domains such as medicine, artificial intelligence and statistics, so that more and more robust learning paradigms be developed.

References

1. S. Belciug, F. Gorunescu, *Intelligent Decision Support Systems—A Journey to Smarter Healthcare* (Springer Nature Switzerland AG, 2020). https://doi.org/10.1007/978-3-030-143 54-1
2. C. Bishop, *Neural Networks for Pattern Recognition* (Oxford University Press, 1996)
3. L.N. Smith, Cyclical learning rates for training neural networks. Comput. Vis. Patter. Recognit. (2015). https://arxiv.org/abs/1506.01186
4. S. Belciug, *Artificial Intelligence in Cancer—Diagnostic to Tailored Treatment* (Elsevier, 2020)
5. J.H. Holland, *Adaptation in Natural and Artificial Systems: An Introductory Analysis with Applications to Biology, Control and Artificial Intelligence* (University of Michigan Press, 1975)
6. N.M. Razali, J. Geraghty, Genetic algorithm performance with different selection strategies in solving TSP, in *Proceedings of the World Congress on Engineering*, vol. II, UK (2011)
7. T. Blickle, K. Thiele, A comparison of selection schemes used in genetic algorithms, TIK-Report, 11 (1995)
8. K. Jebari, M. Madiafi, Selection methods for genetic algorithm. J. Emerg. Sci. **3**(4), 333–344 (2013)
9. A.E. Eiben, Multiparent recombination in evolutionary computing, in *Advances in Evolutionary Computation: Theory and Applications*. ed. by A. Gosh, S. Tsutsui (Springer, Heildelberg, 2003), pp. 175–192
10. A.E. Eiben, J.E. Smith, *Introduction to Evolutionary Computing* (Springer, Heildelberg, 2003)
11. S. Belciug, F. Gorunescu, A hybrid neural network/genetic algorithm system applied to the breast cancer detection and recurrence. Expert Syst. J. Knowl. Eng. **30**(3), 243–254 (2013)
12. M. Mitchell, *An Introduction to Genetic Algorithms* (MIT Press, USA, 1998)
13. D. Whitley, J. Periauxa, G. Winter, *Genetic Algorithms in Engineering and Computer Science* (Wiley, UK, 1995)
14. J.D. Schaffer, et al., Combinations of genetic algorithms and neural networks: a survey of the state-of the-art. In: Whitley, D.L., Schaffer, J.D., (eds.). In: *Proc. Int. Workshop on Combinations of Genetic Algorithms and Neural Networks*, pp. 1–37 (IEEE Computer Society, Los Alamitos, CA, 1992)

15. S. Belciug, F. Gorunescu, Error-correction learning for artificial neural networks using the Bayesian paradigm. Application to automated medical diagnosis. J. Biomed. Inform. **52**, 329–337 (2014)
16. S. Belciug, F. Gorunescu, Learning a single-hidden layer feedforward neural network using rank correlation-based strategy with application to high dimensional gene expression and proteomic spectra datasets in cancer detection. J. Biomed. Inform. **83**, 159–166 (2018)
17. S. Haykin, *Neural Networks. A Comprehensive Foundation*, 2nd edn. (Prentice Hall, 1999)
18. Y. LeCun, L. Bottou, G. Orr, K.-L. Muller, Efficient BackProp. Neural networks: tricks of the trade. Lect. Notes Comput. Sci. **7700**, 9–48 (2012)
19. A. Hajek, Intepretation of probability, in *The Standford Encyclopedia of Philosophy*, ed. by N.Z. Edward (Winter, 2012). http://plato.standford.edu/archives.win2012/entries/probability-interpret/
20. J. Press, *Subjective and Objective Bayesian Statistics: Principles, Models, and Applications*, 2nd edn. (Wiley, 2003). http://onlinelibrary.wiley.com/doi/10.1002/9780470317105.fmatter/pdf
21. E.-J. Wagenmakers, M. Lee, T. Lodewyckx, G. Iverson, Bayesian evaluation of informative hypotheses (statistics for social and behavioral sciences), in *Bayesian Versus Frequentist Inference*. ed. by H. Hoijtink, I. Kulgkist, P. Boelen (Springer, 2008), pp. 181–207
22. F. Gorunescu, S. Belciug, Boosting backpropagation algorithm by stimulus-sampling: application in computer-aided medical diagnosis. J. Biomed. Inform. **63**, 74–81 (2016)
23. S. Belciug, Logistic regression paradigm for training a single-hidden layer feedforward neural network. Application to gene expression datasets for cancer research. J. Biomed. Inform. **102** (2020)
24. S. Belciug, Parallel versus cascaded logistic regression trained single-hidden feedforward neural network for medical data. Exp. Sys. Appl. **170**, 114538 (2021)
25. G.B. Huang, Q.C. Zhu, C.K. Chee-Kheong Siew, Extreme learning machine: a new learning scheme of feedforward neural networks, in *Proceedings of International Joint Conference Neural Networks*, pp. 985–990 (2006)
26. G.B. Huang, X. Ding, H. Zhou, Optimization method based extreme learning machine for classification. Neurocomputing **74**, 155–163 (2010)
27. G.B. Huang, H. Zhou, X. Ding, R. Zhang, Extreme learning machine for regression and multiclass classification. IEEE Trans. Syst. Man Cybernet **42**(2), 513–529 (2011)
28. R. Berglund, S. Belciug, Improving extreme learning machine performance using ant colony optimization feature selection. Application to automated medical diagnosis. Ann. Univ. Craiova, Math. Comput. Sci. Ser. **45**(1), 151–155 (2018)
29. M. Dorigo, V. Maniezzo, A. Colorni, Ant system: optimization by a colony of cooperating agents. IEEE Trans. Syst. Man Cybern. Part B **26**, 29–41 (1996)
30. R.E. Bellman, *Dynamic Programming* (Princeton University Press, Princeton, 1957)

Part III
Advances in Assistive Technologies in Mobility and Navigation

Chapter 8
Smart Shoes for Assisting People: A Short Survey

Nikolaos G. Bourbakis, Iosif Papadakis Ktistakis, and Pulkit Khursija

Summary Lately smart shoes have also become a kind of a fashion for the general public and athletes. However, leggs and feet diseases are more common problems among elderly people as time progresses and to those with certain feet disabilities [1], so smart shoes may offer some comfort to their situation. In particular people with osteoarthritis suffer for movement of the cartilage in a joint wears away over time causing bones to rub against each other [2]. In addition, arthritis is the leading cause of disability for the elderly in the United States of America [2]. Thus, a cause of osteoarthritis is mainly from poor gait, otherwise known as walking patterns, where there is unevenly distributed pressure across the foot while walking. This leads to development of severe pain in the knees over time which progressively gets worse. The uneven wear and tear to the knee cartilage is the specific cause of Osteoarthritis. According to the Centers for Disease Control and Prevention (CDC), more than 50 million adults have been diagnosed with Arthritis, i.e., approximately 1 in 5 adults (C for Disease Control and Prevention, "Arthritis national statistics," 2018. https://www. cdc.gov/arthritis/data_statistics/national-statistics.html). The CDC also predicts that by 2040, more than 78 million people will be diagnosed with Arthritis (C for Disease Control and Prevention, "Arthritis national statistics," 2018. https://www.cdc.gov/art hritis/data_statistics/national-statistics.html).

In the United States of America, the economic impact from Arthritis is troubling. Every year, people with Arthritis and other Rheumatic conditions give up potential income (called "lost wages") due to injury or illness. According to the CDC, in 2013 the total financial cost of Arthritis was $304 Billion [3]. That is, $140 Billion was associated with medical costs while $164 Billion was associated with lost wages due to adults with Arthritis missing work. Furthermore, an Article by Schofield et al. states that in Australia, approximately 54,000 people aged 45–64 lost productive life

N. G. Bourbakis (✉) · I. P. Ktistakis · P. Khursija
CART Center, Wright State University, Dayton, OH, USA
e-mail: nikolaos.bourbakis@wright.edu

N. G. Bourbakis
Technical University of Crete, Chania, Greece

years due to Arthritis in 2015 [4]. Also, agreeing with the CDC, these individuals receive less in income as well. Finally, according to the Arthritis Foundation, 10% of all outpatient visits in 2010 were due to Arthritis and an estimated 706,000–757,000 knee replacement procedures occurred between 2010 and 2011 [5].

In response to these health issues, researchers and practitioners have offered some comfortable solutions in the form of wearable devices that could monitor and help users in their daily activities. The most common category of wearable devices is the smart shoes. Smart shoes have been defined by different ways according to the researchers background. Thus, for computer scientists smart shoes are the ones that use AI algorithms beyond to sensors and a processor that processes data and interacts with the user. Smart shoes are the future of shoes where the shoes would communicate with the user and provide him with vital health data. These shoes could majorly be used for health monitoring, navigation and object detection for the blind and performance enhancement for the athletes. All vital statistics will be provided to the user in real time making the system as a whole efficient and affective.

The design and comfort of these shoes will be the same as regular shoes and will be specially engineered in order to fit in crucial devices. The future of these shoes will have more sensors and algorithms that will enhance the scope for these shoes and will increase the usability. Most of these shoes will generate power (electricity) on their own by converting force exerted by your feet on the ground into electricity to power the device.

Here, a short survey is offered, which does not cover categories and aspects of the importance of the smart shoes defined in various ways, but provides an ideas about the smart shoes and their usefulness. It attempts to address some of their contributions to people in need and those that want to measure certain interests.

8.1 Smart Shoes for People in Need

In this section we provide a brief description of a variety of smart shoes used by different people and for various reasons. This is a short introduction of a small number of smart shoes and their usefulness to people.

8.1.1 Smart Shoes for Visually Impaired People [6]

In this effort the authors have designed a smart shoe for the aid of the visually impaired people. they have used Arduino Uno board along with Ultrasonic Sensors and Water droplet sensor in order to retrieve environmental data. For the power generation they have been using solar panels along with a piezoelectric ring and a battery for uninterrupted power supply. There is also an emergency button on the shoe which when pressed will share details of the person wearing the shoes with his emergency contacts via his cellphone connected to the shoe wirelessly.

In order to detect the path for the blind person, the shoes will gather data from the ultrasonic sensors placed on the shoe. This data will be processed in the Arduino with the help of an algorithm and an output in the form of vibrations and sounds will be produced letting the blind person be aware of any upcoming obstacles.

In rainy conditions the water droplet sensor will be in use. It will see forthcoming patches of water and warn the user in advance for the same. All these sensors will provide data to the Arduino board which will process the data and provide adequate direction and warnings to the user in real time.

Pros-(Good power support; Sensors fitted properly on the shoe will not look odd; The overall system can be fitted in any form of shoe; Multiple ultrasonic sensors are used for accuracy; They require minimum use of power and most of it is generated on its own; Water droplet sensors can also help the user be ready for forthcoming water filled pot holes).

Cons-(The accuracy and integrity of the sensors are not determined; Objects moving towards the person or the person moving towards an object with certain speed will be a dangerous situation; Physical changes for the equipment may not be visible, but they would add a certain weight to the shoe).

8.1.2 Smart Shoes for Blind Individuals [13]

The main objective of this work is to provide an acoustic assistance to the blind people and also to deal with the problems faced by them to walk like the normal human beings. Thus, the authors aim to develop a device that would serve as a guiding assistance to them. One of the biggest problems that the visual impaired people face is during walking in the indoors and outdoors they are not well aware of information about their location and orientation with respect to traffic and obstacles on their way unlike the normal beings. The technology proposed by the authors serves as a potential solution for visual impaired people. Their project consists of smart shoes and a smart cane (stick) that alerts visually-impaired people over obstacles coming between their ways and could help them in walking with less collision. The main aim of this effort is to address a reliable solution encompassing of a cane and a shoe that could communicate with the users through voice alert and pre-recorded messages.

The proposed system uses Arduino Nano and Lily, Bluetooth Module, IR Sensors, Ultrasonic Sensors, Bluetooth Voice Speaker, and a rechargeable battery. The function of the overall system is simple, the person wears the shoes and carries the smart cane along with him/her while walking. As soon as the person wears the shoes, the sensors start to work. The IR sensors and the ultrasonic sensors start to provide data to the Arduino and the Lily for calculation. Based on the results of the calculation objects are identified on the way and a pre-recorded voice action for the same is conveyed to the user with the help of speakers. The shoes also calculate the average length of each step the person takes and provides user the information about an approaching obstacle within certain steps.

Pros-(Smart and intelligent idea to use speakers and prerecorded voice commands; The use of smart cane helps keep the various components out of the shoes making it light and normal to the eyes; The calculation for the number of steps to take in each direction in order to find an obstacle free route is what could help the user maneuver with ease; It has safe control to lower the consumption of the battery when the level drops below 20% in order to maximize the use of the battery; The combination of smart cane and smart shoes together can further improve the accuracy of the overall system).

Cons-(Since both shoes and stick have sensors, both need to be powered and charged; To place a battery in the shoe is not the best of the ideas as batteries tend to become warm after certain duration of use; Batteries usually Li-Ion ones may explode; The length of duration for which the system works is uncertain, i.e. the time or distance for which the system remains in power is undefined).

8.1.3 IoT Based Wireless Smart Shoes and Energy Harvesting System [7]

The author has chosen piezoelectric sensors for the production of electricity that can be used to charge devices such as a smartphone. The shoes are also cable of calculating fitness parameters such as steps taken, calories burned and distance covered through a mobile application.

The use of Node MCU is done in order to process data retrieved by the piezoelectric sensors. The sensors produce electricity in the range of 1 V to 2 V. This electricity is then stored in a cell via the Node MCU chip. This electricity that is stored in the cell can later be used to charge devices via a charging cable that can be plugged into the Node MCU shoe.

Moreover, based on the input voltage by each piezo sensor calculations can be done on the number of steps taken, distance covered and calories burned. For this an intelligent program and an android application was developed. The program is stored in the Node MCU, which later communicates with the android application on the phone.

To reduce the error for step calculation, the user needs to input his BMI info in the app that will later help in accuracy of data, otherwise a standard value will be set for him. Also a certain threshold value of input voltages is set so that the system does not calculates step even if the user is standing. This threshold value is determined after the user takes certain steps while wearing on the shoes.

In addition, the sensors are neatly placed on the insole of the shoe, making it comfortable to wear for the user. Also, this helps the outsole to have adequate grip on the flooring surfaces.

Pros-(It does not require charging; It eliminates the requirement of other sensors for steps and calories calculation; It can be fitted onto any type of shoes; The energy produced although low, however, it can be used to save time for charging otherwise).

Cons-(The accuracy of calculations for the step count has to be readjusted); The components placed inside the shoe, such as a Node MCU module, may make the shoe uncomfortable at times; The batteries kept in the shoe might explode if the person performs heavy outdoor activities).

8.1.4   Smart Shoes for Sensing Force [8]

Here, the authors have created a shoe-patent that uses various pressure sensors in order to calculate the athletic data of the user. This shoe is intended for a person who is deeply into athletics. The shoe comprises of a microcontroller, a battery, pressure sensors, and a battery monitor.

The shoe is neatly designed with all components are placed inside the shoe in such a way that they do not hinder the performance of the athlete. It is simple and easy to use, also visibly it looks like any other ordinary sport shoe. The shoe is also equipped with a reset button on both the shoe of a pair.

The shoe with the help of the sensors will record data and store it on the onboard memory of the microcontroller. The microcontroller will simply store the raw data that it receives from the sensors. Later when the shoes are connected to a PC with the compatible software installed, the software will process the data and produce substantial results that will help the athlete improve. The computer program will recognize, interpret and predict important information based on the results obtained from the pressure sensors. The shoe will just gather information from the sensors and store it on the memory chip inbuilt in the micro controller.

If the user needs to clear the memory, he will be required to either press the reset button on both the shoes or else he will have to reset it via the PC program.

Pros-(Neatly designed; Light weight; Less complex due to lesser number of components used; Comfortable and easy to use; Would be helpful for a person who takes physical activity and cardio activities seriously).

Cons-(Battery isn't a safe thing to be used in a shoe as it might explode; It is inconvenient to connect the shoes to the PC every time activity data is required; It is unconventional and a slow process to charge your shoes before a run/jog).

8.1.5   Smart Shoes for Temperature and Pressure [9]

Here the authors have designed a smart shoe with various different sensors including a pressure sensor, a temperature sensor, an altitude sensor and an acceleration sensor. All these sensors will be centrally controlled by a microcontroller. This microcontroller will be receiving data from at least one sensor during all times.

These sensors such as the altitude sensor and the acceleration sensor have been provided in the shoe so that the calculation of calories burned is accurate. It is observed that walking/ running on an inclined surface requires more energy and the walking pattern for such surface also changes. In order to consider these parameters while making calculations, these sensors have been introduced and activated into the shoe. All the data that is recovered from this shoe is further more accurate and it eliminates certain conditions that might affect acute changes in data resulting in wrong inferences.

A battery is also required for the shoe to be powered on. The user will have to use the compatible application on his/her smart phone in order to see vital statistics that were generated by his/her body during the previous walk/run. The acceleration sensor along with the microcontroller can act as a pair to calculate lap timings and the speed of the person with high accuracy.

Thus, the shoe will be a good product for a sports person or a health athlete, who usually goes out for a jog/run.

Pros-(Smart and intelligent looking shoe; The shoe eliminates certain real-world hindrances to data collection; Well equipped with useful sensors; By far the most accurate and ideal for use).

Cons-(Battery may explode during heavy workout sessions; Complex circuit inside; Placement of sensors inside the shoe would make them uncomfortable at certain instances; The temperature sensor might not be able to perform adequately as most people prefer to wear socks while doing physical workouts such as jogging and running; Shoes require charging and the user may require to charge frequently).

8.1.6   Smart Shoes in IoT [10]

The authors here offer a survey is given that illustrates the use of IOT and Gait analysis, in order to calculate various health disorders and complexities. The shoes are very useful and ideal shoes for someone who requires continuous health monitoring. The shoes are bundled with the latest gizmos that provide complete health monitoring. It also contains wireless tech to transfer data/ information to the cloud, which will intern be sent to a family doctor and to a phone as alert messages.

For the monitoring, the shoes will use gait analysis and calculate the required data based on the gait cycles. For this purpose, various sensors are used. Also, a programmable chip/ microcontroller will be used that will be based on Internet of Things. Thus, a doctor will monitor the health regularly and in case that the user fall sick, an alert message will be sent to the family doctor about the issue user is facing. Also, the health provider may suggest you appropriate treatments for the same.

These shoes may be really beneficial for old people, who stay alone at home. In case of any emergency, appropriate help can be provided to them. If there seems to be a serious health problem recognized by the shoe, the healthcare provider may even send an ambulance for immediate help.

Pros-(Gait analysis is accurate and has wide range of detection of various types of problems; Variety of problems and health deficits can be figured out; Immediate help can be provided as the shoes are always connected wirelessly; Health monitoring systems are always running, except when the user removes the shoe; Both shoes have different functions and gather different data, providing even more varied information to the user; Easy to use and looks almost similar to your regular shoes; May also alert you based on your walking that you are likely to fall).

Cons-(Requires charging in good amount as the sensors are always working; Huge battery required; Good connectivity required).

8.1.7 Smart Shoes for People with Walking Disorders [23]

Smart shoes are designed to help people with a multitude of tasks that include abnormal gait, fall detection, calories burnt, distance traveled, etc. In this system, the authors focus on identifying and providing feedback for abnormal walking patterns. Specifically, by including pressure sensors and vibrators into the sole of the shoe, abnormalities can be detected via the pressure sensors and provide feedback to the user via the vibrators in real-time. Thus, the vibrators activate in the position of the abnormal pressure distribution to physically notify the user that their pressure is unevenly distributed in a certain area. This will allow them to focus on correctly distributing pressure in future steps. To this end, the entire shoe consists of the following items on a per shoe basis.

- Five pressure sensors to record the pressure acting on different areas of the foot.
- Four vibrators to inform the user of the abnormal pattern detected.
- A small memory storage capacity for the pressure sensor data.
- A processor to activate and deactivate the vibrators.
- A Bluetooth transceiver to transmit the data to the Smartphone and to receive instructions for the processor.
- A battery for the power supply.

The pressure sensors were placed in the five most common places that an individual will apply pressure during a gait cycle. Thus, they cover the innermost front side close to the hallux (big toe), the outermost front side, the center, the inner heel, and the outer heel of the foot. In addition, these sensors can detect a multitude of different pressure patterns for abnormalities during a gait cycle. The Smart Software contained within an application on the Smartphone is what houses the Machine Learning algorithm for abnormality predictions.

Pros-(Real time monitoring; Real-time response for correcting walking pattern; Learning capability according to the user walking style).

Cons-(High complexity of the embedded circuit; High cost; Currently not available in the market).

8.2 Special Purpose Smart Shoes

8.2.1 *Smart Shoes with Triboelectric Nanogenerator [11]*

This work discusses the biomechanical energy harvesting systems using triboelectric nanogenerators (TENGS). It is observed that ambulatory footfall generates 20 W of electricity. Also, the traditional use of batteries in wearable devices is discouraged and use of electrostatic induction is encouraged. Thus, this work states that TENGS can efficiently convert irregular and low frequency biomechanical energy from body movements into electrical energy. It also states the limitations of wearable technologies, such as smart bands and watches over the smart shoes. The smart bands and watches can only effectively monitor steps and not other regulatory health. This can only be done by gait analysis, which as of now can only be fitted in smart shoes. But, the limitation of using gait measurement devices is that they are expensive and require comparatively higher amounts of energy. So, for this purpose, TENGS and other piezoelectric devices must be used to provide required and continuous energy to the gait sensors. Moreover, TENGS can easily be mounted into the insole of the shoe while its fabrication process. This will not only help the shoe to look nice but will also make it comfortable to wear and the weight distribution will be equally distributed.

Pros-(No use of batteries; Self-powered).

Cons-(Tangs itself can be dangerous and may produce electric shocks; Fabrication of tangs properly may require the insole to be of specific materials which may incur additional costs; Special materials may make the shoe heavy and might decrease the grip.

8.2.2 *Smart Shoes Gait Analysis [12]*

This work discusses about the gait analysis using smart sensor technology. This plays a vital role in the medical diagnostic process and has several applications in health and fitness monitoring, and rehabilitation and therapy.

In this work a portable system is represented called "smart shoes". They are used to analyze different functionality of gait while walking. The implemented system is based on Bluetooth technology working at 2.4Ghz to collect data using a smartphone in any environment. The system combines different working modules like a step counting (pedometer), obstacle detection, electricity generation, fall detection and health tracking in one system. The system will also be equipped with a rechargeable battery, a buzzer and a GPS module for its better functioning. All these will be connected to Arduino-uno microcontroller that will collect and process data that is incoming from the various other sensor modules. In addition, this work focuses on energy that generated that can be used to power few electronic boards and mobile/hand held devices just by walking. The authors of these smart shoe sensors claim that it can also be helpful for a blind person.

Pros-(Data collection wirelessly to a smartphone; Self-sufficient; Arduino is a reliable chipset; Buzzer will be of great help while navigating rather than vibration motors).

Cons-(Battery may be unsafe to use inside a shoe; Bluetooth working at 2.4Ghz is an old form where transferring speed is pretty low and a lot of noise may hamper the data that is being transferred; It is inconvenient to charge your hand-held devices while walking unless there is a wireless charging system installed).

8.2.3   Smart Shoes for Biomechanical Energy Harvesting [14]

Biomechanical energy harvesting from human motion presents a promising clean alternative to electrical power supplied by batteries for portable electronic devices and for computerized and motorized prosthetics. The authors present the theory of energy harvesting from the human body and describe the amount of energy that can be harvested from body heat and from motions of various parts of the body during walking, such as heel strike; ankle, knee, hip, shoulder, and elbow joint motion; and centre of mass vertical motion.

This work has evaluated biomechanical energy harvesting from human motion presents a promising clean alternative to electrical power supplied by batteries for portable electronic devices and for computerized and motorized prosthetics. It presents the theory of energy harvesting from the human body and describe the amount of energy that can be harvested from body heat and from motions of various parts of the body during walking, such as heel strike; ankle, knee, hip, shoulder, and elbow joint motion; and center of mass vertical motion. Thus, for a device that uses the center of mass motion, the maximum amount of energy that can be harvested is approximately 1 W per kilogram of device weight. For a person weighing 80 kg and walking at approximately 4 km/h, the power generation from the heel strike is approximately 2 W. For a joint-mounted device based on generative braking, the joints generating the most power are the knees (34 W) and the ankles (20 W).

Pros-(Accurately calculates the energy harvested from the entire limb; Calculates the ratio of energy harvested by a 1 W/Kg device to the avg. energy harvested; Compares energy harvested from knees to heel; Provides an alternative (i.e. energy from ankles) to energy being produced from heel strike).

Cons-Does not provide a practical solution to the implementation; Results were provided based on average data of weight and walking speeds, which vary from person to person and from gender to gender; From the limb part it concludes knees produce maximum energy whereas a shoe in the foot does not produce equal energy even if energy harvested from ankles and heel strike are added together.

8.2.4 Smart Shoes with Embedded Piezoelectric Energy Harvesting [15]

Harvesting mechanical energy from human motion is an attractive approach for obtaining clean and sustainable electric energy to power wearable sensors, which are widely used for health monitoring, activity recognition, gait analysis and so on. This work studies a piezoelectric energy harvester for the parasitic mechanical energy in shoes originated from human motion. The harvester is based on a specially designed sandwich structure with a thin thickness, which makes it readily compatible with a shoe. Besides, consideration is given to both high performance and excellent durability. The harvester provides an average output power of 1 mW during a walk at a frequency of roughly 1 Hz. Furthermore, a direct current (DC) power supply is built through integrating the harvester with a power management circuit. The DC power supply is tested by driving a simulated wireless transmitter, which can be activated once every 2–3 steps with an active period lasting 5 ms and a mean power of 50 mW. This work demonstrates the feasibility of applying piezoelectric energy harvesters to power wearable sensors.

A shoe-embedded piezoelectric energy harvester was developed in this work, and it can be integrated in a shoe readily for energy harvesting from human locomotion. Two prototypes with different characteristics are fabricated and tested. One prototype produces more energy, while the other one is more comfortable without creating any inconvenience or discomfort for wearers. The DC power supply system, including the harvester and a power management circuit, is used to collect the mechanical energy dissipated in shoes and power some low-power wearable sensors, such as activity trackers. Even though the harvester is unlikely to replace completely the batteries in all wearable sensors, it is a significant role in reducing the problems related to the use of batteries. The work presents a successful attempt in harnessing the parasitic energy expended during person's everyday actions to produce power for wearable sensors.

Pros-(The paper clearly differentiates between 2 different prototypes one with a better comfort in the shoe and the other with more energy harvesting; The work focuses not only on the performance but also cost and fabrication process for each of the prototype; It provides clear ideas that the prototype is to be used for what type of sort problems).

Cons-(Batteries could not be completely be replaced; The prototype that produces more energy was extremely uncomfortable; The prototype that was comfortable to wear produced considerably lower energy).

8.2.5 Pedestrian Navigation Using Smart Shoes with Markers [16]

The authors propose a novel hybrid inertial sensors-based indoor pedestrian dead reckoning system, which is aided by computer vision-derived position measurements. In contrast to prior vision-based or vision-aided solutions, where environmental markers were used—either deployed in known positions or extracted directly from it by using a shoe-fixed marker. This serves as positional reference to an opposite shoe-mounted camera during foot swing, making the system self-contained. Position measurements can be therefore more reliably fed to a complementary unscented Kalman filter, enhancing the accuracy of the estimated travelled path for 78%, compared to using solely zero velocities as pseudo-measurements.

In addition, their work describes a shoe-fixed marker on the opposite foot, serving as a positional reference to the IMU-camera-compass (IMUCC) unit, mounted onto the other shoe. Using a self-attached visual marker as a positional reference for a combined IMUCC unit is a good approach, since visual markers are usually pre-deployed into the environment. It therefore they propose a self-contained machine vision-aided hybrid PDR aiming to improve foot trajectory estimation in an IMU-based PDR system with ZUPTs. The idea is to minimize foot trajectory error during the swing phase, particularly during slow or disturbed walking, when the swing phase, and thus the error integration time, may last longer. The authors have achieved results by taking advantage of a visual marker-based setup, where traditional environmental markers are being replaced by a user-worn marker, fixed on the user's shoe, while an IMUCC unit is placed on the opposite one. From the time it enters in the camera's field of view, the marker's pose with respect to the camera can be determined by means of augmented reality (AR) machine vision algorithms. Then by using it as a positional reference to the IMU, it can use known spatial relationship between the camera coordinate frame and the IMU coordinate frame. Position measurements can be therefore fed to a complementary unscented Kalman filter (UKF), operating in a unit quaternion space in feedback configuration.

Pros-(It focuses on indoor pedestrian positioning system that helps navigation in indoor areas; A worn marker and camera system are used to make the system self-reliant and sufficient; It has rigidly fixed markers and camera on the shoe unlike other methods that involve gait analysis).

Cons-(Battery powered limitation; The placement of the marker on the shoe is erect and perpendicular to the surface of the shoe, making it look abnormal; Shoe mounted erect marker are difficult to carry and resist the flow of movement of the foot; Difficulties in environment with illumination issues and fuzzy understanding of objects and motion).

8.2.6 Smart Shoes with 3D Tracking Capabilities [17]

Most location-based services are based on a global positioning system (GPS), which only works well in outdoor environments. Compared to outdoor environments, indoor localization has created more buzz in recent years as people spent most of their time indoors working at offices and shopping at malls, etc. Existing solutions mainly rely on inertial sensors (i.e., accelerometer and gyroscope) embedded in mobile devices, which are usually not accurate enough to be useful due to the mobile devices' random movements while people are walking. In this work, the authors propose the use of shoe sensing (i.e., sensors attached to shoes) to achieve 3D indoor positioning. Specifically, a short-time energy-based approach is used to extract the gait pattern. Moreover, in order to improve the accuracy of vertical distance estimation, while the person is climbing upstairs, a state classification is designed to distinguish the walking status including plane motion (i.e., normal walking and jogging horizontally), walking upstairs, and walking downstairs. Furthermore, the authors also report a mechanism to reduce the vertical distance accumulation error. Experimental results show that we can achieve nearly 100% accuracy when extracting gait patterns from walking/jogging with a low-cost shoe sensor, and can also achieve 3D indoor real-time positioning with high accuracy.

The system manages to produce 3D real time positioning with inertial sensor mounted on the shoe. As per the experimental results, the inertial sensor are placed on thigh legs and foot were compared. It was later found that the most accurate results were found when the senor system was placed on the foot. Also, the system uses gait analyses in order to produce its results. So, the foot was a perfect place to place the sensors. The results found were highly accurate and a solution to indoor localisation and 3D positioning were hence found.

Pros-(High accuracy; Gait analysis is reliable and a tested method for calculation through foot swing; Low cost sensors; The sensors can be mounted on any shoe with the help of a strap; Convenient to use; Provides excellent solutions to 3D positioning in indoor platforms such as a mall).

Cons-(Shoe mounted sensors may look odd; Shoe mounted sensors are heavy and will be difficult to use for the elderly and kids).

8.2.7 Pedestrian's Safety with Smart Shoes Sensing [18]

Motivated by safety challenges resulting from distracted pedestrians, this work presents a sensing technology for fine-grained location classification in an urban environment. It seeks to detect the transitions from sidewalk locations to in-street locations, to enable applications such as alerting texting pedestrians when they step into the street. In this work, the authors use shoe-mounted inertial sensors for location classification based on surface gradient profile and step patterns. This approach is different from existing shoe sensing solutions that focus on dead reckoning and inertial navigation. The shoe sensors relay inertial sensor measurements to a smartphone, which extracts the step pattern and the inclination of the ground a pedestrian is walking on. This allows detecting transitions, such as stepping over a curb or walking down sidewalk ramps that lead into the street. The work was carried out walking trials in metropolitan environments in United States (Manhattan) and Europe (Turin). The results from these experiments show that we can accurately deter mine transitions between sidewalk and street locations to identify pedestrian risk.

The system comprises inertial sensor modules on both shoes that share their measurements with a smartphone over a wireless Bluetooth connection. While a sensor on one foot can also achieve most of the ramp detections, sensors on both feet substantially improve the detection of stepping over a curb. It allows the user to detect foot movements over the curb, irrespective of the foot the pedestrian uses for the action. A smartphone can serve as a hub for processing the shoe sensor data and implementing any of the aforementioned applications. The core components of our system are Feature Extraction, Step Profiling and Event Detection. More specifically, the authors proposes a pedestrian safety system that acquires inertial data from a shoe-mounted sensor to detect changes in step pattern and ground patterns caused by ramps and curbs. In particular, a salient feature of this work is that it senses small changes in the inclination of the ground, which are expected due to ramps and the sideways slope of roadways to facilitate water runoff. The shoe-mounted sensor has the capability to measure the foot inclination at any given point in time and reflects the slope of the ground when the foot is flat on the ground. The authors chose inertial sensors because they allow us to infer information about the ground with a very modest power budget, compared to GPS or camera-based approaches.

Pros-(It is accurate when the system placed on the shoe is closest to the ground; Cheaper alternative to camera based or GPS based detection units; Even the smallest changes in inclination of the ground can be calculated).

Cons-(Battery powered device; Magnetic fields or current passing through a street lamp may affect the readings of the magnetometer).

8.2.8 Smart Shoes Insole Tech for Injury Prevention [19]

Impaired walking increases injury risk during locomotion, including falls-related acute injuries and overuse damage to lower limb joints. Gait impairments seriously restrict voluntary, habitual engagement in injury prevention activities, such as recreational walking and exercise. There is, therefore, an urgent need for technology-based interventions for gait disorders that are cost effective, willingly taken-up, and provide immediate positive effects on walking. Gait control using shoe-insoles has potential as an effective population-based intervention, and new sensor technologies will enhance the effectiveness of these devices. Shoe-insole modifications include: (i) ankle joint support for falls prevention; (ii) shock absorption by utilising lower-resilience materials at the heel; (iii) improving reaction speed by stimulating cutaneous receptors; and (iv) preserving dynamic balance via foot centre of pressure control. Using sensor technology, such as in-shoe pressure measurement and motion capture systems, gait can be precisely monitored, allowing the user to visualize the way on how shoe-insoles change walking patterns. In addition, in-shoe systems, such as pressure monitoring and inertial sensors, can be incorporated into the insole to monitor gait in real-time. Inertial sensors coupled with in-shoe foot pressure sensors and global positioning systems (GPS) could be used to monitor spatiotemporal parameters in real-time. Real-time, online data management will enable 'big-data' applications to everyday gait control characteristics.

Pros-(Cleanly helps in fall prevention using gait; Realtime action using cutaneous receptors; Prevents injury by moving the centre of pressure control when a fall is detected).

Cons-(Using various sensors require the shoes to be powered by a battery; Ankle joint support may feel uncomfortable while walking; Big data management for such a data will be waste of resources).

8.2.8.1 Smart Shoes for Women's Safety [20]

This work suggests a smart shoe that not only helps women take care of themselves but also help them be fearless against assaulters. This project makes use of GPS, GSM modules, a shock circuit and camera, that are interfaced with Raspberry Pi board and Arduino. Women facing any troubles or in any kind of danger, can immediately make use of this device, embedded in their shoe to escape from the dangerous situation and even harm the attacker.

This work proposes a new technology to protect women by focusing on their security so that they never feel helpless. The system proposed consists of various modules such as GSM, GPS, shock circuit, camera an Raspberry pi-3. The basic idea of the project is alarming the emergency contacts on pressing the emergency switch located on the side of the shoe. Inside the shoe, there are GPS and GSM modules that combinedly send the location to the emergency contact. The shock circuit generates a shock of 400 kV that is sufficient to buy the victim enough time to escape. Meanwhile, on pressing the switch, message will be sent asking for help containing a link that directs the guardian to Amazon Web Services Kinesis that offers live streaming or to a google drive link (by MotionEyeOS), where 30 s short clips would be saved. The live video is recorded with the help of camera. The 5 V battery used in the shock circuit is powered by piezoelectric sensors connected in series at the sole of the shoe. If a woman is walking alone or faces any danger, she should immediately press the emergency button present on the shoe. As soon as the button is pressed, electric teaser is activated that can be used to give shock to the attacker, the electric teaser produces 400 kV shock, simultaneously, GPS and GSM modules are activated with the help of Arduino Uno.

Pros-(Self-sustainable system, needs not to be charged frequently; AWS is pretty quick and live telecasting of the 30 s videos will be smooth; Electric taser will be a good option in order to get adequate time).

Cons-(Once the taser is used the system will be out of power as the taser requires a lot of supply of voltage; Using the taser and video telecasting will not be a practical thing with that size of battery; GSM module fitted inside the shoe may experience delayed or weak connectivity).

8.2.8.2 Novel Pedestrian Navigation Algorithm for a Foot-Mounted Inertial-Sensor-Based System [21]

This work proposes a zero velocity update (ZUPT) method for a foot-mounted pedestrian navigation system (PNS). First, the error model of the PNS is developed and a Kalman filter is built based on the error model. Second, a zero velocity detection algorithm based on the variations in speed over a gait cycle is proposed. A finite state machine including three states is employed to model a gait cycle. The state transition conditions are determined based on speed using a sliding window. Third, the ZUPT software flow is illustrated and has been described. Finally, the performances of the proposed method and other methods are examined and compared experimentally.

The experimental results show that the mean relative accuracy of the proposed method is 0.89% under various motion modes. PNS' that employ MEMS inertial sensors are proposed in. In this effort, a strapdown inertial navigation algorithm is applied to a PNS. Such a PNS is useful for locating and guiding emergency first responders, blind individuals and security personnel.

The experimental results show that the algorithm proposed in this work can correctly detect zero velocity intervals under various motion modes. The algorithm is insensitive to the mounting position of the IMU and differences in pedestrians. They have also conducted experiments under an elevator environment. Because the acceleration of the elevator was lower than the threshold Th, the ZUPT state machine remained in the zero velocity state regardless of whether the elevator was rising or descending. They will attempt to solve this problem in the future. ZUPT cannot estimate or compensate for heading errors, which are a primary error source in PNS.

Pros-(Zero velocity intervals could be detected properly; Mounting position is insensitive and same results can be found every time; Results show that the mean relative accuracy of the proposed method is 0.89% under various motion modes; PNS is useful for locating and guiding emergency first responders, blind individuals and security personnel).

Cons-(Heading errors could not be compensated; Relative accuracy is low; Rapid accumulation of navigation errors).

8.2.8.3 Stride Counting Using Insole Sensors [22]

This effort proposes a method of estimating walking distance based on a precise counting of walking strides using insole sensors. We use an inertial triaxial accelerometer and eight pressure sensors installed in the insole of a shoe to record walkers' movement data. The data is then transmitted to a smartphone to filter out noise and determine stance and swing phases. Based on phase information, they count the number of strides travelled and estimate the movement distance. To evaluate the accuracy of the proposed method, they have created two walking databases on seven healthy participants and tested the proposed method. The first database, which is called the short distance database, consists of collected data from all seven healthy subjects walking on a 16 m distance. The second one, named the long-distance database, is constructed from walking data of three healthy subjects who have participated in the short database for an 89 m distance. The experimental results show that the proposed method performs walking distance estimation accurately with the mean error rates of 4.8 and 3.1% for the short and long-distance databases, respectively. Moreover, the maximum difference of the swing phase determination with respect to time is 0.08 and 0.06 s for starting and stopping points of swing phases, respectively. Therefore, the stride counting method provides a highly precise result when subjects walk.

The insole sensor module is commercially called the "Footlogger". The insole sensor module includes a triaxial accelerometer, eight pressure sensors attached to the insole of the shoe, and a microcontroller (MCU) kit supporting Bluetooth connection to transmit the movement information. The sampling frequency of the insole sensor module is set to 50 Hz. Commonly, these sensors are sampled at frequency range of 20 Hz to 200 Hz. In posture and activity classification, a low sampling rate of accelerometer and pressure sensor possibly produces excellent recognition accuracy. While, pressure and acceleration data were sampled at 25 Hz by a 12-bit analogue-to-digital converter to accurately identify sitting, standing and walking postures. The authors have stated that reduction of sampling frequency of accelerometer and pressure sensor from 25 to 1 Hz does not create significant lost of accuracy (98% to 93%). In estimation applications, Aminian et al. record body acceleration at a sampled rate of 40 Hz to estimate walking speed. The experimental results showed that the maximum of speed-predicted error is 16%.

Pros-(8 sensors provide overall data; Due to the presence of so many sensors, accuracy of data is high; Error is low at 16%; Highly precise results are obtained; Wireless Bluetooth connectivity).

Cons-(Power consumption by the sensors is very high; Noise filtration could possibly delete vital data as well; Sampling size of patients is very small).

8.3 Maturity Evaluation of the Smart Shoes

At this point, we attempt to quantitatively evaluate the systems' maturity in order to offer some reference work for them rather a typical comparison. For this purpose we have selected a number of features (Table 8.1) and for every feature we assign a weight W_{ik}, which reflects its importance for each of the evaluators (Developer, User, Medical Doctor). Thus, every feature is expressed as

Table 8.1 Features

Features	Symbol	Defination
Cost	F1	Smart shoes must have a reasonably low cost
Friendliness	F2	Smart shoes must be use for use
Range	F3	Smart shoes must have power lasting long enough
Calibration	F4	Smart shoes must be easy to calibrate them
System complexity	F5	The hardware complexity of the wearable system must be reasonably low
Software complexity	F6	The software complexity of the wearable system must be reasonably low
Robustness	F7	Smart shoes must be highly robust
Scalability	F8	Smart shoes must be use to upgrade them
Lifetime	F9	Wear and tear should be less increasing life
Realtime	F10	Smart shoes must respond in real time to needs
Reliability	F11	Smart shoes must respond reliably
Always on	F12	Smart shoes must work when the user needs them
Alternative power	F13	Smart shoes must have an alternative power
Intelligent interaction	F14	Smart shoes must have intelligent behaviour

$$F_k = \sum_{i=1}^{3} (gik*Wik)/3$$

where, gik represents the value of each evaluator for the F_k feature. For instance, by using the simple formula above the feature F_1 has value 7.0 for the system reported in [9]. In addition, for every feature (F_k) and for every system, we assign a score based on the average information we have selected for each system, which most of the times is solely got from the literature and our discussion with our groups of users and engineers.

As a result of this effort, we have produced the following tables (Table 8.2a and b) that provide for each system under consideration a number of values for the 14 features selected for this evaluation. The final scores at the bottom of these tables represents the overall "performance" of each system based on the evaluation coming from the experts. Also, the last column of the Table 2b provides the average value for each feature evaluated for all the systems, and at the bottom offers an average score from all the features and systems together.

In addition, the graphical representation of the scores of the systems under evaluation are shown in Table 8.3.

Where, AVG is the average value of each feature coming from each system under evaluation. Also, the work in [23] offers a unique feature of machine learning for adjusting the walking style based on the inputs from the sensors. The adjustment is coming through a set of vibrators that indicate to the user to change the walking style to the level that no more vibrations occur.

Table 8.2 Evaluation of systems and devices

(a)

	[6]	[7]	[8]	[9]	[10]	[11]	[12]	[13]
F1	6	6	7	7	5.25	6	5.25	5.5
F2	7	6.75	6.75	6.25	6.25	6.75	6	4
F3	8.75	8.75	8.75	7.75	7.5	8.75	5.75	8
F4	6.5	6	6	6	6	6	5.5	6
F5	3.75	5	7.75	6	6.5	5	5.5	5
F6	3.75	8.5	8.5	7.25	8.75	8.5	4	8.5
F7	5.75	6.75	6.75	6.75	6.75	6.75	6.75	6.75
F8	4.75	9	9	9	9	9	7	9
F9	5	5.75	5.75	5.75	5	6.66	5.33	5.33
F10	7.5	6	6	6	6	6.75	6.5	6
F11	5.25	4.5	4.5	5.25	4.5	5.75	5.5	4.5
F12	7	7.25	7.25	8.25	6	7.75	6.75	7.25
F13	6.25	3	9.75	2.25	2.25	9.75	3	7.25
F14	NA	NA	NA	NA	NA	NA	NA	NA
Score	77.25	82.4	93.75	83.5	78.05	93.41	72.83	83.08

(b)

	[14]	[15]	[16]	[17]	[18]	[19]	[20]	[21]	[22]	[23]	AVG
F1	6.25	5	6	4.25	7	6	5.67	8	3	4.75	5.72
F2	4.5	6.75	6.75	5.5	6.75	6.75	6.75	7.2	7	8.5	6.4
F3	8.75	8.75	8.75	8.75	4.75	8.75	6.25	6.5	6.75	7.25	7.53
F4	6	6	6	6	6.25	6	6.75	7.25	5.25	7.8	6.15
F5	4.75	5	5	3	7.25	5	7.5	8.15	6.25	4.6	5.62
F6	5.75	8.5	8.5	5	6.5	8.5	5.75	4.25	3	7.1	6.7
F7	6.25	6.75	6.75	6.75	6.75	6.75	7.25	6	4	4.6	6.77
F8	8.25	9	9	7.5	8.25	9	8.5	6.8	5.25	7.8	8.06
F9	5.33	5.33	5.33	6.75	6.25	5.33	8.25	7.15	8.25	5.88	6.2
F10	6	6	6	6.75	6	6	8	5.5	8.5	8.23	6.54
F11	5.5	4.5	4.5	6.5	6.5	4.5	6.5	4.3	5	6.98	5.25
F12	5.5	7.25	7.25	6.75	6.75	7.25	6.5	7.75	6.5	7.66	7.01
F13	6.25	3	9.75	2.25	2.25	9.75	3	5.15	4.75	3.75	5.19
F14	NA	NA	NA	NA	NA	NA	NA	NA	NA	8.50*	--
Score	79.08	81.8	86.58	75.75	81.25	89.58	86.67	84	73.5	93.40*	83.14*

Table 8.3 Graphical representation of the scores of the systems under evaluation

8.4   Conclusion

In this rapidly changing and fast-growing world, people are giving less attention to their personal health, as they are more involved in their work. Visiting clinics and hospitals for health monitoring is not a feasible option as the process takes a lot of time. At the same time older people are more prone to heart diseases, oxygen blood level dip and are more prone to fall as well. In addition, blind and visually impaired people need navigation and obstacle avoidance for going travelling. For athletes and sports persons, tracking their health becomes easy and hassle-free via smart-shoes. The universality of smart-shoes makes conceivable a nonstop medical supervision, harder to accomplish with different gadgets. Besides, the flexibility of smart-shoes gives a proper ground to execute a brought together stage giving products administrations working with elderly and blind people.

Smart-shoes have authoritatively the possibility to turn out to be close aide to help blind and those in need in their everyday life. People may track their health and their sleep, in order to stay healthy and fit.

Thus, the short survey is coming out to address and make visible the potential benefits of the smart shoes in both ways humane and economy. In addition, the incredible evolution of AI, the great potential of smart sensors and the advanced algorithms will come into effect which will enhance the usability of these shoes by making the lives of people more enjoyable and productive.

References

1. C for disease control and prevention arthritis national statistics (2018)https://www.cdc.gov/art hritis/data_statistics/national-statistics.html
2. A foundation what is arthritis? [Online] https://www.arthritis.org/health-wellness/about-arthritis/understanding-arthritis/what-is-arthritis

3. L.B. Murphy, M.G. Cisternas, D.J. Pasta, C.G. Helmick, E.H. Yelin, Medical expenditures and earnings losses among us adults with arthritis in 2013. Arthritis Care Res. **70**(6), 869–876 (2018)

4. D.M. Schofield et.al., The long-term economic impacts of arthritis through lost productive life years: results from an Australian microsimulation model. *BMC public health*, **18**(1), 1–10 (2018)

5. A. foundation arthritis by the numbers. [Online]. (2019)https://www.arthritis.org/getmedia/e12 56607-fa87-4593-aa8a-8db4f291072a/2019-abtn-final-march-2019.pdf

6. N. Vignesh, et al., Smart shoe for visually impaired person. Int. J. Eng. Technol. **7**(3.12), 116–119 (2018)

7. P. Vijayakumar, et al., IOT based wireless smart shoe and energy harvesting system. Int. J. Innov. Technol. Explor. Eng. (IJITEE) **8**(7) May, ISSN: 2278–3075 (2019)

8. T. Wood, U.S.A. patent smart shoes, sensing force, U.S. 5373651 (1994)

9. Yong Won Yang, U.S.A. patent smart shoes, temperature and pressure measurement, U.S.104763483 (2019)

10. B.M. Eskofier, et al. An overview of smart shoes in the internet of health things: gait and mobility assessment in health promotion and disease monitoring. Appl. Sci. (2017)

11. Yongjiu Zou, et al. Triboeletric Nanogenerator enable smart shoes for wearable electricity generation, AAAS Research, ID 7158953 (2020)

12. R. Anil Shinde, Smart shoes: walking towards a better future. Int. J. Eng. Res. Technol. (IJERT) **8**(07) (2019)http://www.ijert.org ISSN: 2278–0181 IJERTV8IS070167

13. S. Rastogi, Smart assistive shoes and cane: solemates for the blind people. Int. J. Adv. Res. Electron. Commun. Eng. (IJARECE) **6**(4), (2017)

14. R. Riemer, A. Shapiro, Biomechanical energy harvesting from human motion: theory, state of the art, design guidelines, and future directions. J. Neuro-Eng. Rehabil. **8**(22), (2011)

15. J. Zhao, Z. You, A Shoe-embedded piezoelectric energy harvester for wearable sensors. Sens.S. **14**, 12497–12510 (2014). https://doi.org/10.3390/s140712497

16. M. Placer, S. Kovacic, Enhancing indoor inertial pedestrian navigation using a shoe-worn marker. Sens.S. **13**, 9836–9859, (2013). https://doi.org/10.3390/s130809836.

17. Fangmin Li, 3D Tracking via shoe sensing, Sens.S. **16**, 1809 (2016). https://doi.org/10.3390/s16111809

18. S. Jain, et al. Look-up: enabling pedestrian safety services via shoe sensing. (MobiSys, Florence, Italy, 2015). https://doi.org/10.1145/2742647.2742669.

19. H. Nagano, R.K. Begg, Shoe-insole technology for Injury prevention in walking. Sens.S. **18**, 1468 (2018). https://doi.org/10.3390/s18051468

20. V. Sharma, Y. Tomar, D. Vydeki, Smart shoe for women's safety. (School of Engineering, VIT, Chennai, IN, 2016), Vishesh.sharma2016@vitstudent.ac.in.

21. M. Ren et al., A novel pedestrian navigation algorithm for a foot-mounted inertial-sensor-based system. Sens.S. **16**, 139 (2016). https://doi.org/10.3390/s16010139

22. P.H. Truong, et al., Stride counting in human walking and walking distance estimation using insole sensors. Sens.S. **16** (2016).https://doi.org/10.3390/s16060823.

23. J. Sudharshan, G. Goodman, N. Bourbakis, Smart shoes for temporal identification and corrections to assist people with abnormal walking patterns, in *Proceedings of IEEE Conference on BIBE-2020,* (USA, 2020)

Chapter 9
Re-Examining the Optimal Routing Problem from the Perspective of Mobility Impaired Individuals

K. Liagkouras and K. Metaxiotis

Abstract The optimal routing problem is a well-known problem at the intersection of operations research and artificial intelligence. The optimal routing problem has been successfully applied in many real-life situations. In this paper, we re-examine the optimal routing problem from the perspective of mobility impaired individuals. As mobility impaired individuals face a number of additional constraints and transportation challenges in their daily lives, this study attempts to address these challenges by adjusting the classical routing problem to facilitate the needs of mobility impaired individuals. In order to solve the problem, we propose a routing recommendation algorithm, which takes into consideration the level of accessibility of the alternative routes. Numerical experiments are reported to examine the performance of the routing algorithm.

Keywords Optimal routing problem · Mobility impaired individuals · Commuters · Evolutionary algorithms

9.1 Introduction

Society today is more concerned with the quality of life. At the same time life expectancy today is much higher than ever before. As a result, today there are more people that retire from driving and thus require improved journey planner options which may combine accessible public transport and walking routes to meet their mobility needs [31]. For example, they will require mobility to access health care services, various social activities, shopping, and simply maintain community connections [26]. However, there are many perceived barriers, which limit the usual mobility requirements of elderly people and those with special needs.

K. Liagkouras (✉) · K. Metaxiotis
Department of Informatics, University of Piraeus, 80, Karaoli & Dimitriou Str, 18534 Piraeus, Greece
e-mail: kliagk@unipi.gr

K. Metaxiotis
e-mail: kmetax@unipi.gr

G. A. Tsihrintzis et al. (eds.), *Advances in Assistive Technologies*, Learning and Analytics in Intelligent Systems 28, https://doi.org/10.1007/978-3-030-87132-1_9

Public transport [10] has a significant influence on access to various health services for elderly people and for mobility impaired individuals. Nowadays, society has become more aware about the needs of mobility impaired individuals. As a result, special attention is given while constructing or upgrading road and footpath infrastructure [32], as this can have impact on mobility impaired individuals. Meeting everyone's mobility needs can be challenging for public infrastructure, but it is a necessity to ensure that every individual, independently from his special circumstances, do not become isolated from society [22]. Despite the undeniable progress that has been made with respect to public infrastructure for mobility impaired individuals, the following question still remains: "Which is the most accessible route to take between two points within walking distance?".

By point, we mean a place where a journey starts or finishes (e.g. home, hospital, public transport station, or community place). Nowadays, it has become easier to go from one place to another by using navigation systems. Modern navigation systems are able of computing different routes based on various criteria such as shortest route and fastest route. Although these systems are built to help people to be mobile, they cannot always satisfy every type of user. For example, a person with a manual wheelchair, who may be querying a navigation system to travel between two locations, may not be satisfied with the outcome of his query. He may be directed to a path which is inaccessible or too steep and risky [14–16] for him.

This happens as navigation systems only consider paths that are shortest and fastest. But, for mobility impaired individuals, route accessibility is the key factor that needs to be considered. It is very important the system to have the ability to make recommendations for a route based on the person's physical capability. A navigation system is of less use to a daily commuter, as he would be well aware of the environment. However, it is a completely different case when the user wants to visit a new or unknown place. It is absolutely necessary to design an accessible path recommendation for mobility impaired individuals [2] to fit their physical abilities.

It is a challenging task to assess the level of accessibility of a particular route, as there are many factors that can be taken into consideration, such as the gradient of the route, the existence of access ramps for disabled people, and disabled parking spots. For example, people with a wheelchair can comfortably wheel themselves up a specific gradient, but not beyond a slope of one-in-fourteen [29]. The challenge is to pick the best route from all alternative routes based on the pre-selected criteria. The existing path routing algorithms try to minimize the total travel distance or travel time. However, the existing implementations do not evaluate the accessibility of a particular route.

In this paper, we propose a routing algorithm that takes into consideration the accessibility of the alternative routes. The user can choose the most suitable path according to his accessibility requirements. The rest of the paper is organized as follows. Section 9.2 provides a literature study on route recommendation systems. Section 9.3 presents related research in the field. Section 9.4 presents the proposed route recommendation algorithm, which takes into consideration the level of accessibility of the alternative routes. The experimental results and discussion are presented in Sect. 9.5. Finally, Sect. 9.6 concludes the paper and presents some directions for future work.

9.2 Literature Review

Dijkstra [9] and A*[11] search algorithms, are mostly used to find the shortest path in terms of distance or travel time. These search algorithms have been proved very effective in finding the optimal route. However, these approaches do not take into consideration the special requirements of the mobility impaired individuals. For example, a route segment may contain a very small portion which is wheelchair inaccessible due to a steep slope or steps. In this paper, we propose a routing algorithm that takes into consideration the accessibility of the alternative routes. The user can choose the most suitable path according to his accessibility requirements.

9.2.1 Mobility Aspects for People with Special Needs

Modern navigation systems are able of computing different routes based on various criteria such as shortest route and fastest route [21]. However, none of these systems offers specific support for mobility impaired individuals. For example, a person with a manual wheelchair, who may be querying a navigation system to travel between two locations, may not be satisfied with the outcome of his query. He may be directed to a path which is inaccessible [28] or too steep and risky [17–19] for him. Because of these accessibility issues, wheelchair users often become reluctant to move along unknown routes.

Scandinavian countries were the first to design public spaces and transportation systems accessible to wheelchair users [6, 27]. The plan was to provide barrier-free access to the public infrastructure for all members of the community [7, 13, 24, 30]. The purpose of this initiative was to integrate the physically disabled into mainstream society [5, 20, 25, 27]. Later on other areas of the world also adopted the concept of barrier-free accessibility, including other European countries, the United States and Japan. At the same chronological period a series of corresponding laws, regulations, and guidelines were developed and implemented [1, 13, 27] regarding the creation of a barrier-free society.

Accessibility is crucial for senior citizens and mobility impaired individuals alike. These two categories of citizens are confronted, in their everyday lives with many impediments on their routes such as stairs or steep slopes. As a result, this kind of people needs to conscientiously choose their route according to those constraining characteristics [4]. Similarly, people with some mobility reduction, such as pregnant women, or people that want to walk with small kids may be also affected by the same constraints. Based on the aforementioned justification, it is essential the proper design of a route according to user's preferences for promoting the equal participation of every citizen in the economic and social life of the society.

Over the last years, there is some ongoing research towards the incorporation of user's preferences regarding the accessibility of a particular route. Balstrøm [3] proposed a method to find the faster walking routes in open field and the criteria were defined according to the surface's friction. The accessibility of routes for people with reduced mobility such as elderly people was determined by Borst et al. [4]. The authors [4] used up to twenty-three physical different criteria such as stair, street elevation level, shops and zebra crossings for evaluating the level of accessibility of a particular route. In another study, Hochmair [12] uses the waiting time, turns and traffic lights as parameters for finding the optimal route.

9.3 Related Work

Our work has many analogies with the classical traveling salesman problem (TSP), for that reason we believe that it is worthwhile to present the relevant research in the field. The TSP can be formulated as an integer linear program. The two most well-known formulations of the TSP are the following: a) Miller–Tucker–Zemlin (MTZ) formulation and b) Dantzig–Fulkerson–Johnson (DFJ) formulation. Below, we provide a presentation of these two competing formulations.

9.3.1 Miller–Tucker–Zemlin Formulation of the Traveling Salesman Problem

We begin with the formulation of Miller et al. [23]. For all pairs (i, j), let x_{ij} be a binary variable, taking the value 1, if and only if the salesman travels from node i to node j. Also, for $i = 2, \ldots, n$ let u_i be a continuous variable representing the position of node i in the route.

$$x_{ij} = \begin{cases} 1 & \text{if} (i, j) \text{ belongs to a tour} \\ 0 & \text{otherwise} \end{cases}$$

Finally, $c_{ij} > 0$ represents the distance (cost) from node i to node j.
Then the TSP can be written as the following integer linear programming problem: Minimize the total cost of traveled distance.

$$min \sum_{i=1}^{n} \sum_{j=1}^{n} c_{ij} x_{ij} \tag{9.1}$$

$$\sum_{i=1, \ i \neq j}^{n} x_{ij} = 1 , \quad j = 1, \ldots, n \tag{9.2}$$

$$\sum_{j=1, \ j \neq i}^{n} x_{ij} = 1, \qquad i = 1, \ldots, n \tag{9.3}$$

$$x_{ij} \in \{0, 1\}, \quad i, j = 1, \ldots, n \tag{9.4}$$

$$u_i \in Z, \quad i = 2, \ldots, n \tag{9.5}$$

$$u_i - u_j + n x_{ij} \leq n - 1, \quad 2 \leq i, j \leq n; i \neq j \tag{9.6}$$

–

$$0 \leq u_i \leq n - 1, \quad 2 \leq i \leq n \tag{9.7}$$

The constraints (2) and (3) ensure that the citizen arrives at and departs from each node exactly once. The constraints (6) and (7) ensure that there is only a single tour covering all cities and not two or more disjointed tours that only collectively cover all cities.

9.3.2 Dantzig–Fulkerson–Johnson Formulation of the Traveling Salesman Problem

Dantzig et al. [8] proposed the following problem formulation. The cities are labeled with the numbers $1, \ldots, n$ and define:

$$x_{ij} = \begin{cases} 1 & \text{if } (i, j) \text{ belongs to a tour} \\ 0 & \text{otherwise} \end{cases}$$

Respectively, $c_{ij} > 0$ represents the distance from node i to node j.
Then the TSP can be written as the following integer linear programming problem: Minimize the total cost of traveled distance:

$$min \sum_{i=1}^{n} \sum_{j=1}^{n} c_{ij} x_{ij} \tag{9.1}$$

$$\sum_{i=1, i \neq j}^{n} x_{ij} = 1, \quad j = 1, \ldots, n \tag{9.2}$$

$$\sum_{j=1,\, j\neq i}^{n} x_{ij} = 1, \quad i = 1, \ldots\ldots\ldots, n \tag{9.3}$$

$$x_{ij} \in \{0,\ 1\}, \qquad i, j = 1, \ldots\ldots, n \tag{9.4}$$

$$\sum_{i\in Q}\sum_{j\in Q} x_{ij} \leq |Q| - 1, \qquad \forall Q \subseteq \{1, \ldots, n\}, \ |Q| \geq 2 \tag{9.5}$$

The constraints (2) and (3) ensure that the citizen arrives at and departs from each node exactly once. The constraint (5) ensures that there are no sub-tours among the non-starting vertices, so the solution returned is a single tour and not the union of smaller tours.

9.4 The Optimal Routing Problem from the Perspective of Mobility Impaired Individuals

In Sect. 9.3, we presented two formulations of the classical traveling salesman problem (TSP). In this section we borrow elements from the classical TSP formulation, specially adapted to address the needs of mobility impaired individuals.

9.4.1 Measuring Route Scores Based on the Degree of Accessibility

Nowadays, navigation systems are able of computing different routes based on various criteria such as shortest route and fastest route. However, the majority of the existing implementations do not consider the issue of accessibility of alternative routes. Without that functionality the usefulness of such systems is questionable for mobility impaired individuals. For example, a person with a manual wheelchair, who may be querying a navigation system, may be directed to a path which is inaccessible or too steep. Mobility impaired individuals, like everybody else, require mobility to access health care services, various social activities, shopping, and maintaining relationships with the community. However, there are many perceived barriers which limit their ability to access different locations.

The route problem for mobility impaired individuals can be stated as follows: given origin (start) and destination (goal) points, and a set of user preferences and constraints, find the best route from start to goal according to the user's requirements. To solve the problem, two issues should be considered.

(a) Map Construction.

The first issue is the construction of the streets' map, which requires gathering information from maps repositories. At the end of the process, we will end up with an undirected graph $G = (V, E)$ where each edge in E represents a street segment and V represents the set of vertices (or nodes). Both edges and nodes have a set of features F that could be the street elevation level and issues of accessibility, such as potential barriers.

(b) Routes Calculation.

From the aforementioned process we obtain a new graph, where every edge has a cost assigned, based on its distance and the degree of accessibility of the particular street segment. Finally, a routing algorithm is applied to obtain the optimal and (possibly) some alternatives routes.

9.4.2 Problem Statement: The Optimal Routing Problem from the Perspective of Mobility Impaired Individuals

Below, we present the proposed model formulation for solving the optimal routing problem from the perspective of mobility impaired individuals. The staring point is denoted as point 0, to a given set of n other points, typically referred to as destinations (e.g. supermarker, bakery, pharmacy) $N = \{0, 1, \ldots, n\}$.

Let $V = \{0\} \bigcup N = \{0, 1, \ldots, n\}$ be the set of vertices (or nodes). In the symmetric case i.e. when the cost for moving between i and j does not depend on the direction, *i.e.* either from i to j or from j to i, the underlying graph $G = (V, E)$ is complete and undirected with edge set $E = \{e = \{i, j\} = \{j, i\} : i, j \in V, i \neq j\}$ and edge costs c_{ij} for $\{i, j\} \in E$. The edge cost c_{ij} is not determined only by its distance, but also by its accessibility. Thus, the final cost c_{ij} of an edge is calculated by dividing its distance by the its level of accessibility. The level of accessibility takes values in the interval $[0, 1]$, where, a value of 0 is assigned to an inaccessible street and respectively a value of 1 is assigned to a street with ideal conditions for mobility impaired individuals, such as streets with no elevation level, access ramps, and disabled parking spots. Obviously, a level of accessibility equal to 0 indicates an inaccessible edge (i.e. street) and as a result re-routing is required.

Thus, the underlying problem is defined by a complete weighted graph $G = (V, E, c_{ij})$. A route is a sequence let $r = \{i_0, i_1, i_2, \ldots, i_s, i_{s+1}\}$ with $i_0 = i_{s+1} = 0$ in which the set $S = \{i_1, \ldots, i_s\} \subseteq N$ of destinations is visited. The route r has cost $\sum_{p=0}^{s} c_{i_p, i_{p+1}}$. It is feasible if no destination is visited more than once, i.e. $i_j \neq i_k$ for all $1 \leq j \leq k \leq s$ and also satisfies the degree of accessibility that is a user specific parameter. In this case, one says that $S \subseteq N$ is a feasible route.

Basic Notation.

a. The problem is given by a set of destinations $N = \{0, 1, \ldots, n\}$, that are positioned at n different locations.
b. Let $S \subseteq N$ be an arbitrary subset of vertices. For undirected graphs the set $E(S) = \{\{i, j\} \in E : i, j \in S\}$ is the set of edges with both endpoints in S.
c. The goal is to find a tour of all n points, starting and ending at point 0 with the cheapest cost.

Assumptions.

a. The cost is symmetric $c_{ij} = c_{ji}$ and direction of the tour does not matter.
b. Triangle inequality: $c_{ij} + c_{jk} \geq c_{ik}$
c. The destinations are points in the plane.
d. The cost c_{ij} from i to j is not determined only by its distance, but also by its accessibility.

A tour is described by variables $x_{ij} \in \{0, 1\} : x_{ij} = 1$, if tour goes from i to j, $x_{ij} = 0$ otherwise.

Then the optimal routing problem from the perspective of mobility impaired individuals can be written as follows:

$$min \sum_{i=1}^{n} \sum_{j=1}^{n} c_{ij} x_{ij} \tag{9.1}$$

$$\sum_{i=1, i \neq j}^{n} x_{ij}, \quad j = 1, \ldots\ldots, n \tag{9.2}$$

$$\sum_{j=1, j \neq 1}^{n} x_{ij} = 1 \quad i = 1, \ldots\ldots, n \tag{9.3}$$

$$x_{ij} \in \{0, 1\} \quad i, j = 1, \ldots\ldots, n \tag{9.4}$$

$$c_{ij} = \frac{d_{ij}}{a_{ij}} \tag{9.5}$$

where d_{ij} is the distance from i to j and $a_{ij} \in [0, 1]$ is the accessibility of edge from i to j

$$\sum_{i \in Q} \sum_{j \in Q} x_{ij} \leq |Q| - 1 \; \forall Q \subseteq \{1, \ldots, n\}, |Q| \geq 2 \tag{9.6}$$

Equation 9.1 minimizes the total cost of traveled distance. The constraints (2) and (3) ensure that each destination is visited exactly once. The constraint (5) ensures that cost c_{ij} is not determined only by its distance, but also by its accessibility. The constraint (6) ensures that there are no sub-tours among the non-starting vertes, so the solution returned is a single tour and not the union of smaller tours.

9.4.3 The Proposed Solution Approach

Below, we analyze the solution approach for solving the optimal routing problem from the perspective of mobility impaired individuals. For solving the aforementioned problem we introduce the following solution representation.

(i)   A binary-valued vector (E) is used to represent the set of edges $(e_1, e_2, \ldots, e_n)$ of an undirected graph. Each edge $e_i = \{0, 1\}, i = 1, \ldots, n$ represents a segment of the route $E = \{e_1, e_2, \ldots, e_n\}$. If $e_i = 1$ the edge e_i belongs to a tour and $e_i = 0$ otherwise.

(ii) A real-valued vector (A) is used to perform the screening based on the level of accessibility of each edge to eliminate the edges that do not satisfy the accessibility level specified by the user from the feasible route. The level of accessibility $A = \{a_1, a_2, \ldots, a_n\}$ takes values in the interval $a_i \in [0, 1], i = 1, \ldots, n$, where a value of 0 is assigned to an inaccessible edge and respectively a value of 1 is assigned to an edge with ideal conditions for mobility impaired individuals.

(iii) Finally, a real-valued vector (D) is used to represent the distance of graph's edges. Thus $D = \{d_1, \ldots, d_n\}, 1 \le d_i \le 10, i = 1, \ldots, n$.

Notice that the three different $(E, A$ and $D)$ arrays of the proposed solution representation are of equal size n. This is happening because the imposed accessibility constraint is a rigid constraint that eliminates the non eligible segments of the street from the optimal route. Therefore, only the eligible segments of the route that satisfy the imposed accessibility constraint are taken into consideration during the optimization process.

The proposed solution representation allows the comparison of alternative routes by simply updating the binary-valued vector (E).

9.5 The Experimental Results and Discussion

In this section we present experimentally the optimal routing problem from the perspective of mobility impaired individuals. Table 9.1 provides an example for demonstrating the efficiency of the proposed approach. In Table 9.1 are given the edges $E = \{e_1, e_2, \ldots, e_{10}\}$ of an undirected graph. Each edge corresponds to a pair of nodes. Furthermore, Table 9.1 provides the distance and the level of accessibility of graph's edges.

As shown in Fig. 9.1 the resulting graph obeys the following rules: (i) it is an undirected graph that starts and finishes at point a, (ii) each destination is visited exactly once, (iii) the solution returned is a single tour and not the union of smaller tours.

Based on the information that is provided above the following two (2) alternative routes are formed: (a) route A $= \{e_1, e_2, e_3, e_4, e_5, e_6\}$ as shown in Fig. 9.2 and (b)

Table 9.1 An undirected graph having $N = 6$ vertices and $M = 10$ edges and each vertex is associated with a cost and a level of accessibility

Edge	Nodes (i, j) – undirected graph	Distance (i, j)	Level of accessibility (i, j)
e_1	(a, b)	2	0.9
e_2	(b, c)	4	0.3
e_3	(c, f)	2	0.8
e_4	(e, f)	2	0.7
e_5	(d, e)	4	0.8
e_6	(a, d)	2	1.0
e_7	(b, d)	3	0.2
e_8	(c, e)	3	0.9
e_9	(b, e)	5	0.9
e_{10}	(c, d)	5	0.8

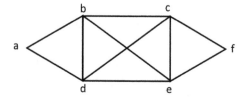

Fig. 9.1 The derived graph

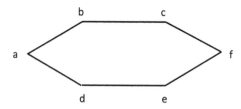

Fig. 9.2 Route A

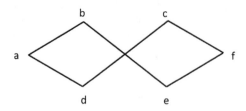

Fig. 9.3 Route B

route B $= \{e_1, e_9, e_4, e_3, e_{10}, e_6\}$ as shown in Fig. 9.3.

The solution representation for route A as shown in Fig. 9.2 is given as follows:

E $= \{1, 1, 1, 1, 1, 1, 0, 0, 0, 0\}$, represents the set of edges of route A.

A $= \{0.9, 0.3, 0.8, 0.7, 0.8, 1.0, 0.2, 0.9, 0.9, 0.8\}$, represents the level of accessibility of edges.

D $= \{2, 4, 2, 2, 4, 2, 3, 3, 5, 5\}$, represents the distance of graph's edges.

The solution representation for route B as shown in Fig. 9.3 is given as follows:

E $= \{1, 0, 1, 1, 0, 1, 0, 0, 1, 1\}$, represents the set of edges of route B.

A $= \{0.9, 0.3, 0.8, 0.7, 0.8, 1.0, 0.2, 0.9, 0.9, 0.8\}$, represents the level of accessibility of edges.

D $= \{2, 4, 2, 2, 4, 2, 3, 3, 5, 5\}$, represents the distance of graph's edges.

As soon as we determine the alternative routes, the proposed solution representation can be easily combined with any single-objective optimization algorithm for finding the optimal solution to the examined problem.

In this case that we only have two (2) alternative routes we can easily calculate the alternative solutions.

Total Cost of Route A:

$$\sum_{i=1}^{n} e_i\, c_i = 1\frac{2}{0.9} + 1\frac{4}{0.3} + 1\frac{2}{0.8} + 1\frac{2}{0.7} + 1\frac{4}{0.8} + 1\frac{2}{1.0} + 0\frac{3}{0.2} + 0\frac{3}{0.9} + 0\frac{5}{0.9}$$
$$+ \; 0\frac{5}{0.8} = 27.91$$

Total Cost of Route B:

$$\sum_{i=1}^{n} e_i\, c_i = 1\frac{2}{0.9} + 0\frac{4}{0.3} + 1\frac{2}{0.8} + 1\frac{2}{0.7} + 0\frac{4}{0.8} + 1\frac{2}{1.0} + 0\frac{3}{0.2} + 0\frac{3}{0.9} + 1\frac{5}{0.9}$$
$$+ \; 1\frac{5}{0.8} = 21.38$$

Based on the calculations presented above results that when we take into consideration the level of accessibility of the alternative routes, *route B* appears to be the most efficient route as its total cost is only 21.38 compared to 27.91 of *route A*.

Please note that if we do not take into consideration in our calculations the level of accessibility, the total cost of *route A* becomes 16 compared to 18 of *route B*. From the calculations presented above, becomes clear that the proposed approach can help mobility impaired individuals to select the most suitable route for their mobility requirements, not only in terms of travelled distance, but also in terms of accessibility.

The proposed solution approach also supports a user determined parameter that can be used for adjusting the minimum allowable level of accessibility of the alternative routes. Thus the user, based on his accessibility requirements can easily set

Fig. 9.4 A segment of *route*
A (bc) is not eligible for
accessibility level 0.4

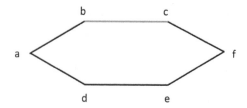

the minimum allowed level of accessibility for each route. With reference to our previous example, we assume that the user sets the minimum accessibility level for any segment of the route equal to 0.4 where, a value of 0 is assigned to an inaccessible edge and respectively a value of *1* is assigned to an edge with ideal conditions for mobility impaired individuals. In this case *route A* becomes not eligible as a segment (see Fig. 9.4) of *route A* does not satisfy the accessibility constraint 0.4 and *route B* becomes the only feasible option.

9.6 Conclusions

This paper examines the optimal routing problem from the perspective of mobility impaired individuals. As mobility impaired individuals face a number of additional constraints and transportation challenges in their daily lives, this study attempts to address these challenges by adjusting the classical routing problem to facilitate the needs of mobility impaired individuals. In order to solve the problem, we propose a routing recommendation algorithm, which takes into consideration the level of accessibility of the alternative routes. Numerical experiments are reported to examine the performance of the routing algorithm.

In our future work, we plan to further elaborate the concept of accessibility. That is, depending on the particular circumstances of each mobility impaired individual, different scores will be assigned to each segment of the route, providing individualized level of accessibility, making the choice of optimal route different for each user.

References

1. T. Akiyama, Japan's transportation policies for the elderly and disabled, in: *Workshop on Implementing Sustainable Urban Travel Policies in Japan and Other Asia-Pacific Countries. European Conference of Ministers of Transport,* (Tokyo 2005), pp. 1–13
2. S. Azenkot, S. Prasain, A. Borning, E. Fortuna, R. Ladner, J. Wobbrock, Enhancing independence and safety for blind and deaf-blind public transit riders. Proc. SIGCHI Conf. Hum. Factors Comput. Syst. ACM **2011**, 3247–3256 (2011)
3. T. Balstrøm, On identifying the most time-saving walking route in a trackless mountainous terrain. Dan. J. Geogr. **102**(October), 51–58 (2002). https://doi.org/10.1080/00167223.2002.

10649465

4. H.C. Borst, S.I. de Vries, J.M. Graham, J.E. van Dongen, I. Bakker, H.M. Miedema, Influence of environmental street characteristics on walking route choice of elderly people. J. Environ. Psychol. **29**(4), 477–484 (2009) (ISSN: 02724944), http://dx.doi.org/https://doi.org/10.1016/j.jenvp.2009.08.002

5. B. Cheng, X. Zhao, Summary of the international legislative process on the construction of barrier-free environment for the past 20 years. Build. Sci. **24**(3), 157–159 (2008). https://doi.org/10.3969/j.issn.1002-8528.2008.03.042

6. T.S. Chiu, Investigating user problems in the construction of barrier-free environments in Taiwan. Taiwan. J. Soc. Welf. **7**(2), 19–46 (2009)

7. B.A. Cooper, U. Cohen, B.R. Hasselkus, Barrier-free design: a review and critique of the occupational therapy perspective. Am. J. Occup. Ther. **45**(4), 344–350 (1991). https://doi.org/10.5014/ajot.45.4.344

8. G.B. Dantzig, D.R. Fulkerson, S.M. Johnson, Solution of a large-scale travelingsalesman problem. Operational Research **2**(1954), 363–410 (1954)

9. E. Dijkstra, (1959) A note on two problems in connexion with graphs. Numer. Math. **1**(1), 269–271 (1959)

10. L. Ferrari, M. Berlingerio, F. Calabrese, B. Curtis-Davidson, Measuring public transport accessibility using pervasive mobility data. IEEE Pervasive Comput. **2013**, 26–33 (2013)

11. P. Hart, N. Nilsson, B. Raphael, (1968) A formal basis for the heuristic determination of minimum cost paths. IEEE Transactions on Systems Science and Cybernetics **4**(2), 100–107 (1968)

12. H.H. Hochmair, Grouping of optimized pedestrian routes for multi-modal route planning: a comparison of two cities, in *The European Information Society*, in *Lecture Notes in Geoinformation and Cartography*, (Springer, Heidelberg, Berlin 2008), pp. 339–358, ISBN: 978–3–540–78945–1, https://doi.org/10.1007/978-3-540-78946-8

13. S. Kose, From barrier-free to universal design: an international perspective. Assist. Technol. **10**(1), 44–50 (1998). https://doi.org/10.1080/10400435.1998.10131960

14. K. Liagkouras, K. Metaxiotis, Multi-period mean–variance fuzzy portfolio optimization model with transaction costs. Eng. Appl. Artif. Intell. **67**, 260–269 (2018)

15. K. Liagkouras, K. Metaxiotis, A new efficiently encoded multiobjective algorithm for the solution of the cardinality constrained portfolio optimization problem. Ann. Oper. Res. **267**(1–2), 281–319 (2018)

16. K. Liagkouras, A new three-dimensional encoding multiobjective evolutionary algorithm with application to the portfolio optimization problem. Knowledge Based System **163**, 186–203 (2019)

17. K. Liagkouras, K. Metaxiotis, Handling the complexities of the multi-constrained portfolio optimization problem with the support of a novel MOEA. J. Oper. Res. Soc. **69**(10), 1609–1627 (2018)

18. K. Liagkouras, K. Metaxiotis, Examining the effect of different configuration issues of the multiobjective evolutionary algorithms on the efficient frontier formulation for the constrained portfolio optimization problem. J. Oper. Res. Soc. **69**(3), 416–438 (2018)

19. K. Liagkouras, K. Metaxiotis, G. Tsihrintzis, Incorporating environmental and social considerations into the portfolio optimization process. Ann. Oper. Res., 1–26 (2020)

20. S.M. Lv, Current situation and outlook on barrier-free environment construction in China. Disability Research **2**, 3–8 (2013)

21. S. Mascetti, G. Civitarese, O.E. Malak, C. Bettini SmartWheels: detecting urban features for wheelchair users' navigation, Pervasive Mob. Comput. **62**, 101115 (2020)

22. J. Meurer, M. Stein, D. Randall, M. Rohde, V. Wulf, Social dependency and mobile autonomy: supporting older adults' mobility with ridesharing ICT, in *Proceedings of The 32nd Annual ACM Conference on Human Factors in Computing Systems*, ACM, 2014, pp. 1923–1932

23. C.E. Miller, A.W. Tucker, R.A. Zemlin, Integer programming formulations and travelling salesman problems. J. Assoc. Comput. Mach. **7**(1960), 326–329 (1960)

24. H Oliver, D. Christine, E. Johann, Detail practice: barrier-free design. Birkhäuser GmbH, Basel, Berlin, (2010)
25. H.X. Pan, J.Y. Xiong, B. Liu, The tendency of integral concept of barrier-free environment construction. Urban Plan. Forum **168**(2), 42–46 (2007). https://doi.org/10.3969/j.issn.1000-3363.2007.02.008
26. M.S. Rahaman, Y. Mei, M. Hamilton, F.D. Salim, CAPRA: A contour-based accessible path routing algorithm. Inf. Sci. **385–386**(2017), 157–173 (2017)
27. S.Y. Tzeng, From barrier-free design to Universal design-comparisons the concept transition and development process of barrier-free design between America and Japan. J Des **8**(2), 57–76 (2003)
28. R. Velho, Transport accessibility for wheelchair users: a qualitative analysis of inclusion and health. Int. J. Transp. Sci. Technol. (2018)
29. T. Völkel, G. Weber, RouteCheckr: personalized multicriteria routing for mobility impaired pedestrians, in: *Proceedings of the 10th International ACM SIGACCESS Conference on Computers and Accessibility,ACM*, pp. 185–192 (2008)
30. G.Y. Wang, Studies on disability and social participation: barrier free, universal design, capability and difference. Chin. J. Sociol. **35**(6), 133–152 (2015) https://doi.org/10.15992/j.cnki.31-1123/c.2015.06.015
31. H. Wennberg, A. Ståhl, C. Hydén, (2009) Older pedestrians' perceptions of the outdoor environment in a year-round perspective. Eur. J. Ageing **6**(4), 277–290 (2009)
32. M. Whelan, J. Langford, J. Oxley, S. Koppel, J. Charlton, The elderly and mobility: a review of the literature, Monash University Accident Research Centre, Australia (2006)

Chapter 10
Human Fall Detection in Depth-Videos Using Temporal Templates and Convolutional Neural Networks

Earnest Paul Ijjina

Abstract The task of Human Fall detection is taken up to develop efficient technology to assist humans in activities of their daily living. In homes, the aged people are more prone to fall with improper gait, due to weakened muscles, lack of balance, sensation and co-ordination. The toddlers frequently fall down as the above abilities are not yet well developed in them. The elderly people and the toddlers need constant watch in homes by a system to alert the care-takers immediately after a fall, that makes our homes smart and safe. This research addresses this problem. In this work, we use motion information obtained from video to compute new temporal templates, which are in turn used by a convolutional neural network to recognize the human actions. A new representation capturing the pose of the subject over a period of time is proposed for human action recognition. This temporal template representation computed at the beginning and at the end of the fall-event is explored in this work for fall detection. The ConvNet feature extracted from these temporal templates are used by an extreme learning machine (ELM) for action recognition. The efficacy of the proposed approach is demonstrated on SDU Fall detection, UP-Fall, UR Fall, and MIVIA action datasets.

Keywords Fall detection · Depth video · Temporal template · Convolutional neural network · Extreme learning machine · SDU Fall dataset · UP-Fall dataset · UR Fall detection dataset · MIVIA Action dataset

10.1 Introduction

The topic of behaviour analysis is a major field of research in academy and industry due to its broad spectrum of applications in human-computer interaction (HCI), automatic video surveillance, robotics, gaming, entertainment, video indexing, search and retrieval. The analysis of human behaviour may consider a specific region of

E. Paul Ijjina (✉)
Department of Computer Science and Engineering, National Institute of Technology Warangal, Hanamkonda 506004, Telangana, India
e-mail: iep@nitw.ac.in

© The Author(s), under exclusive license to Springer Nature Switzerland AG 2022
G. A. Tsihrintzis et al. (eds.), *Advances in Assistive Technologies*, Learning and Analytics in Intelligent Systems 28, https://doi.org/10.1007/978-3-030-87132-1_10

217

a subject like hand gesture recognition, a full body subjects action, the interaction between subjects (or) subject-object, or a crowd-behaviour based on the number of subjects considered. The duration of analysis differs for motion, action and activity, which is another classification of human behaviour analysis. This work focuses on the actions/interactions of a single person, as a part of human behaviour analysis.

We aim to devise a system that recognizes human fall which would typically last around a second, from visual observations of these actions. Due to the rapid increase in elderly population, automated recognition of human fall is needed to provide in-time critical medical assistance to the victims. This work contributes to these efforts by designing a human fall recognition system using the information of the subject computed from video observations. A convolutional neural network (CNN) architecture trained by extreme learning machines (ELM) is used in this work to minimize the overall recognition time. The streaming video captured by surveillance cameras (RGB—during daytime and Infrared—during the night time) with well-defined background information can also be processed by the proposed approach even without pre-defined observation boundaries, due to the use of foreground information in video frames selected by a sliding temporal window for input representation. The next section outlines some of the existing approaches to human fall detection.

Some of the existing approaches to human fall detection use sensory information like acceleration, captured by internal-measurement-units (IMUs)/smart-phones. A smart-phone audio based fall detection system using MFCC and spectrogram features is proposed in [1]. The need to carry and position the phone for noise-free audio, limits its utility. Another smartphone based fall monitoring system is proposed in [2] that uses compass and accelerometer data to detect a fall event. Signal magnitude vector and area features computed on accelerometer data are used to detect fall in [3]. The change in acceleration from subject-to-subject and the position (location) of the accelerometer, makes its ineffective for common use.

In literature, action recognition approaches using specialized sensory setup were also proposed. An approach for recognizing activities of daily living (ADL) using an accelerometer and a surface electromyography, utilizing the angle and amplitude of acceleration data along the three axis for fall detection is given in [4]. The need for specialized sensory setup limits its wider use. A radar based telehealth fall detection system deployed on an embedded system was proposed in [5]. A multi-sensor approach using fusion of heterogenous data with evidential network was proposed in [6] for action recognition. The special nature of sensors and their sensitivity to capturing conditions limits its use. Body sensor network is used to monitor the 3D motion of a subject, along with a cloud computing recognition model for fall detection [7]. The need to wear subject-specific design with specialized sensory setup, affects its prolonged usage. An approach for fall detection using wireless body network with sensors for air pressure was proposed in [8]. The use of pressure sensors and the need to wear it during a fall makes it impractical for regular use.

The recent technological advancements led to the use of video-surveillance in homes, office spaces and even for public places. As a result, action recognition approaches based on video surveillance became popular over the last decade. A multi-view video-based fall detection approach utilizing the different camera-views to detect the fall-pose and estimate the contact area between the subject and surface is proposed in [9]. One of the major limitation is the possibility of misclassification of a subject on the floor. A 2-stage fall recognition based on Microsoft Kinect was proposed in [10], that obtains the vertical state of the subject by segmentation of depth video in the first stage and with recognition and tracking in the 2nd stage. A human action recognition approach based on depth motion maps [11], obtained by projecting the depth video sequence onto the Cartesian planes, is used for action recognition. A gradient boundary convolutional network utilizing a CNN and gradient boundaries for action recognition was proposed in [12].

In the recent years, many variations of extreme learning machines (ELM) were proposed for various domains due to their low computation cost and better generalization capabilities. An approach for predicting the secondary structure of protein using a binary ELM classifier to predict the three secondary structures of the protein is proposed in [13]. The structural information of XML document is captured by a structured link vector model, to generate the feature vector used by an ELM for document classification [14]. A distributed ensemble classification framework based on ELM in a hierarchical P2P network was proposed in [15] through incremental learning of online sequential training, to decrease the overall prediction error with marginal network overhead. The low computation cost and neural network implementation of these models resulted in their increased use in the last decade. This work exploits these characteristics to design an effective fall detection framework.

The above discussion suggests the existing approaches have limitations like: (1) the need to wear the sensors for monitoring, (2) the use of specialized sensors for fall detection, (3) the assumption of the availability of pre-defined observation boundaries for action recognition, (4) the use of computationally expensive features, and (5) the use of pose of the subject after a fall to recognize fall event. To overcome these limitations, we consider the widely available visual information (that can be obtained from either depth/RGB/IR video) to compute a simple temporal template representation from a fixed number of video frames (that are selected by a sliding temporal window on the video/streaming-video) for recognition. Due to the sliding nature of the temporal window, at some position of the window, the frames corresponding to the falling action are selected, thereby obtain the most discriminative data to recognize human fall. The effectiveness is further improved by utilizing the robustness of ConvNet features and the generalization capability of extreme learning machines (ELM) for recognition.

The novelty of this work lies in (a) the use of a temporal template representation to effectively recognize human actions, and (b) the utilization of change in pose at the start and at the end of a fall-event, for better recognition. In this work, an efficient recognition framework using convolutional neural networks is proposed, that can be extended to streaming video using a sliding temporal window. Thus, this approach is suitable for surveillance videos without observation boundaries which require real-time processing. The organization of this chapter is as follows: Sect. 10.2 presents the process of computing temporal template and its use for recognizing human actions in the proposed approach. Section 10.3 details the experimental study of the proposed approach on two video datasets and, Sect. 10.4 gives the future work and concluding remarks.

10.2 Proposed Method

In this work, a fall detection system utilizing a new temporal template representation based on depth information is proposed. The ConvNet features extracted from this temporal template is used by an extreme learning machine (ELM) for detecting human actions. The block diagram of the proposed approach is shown in Fig. 10.1. The temporal templates of observations are used in this work for their capability to capture the entire observation as a single image. The ConvNet features on temporal template representation and the generalization capabilities of extreme learning machines are utilized to develop an effective human action detection framework.

The change in frame difference for human actions considered in this work are shown in Fig. 10.2. Some of the observations from this figure are, there is no/low value in the start of the observation as subject is not in the field-of-view of the camera, and there is no/low value at the end of the observation due to lack of subjects motion. Hence, traditional temporal templates that assign higher significance (i.e., set higher gray-value) to recent motion may not result in an efficient discriminative representation for recognizing these human actions.

To overcome this drawback of traditional temporal templates, we use foreground information in video frames to compute a new temporal template suitable for action recognition. The foreground information in n video frames is scaled by $\frac{1}{n}$ and aggregated to compute this new temporal template representation. As a result, this representation captures the subject spatial location in given duration. Lets assume that $F(t)$, $FG(t)$ denotes the video frame, foreground respectively of the tth frame of the observation with a background B. The computation of $FG(t)$ is given in Eq. 10.1.

$$FG(t) = |F(t) - B| \tag{10.1}$$

This binarized foreground information (BFG) is computed by thresholding BFG with (α), as stated in Eq. 10.2.

$$BFG(t) = FG(t) < \alpha \tag{10.2}$$

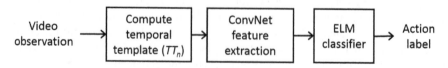

Fig. 10.1 Block diagram of the proposed approach

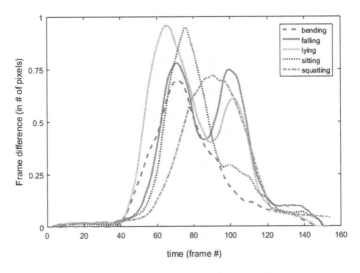

Fig. 10.2 Typical variation in frame difference for some human actions

The new temporal template (TT_n) for an observation with n frames is computed by scaling the sum of BFG of these n frame, as shown in Eq. 10.3.

$$TT_n = \frac{1}{n} \sum_{t=1}^{n} BFG(t) \qquad (10.3)$$

Algorithm 1 Compute temporal template, $TT(F, n, \alpha, B)$

Require: F, n, α, B {video F with n frames and background B}
  $TT \leftarrow 0$ {Initialize result to zero array}
  for t = 1 to n **do**
    $FG \leftarrow |F(t) - B|$ {Compute foreground of current frame, t}
    $BFG \leftarrow FG < \alpha$ {Compute binarized foreground of current frame, t}
    $TT \leftarrow TT + \frac{1}{n}BFG$ {Update TT with current frames BFG information}
  end for
  return TT

It can be observed that in a video with no/low motion of subject, this will capture the subjects pose. Due to the scaling of foreground information, the intensity of noise in depth video stream also decreases proportionally.

For an observation with a *fall* event, the frames before the event contain *non-fall* actions like *walking* and the frames after the *fall* event will contain the subject in fallen pose without any motion. An illustration of temporal template (TT_n) computation from depth video for a *fall* event is shown in Fig. 10.3. The illustration depicts the depth video frames that are 10 frames apart, a sliding window of 20 frames width

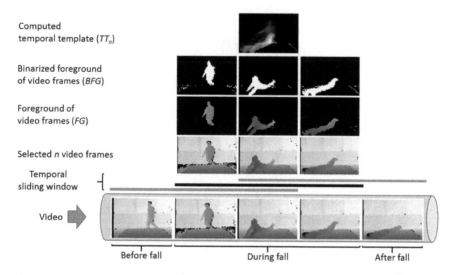

Fig. 10.3 Illustration demonstrating the computation of the proposed temporal template TT_n for an observation, using Eqs. 10.1 to 10.3

shifted by 10 frames. This illustration considers the window location that overlaps with the frames containing the fall action (i.e., capturing the subjects fall). The frames selected using the sliding window are processed to compute the proposed temporal template (TT_n) shown in the figure, using Eqs. 10.1 to 10.3. The pseudo-code for computing the temporal template for a video is given in Algorithm 1. One can observe that, even when there is noise in the binarized foreground information at the edges of the mattress, scaling the foreground information reduced its intensity in the computed temporal template.

From Fig. 10.2, it can be observed that this representation is used for ConvNet [16] feature extraction that is in turn used by an ELM classifier [17] for recognizing actions. This work considers the CNN architecture given in [18] that is trained by extreme learning machines (ELM) [19] to minimize the computation time needed for action recognition.

The overall block-diagram of the proposed fall-detection framework is given in Fig. 10.4. As demonstrated in the figure, the region of the observation where the frame-difference is above the given threshold (of around 0.2 of the maximum frame-difference in the observation) is processed using a sliding window of fixed-size (w) of 20 frames for a 20 fps video) and sliding-length ($\frac{w}{2}$). The first location of the sliding-window in this region (marked 2 in the figure) indicates the temporal template computed for the beginning of the fall-event and the last location of the sliding-window in this region (marked 6 in the figure) indicates the temporal template computed at the end of the fall-event. The ConvNet features extracted from these two templates for an observation are used to recognize the action associated with it, using ELM classifier on the individual set of features and by combining the results using early and late fusion, as illustrated in the Fig. 10.4.

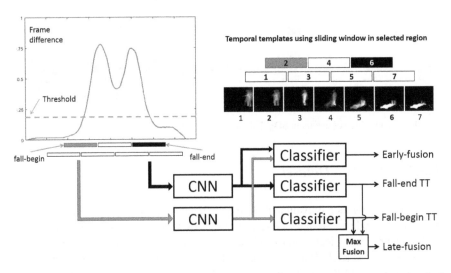

Fig. 10.4 Illustration of fall-detection using temporal-templates computed at the begin and end of fall event with various fusion rules

The next section presents the experiment evaluation of the proposed fall-detection approach on SDU Fall and UP-Fall datasets. The feasibility to utilize this representation for action recognition is demonstrated on UR Fall and MIVIA action datasets.

10.3 Experiments, Results and Discussion

In this section, we will evaluate the proposed approach on four human action recognition datasets, namely, SDU Fall dataset [20], UP-Fall dataset [21], UR Fall dataset [22] and MIVIA action dataset [23, 24]. The temporal template representation proposed in this work is computed using the Algorithm 1 with a suitable α value. The value of α is dependent on the dataset (200 for SDU Fall, 220 for UR Fall and 200 for MIVIA) due to variation in experimental setup in each dataset. A suitable constant value to remove the background information of the depth video observations is chosen to select the foreground information. The typical depth video for the depth-video datasets is shown in Fig. 10.5. For UP-Fall dataset, binarized frame difference is used as the binarized fore-ground information, to compute the temporal template.

The input representation computed for the observations in this dataset are downsampled to generate a 88×88 matrix that is used for action recognition using a convolutional neural network (CNN) and an extreme learning machine (ELM) classifier. The number of convolution masks in CNN is set to the number of action types being recognized. The study was carried in Matlab R2019a on a PC with i5-3210M CPU @ 2.50GHz and 8GB RAM. The experimental set-up corresponding to each dataset is discussed in the following sub-sections.

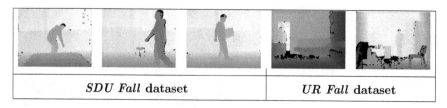

| SDU *Fall* dataset | UR *Fall* dataset |

Fig. 10.5 Depth video of events in SDU Fall dataset and UR Fall dataset

10.3.1 SDU Fall Dataset

The proposed approach is evaluated on SDU Fall dataset,[1] consisting of 6 actions performed by 10 male and female subjects 10 times by varying the lighting & capturing conditions and the objects carried while performing the actions. The modalities of the observations is RGB-D and skeletal tracking information is evaluated using leave one subject out (LOSO) evaluation scheme [20]. Hence, LOSO is used in this study for this dataset.

A temporal window of size 20 which shifts by 10 frames is used to collect the video samples from the video observation, which are in turn used in computing temporal templates as discussed in Sect. 10.2. The typical templates obtained for the six actions are shown in Fig. 10.6. It can be observed that the templates obtained for last temporal window location for *fall* and *lying* are identical. Hence, the last temporal window of the observation cannot provide discriminative information for these actions. For better discrimination, we consider the template for a window location closer to the end, using a threshold on frame difference. This template is used as the input for ConvNet feature extraction for the video. The t-SNE visualization [25] of this input representation for this dataset is shown in Fig. 10.7. From the figure, it can be observed that (a) *walking* action can be well discriminated from the rest of the actions, (b) there might be misclassification between *falling* and *lying* actions due to their high similarity, and (c) misclassification across *bending*, *sitting*, and *squatting* is expected due to their close similarity in motion. This template representation of observations is given as input to a CNN for ConvNet feature extraction that is in turn used by an ELM for action recognition, as discussed in the previous section. The proposed approach achieved an accuracy of 91.58% for discriminating the 6 actions. The associated confusion matrix is given in Fig. 10.8. As the vertical labels correspond to the actual class and the horizontal labels represent the predicted class, the non-diagonal elements represent the percentage of misclassified observations. It can be observed that the misclassification follow the observations from the t-SNE visualization. The performance comparison of the proposed approach for recognizing

---

[1] http://www.sucro.org/homepage/wanghaibo/SDUFall.html.

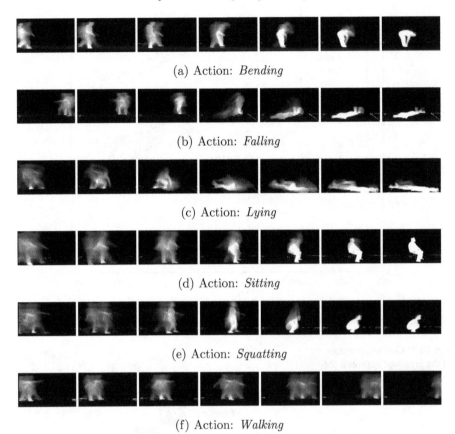

(a) Action: *Bending*

(b) Action: *Falling*

(c) Action: *Lying*

(d) Action: *Sitting*

(e) Action: *Squatting*

(f) Action: *Walking*

Fig. 10.6 Plots of typical temporal templates computed for the six actions in SDU Fall dataset, using a temporal window length of 20 frames and a shift width of 10 frames in the selected region of the observation

the six human actions is given in Table 10.1. The comparative study of the existing and proposed approaches for fall/non-fall action detection is given in Table 10.2. It can be observed that the proposed approach performs better than the existing approaches for the above two scenarios. This could be due to the robustness of ConvNet features extracted from the proposed temporal template representation of video observations.

The use of background subtraction for preprocessing eliminated the use of chair and mattress for recognizing the actions in this approach. The variations in execution of the action in the data shown in Fig. 10.9 suggests that realistic variants of these action were considered in this dataset, making it a challenging dataset to work on. The average duration of observation in SDU Fall dataset is given in Table 10.3. From the table, it can be observed that the average duration of an observation is 5.7 s. The performance comparison on this dataset is given in Table 10.4. It can be observed that our approach can process 20 frames in 0.47 s for human action

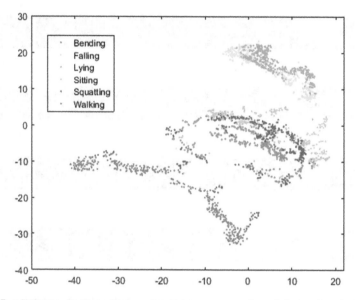

Fig. 10.7 t-SNE visualization of temporal template representation of observations in SDU Fall detection dataset. (Best viewed in colour)

Fig. 10.8 Confusion matrix of the proposed approach for SDU Fall dataset

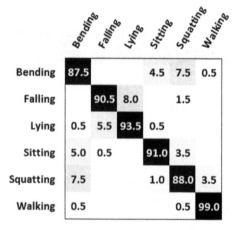

recognition. Since the video frame rate is 30 fps and the proposed approach can process 20 frames (corresponding to $\frac{2}{3} = 0.67$ s) in 0.47 s, this approach can be used for real-time monitoring. The best among the existing approaches requires 1.067 s to process a 5.7 s observation, and so it is not suitable for real-time action recognition. In addition to real-time processing, the proposed approach takes only $\frac{1}{3}$ seconds to recognize the action after its occurrence, due to the use of a sliding window of length $\frac{2}{3}$ seconds (20 frames of a video @30fps) to select the frames used in computing the input representation. The ability to process streaming video (without

Table 10.1 Performance (accuracy in %) comparison for recognizing the six actions in SDU Fall dataset

Approach	Acc.
Shape features + SVM classifier [20]	32.4
Shape features + ELM classifier [20]	51.31
Shape features + PSO-ELM classification [20]	51.94
Shape features + VPSO-ELM classification [20]	54.16
FV-SVM [26]	64.67
Fisher Vector-DVM [27]	64.67
Silhouette orientation volumes [28]	89.63
Proposed TT with LRF-ELM [29]	91.50
Proposed TT with architecture in Fig. 10.3	36.58
Motion Energy image with proposed architecture in Fig. 10.3	64.42
Motion History image with proposed architecture in Fig. 10.3	70.50
Recognition framework in Fig. 10.4 for Early fusion	**91.58**

Table 10.2 Performance (accuracy in %) comparison for recognizing Fall versus Non-Fall actions in SDU Fall dataset

Approach	Acc.
Shape features + SVM classifier [20]	63.12
DMM classifier [30]	75.50
Shape features + ELM classifier [20]	84.36
Shape features + PSO-ELM classification [20]	85.34
Shape features + VPSO-ELM classification [20]	86.83
FV-SVM [26]	88.83
Fisher Vector-DVM [27]	89.17
Proposed TT with LRF-ELM [29]	91.0
Silhouette orientation Volumes [28]	91.89
Proposed fall-detection approach with Early fusion	**92.25**

action boundaries) in real-time, with minimum recognition delay, makes this an ideal approach for real-time monitoring. As ConvNet feature extraction takes more time in the proposed approach, the overall computation time can be further reduced by considering hardware implementations of CNN for ConvNet feature extraction. It is observed that computation of the temporal template on the entire video won't yield good results. It is observed that the best results were obtained when fall event

Fig. 10.9 Temporal templates corresponding to some of the variations in executing the **a** *Bending*, **b** *Falling*, **c** *Lying*, **d** *Sitting*, and **e** *Squatting* actions in SDU Fall dataset. (Here, ext. represents extended arms and clo. represents closed arms)

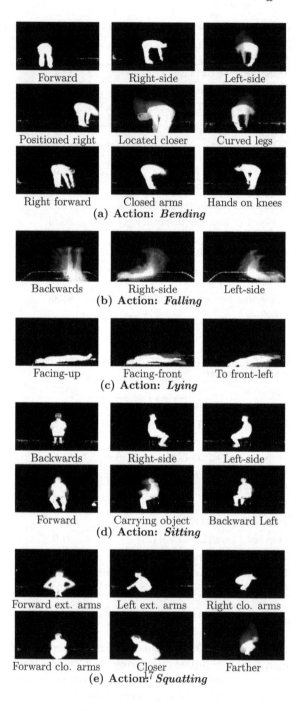

Forward   Right-side   Left-side

Positioned right   Located closer   Curved legs

Right forward   Closed arms   Hands on knees

(a) **Action:** *Bending*

Backwards   Right-side   Left-side

(b) **Action:** *Falling*

Facing-up   Facing-front   To front-left

(c) **Action:** *Lying*

Backwards   Right-side   Left-side

Forward   Carrying object   Backward Left

(d) **Action:** *Sitting*

Forward ext. arms   Left ext. arms   Right clo. arms

Forward clo. arms   Closer   Farther

(e) Action: *Squatting*

Table 10.3 Average duration of observations per each action-type in SDU Fall dataset

Action	Average duration (in sec)
Bending	5.336667
Falling	5.646500
Lying	6.844167
Sitting	5.790000
Squatting	5.316167
Walking	5.235167

Table 10.4 Comparison of computation time (in seconds) of the proposed (for 20 frames) and existing approaches on SDU Fall dataset

Method →	Proposed approach				BoCSS +	DMM +
Time ↓	InpRep	ConvNet	ELM	Total	Classifier [20]	Classifier
Training	–	5.3600	0.2292	**5.5892**	≈ 8	≈ 1.067
Testing	0.0128	0.4575	0.00076	**0.4711**		

is detected from the temporal template computed from the beginning and from the ending of the selected region, which corresponds to the start and end of a fall-event. The next section discusses the experimental study conducted on UP-Fall dataset.

10.3.2 UP-Fall Detection Dataset

The UP-Fall detection dataset was proposed in [21], in which 5 types of fall-events, namely, *Falling forward using hands*, *Falling forward using knees*, *Falling backwards*, *Falling side-ward* and *Falling sitting in empty chair* were performed by 17 subjects. The RGB observations were captured at 18 fps and the observations are approximately 10 s duration. These RGB observations are publicly available for academic research. This dataset is considered in this study to demonstrate the feasibility of using this approach on the regular colour video, by using the binarized frame difference as the binary foreground information (BFG) in the computation of the proposed temporal templates. A window length of 40 frames, with 20 frames sliding is used for experimental evaluation as suggested in the dataset. The experimental set-up for this dataset is similar to the previous dataset. A recognition accuracy of 68.24% is achieved by the proposed approach, whose confusion matrix is shown in Fig. 10.10. The next section discusses the experimental study conducted on MIVIA action dataset.

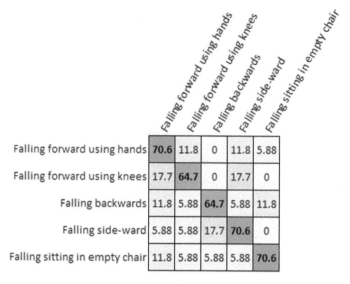

Fig. 10.10 Confusion matrix of the proposed approach for UP-Fall dataset

| | **Fall** Action | **ADL** Action |

Fig. 10.11 Temporal templates corresponding to some of the variations of *Fall* and *ADL* actions in UR Fall dataset

10.3.3   UR Fall Detection Dataset

The UR Fall detection dataset [22][2] consists of *Fall* events and activities of daily living (*ADL*) performed by 5 subjects captured using Kinect depth camera and accelerometer. There are 30 fall events and 40 activities of daily living. This multimodal dataset is generally used to recognize fall from the sudeen change in acceleration information during fall and the fall pose from camera after fall. The proposed approach is evaluated on this dataset, by computing temporal templates from depth infromation, without background subtraction (due to the complexity of the background). Some of the typical temporal templates geneated for *Fall* and *ADL* are shown in Fig. 10.11. The performance camparison of the proposed approach against the existing approaches is given in Table 10.5.

---

[2] http://fenix.univ.rzeszow.pl/~mkepski/ds/uf.html.

Table 10.5 Performance of existing and proposed approaches on UR Fall detection dataset

Approach	Accuracy (in %)
Curvelets features + HMM [31]	96.88
Two-stream CNN [32]	98.0
Proposed TT with LRF-ELM [29]	100
Proposed TT with architecture in Fig. 10.3	**100**

10.3.4   MIVIA Action Dataset

The MIVIA actions dataset[3] consists of RGB-D observations of 7 human actions performed by 7 male and 7 female subjects. Each action is performed twice by each subject and the average duration of observation for each action type is given in Table 10.6. In contrast to the previous dataset where each observation is of similar length, the observations of different classes in this dataset have different durations. A sliding window approach cannot be used for selecting the frames used for computing the temporal templates due to the nature of the actions and the variations in their duration. Hence, all the frames in the observation are used to compute the template, that is used for input representation. The ConvNet features extracted on this template representation is used by an ELM classifier for action detection. The t-SNE visualization [25] of the observations using the proposed template representation is shown in Fig. 10.12. From this visualization, it can be observed that a) *opening a jar* and *sitting* actions are well separated from other classes, b) misclassification between *drinking* and *random motion* is likely and c) misclassification between *sleeping* and *interacting with a table* is expected. The leave one subject-out (LOSO) evaluation protocol proposed for this dataset in [23] and [24] is used to evaluate the performance of the proposed approach on this dataset. An accuracy of 86.85% is obtained by the proposed approach whose confusion matrix is given in Fig. 10.13. Here, the non-diagonal elements represent the percentage of misclassified observations. It can be observed that the misclassification follow the observations from the t-SNE visualization. The performance comparison on this dataset is given in Table 10.7. It can be observed that our approach performs better than the existing approaches, which could be due to the use of ConvNet features for recognizing human actions from the proposed temporal template representation.

As temporal templates are computed from raw video, a real-time action recognition system could be developed by considering a parallel implementation using GPU, for action detection. The next section presents the conclusions of this work.

---

[3] http://mivia.unisa.it/datasets/video-analysis-datasets/mivia-action-dataset/.

Table 10.6 Average duration of observations per each action-type in MIVIA action dataset

Action	Average duration (in sec)
Opening a jar	2
Drinking	3
Sleeping	3
Random motion	11
Stopping	7
Interacting with a table	3
Sitting	3

Fig. 10.12 visualization of temporal template representation of observations in MIVIA actions dataset. (Best viewed in colour)

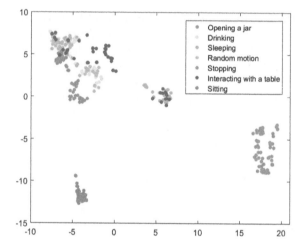

Fig. 10.13 Confusion matrix of the proposed approach for MIVIA action dataset

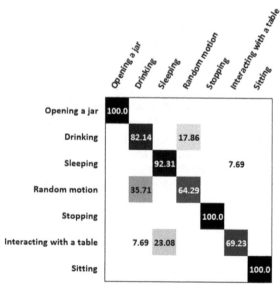

Table 10.7 Performance of existing and proposed approaches on MIVIA actions dataset

Approach	Accuracy (in %)
Proposed TT with LRF-ELM [29]	70.86
Hu moments etc.,+ GMM (Reject mechanism) [23]	79.8
Kernels of visual strings (HaCK) [33]	80.1
Bag of words (BoW) [24]	84.1
Deep learning [34]	84.7
Aclet strings + Edit distance [35]	85.2
Proposed TT with architecture in Fig. 10.3	**86.85**

10.4   Conclusions and Future Work

In this work, we propose an action recognition framework using a new temporal template representation assigning equal weight to foreground information in the selected video frames. A CNN architecture trained by extreme learning machines is used for ConvNet feature extraction for action detection. The proposed templates can recognize actions even in the absence of motion, thereby making it suitable for recognizing activities of daily living including fall event. The simple arithmetic for computing the templates and the existence of parallel implementations of ConvNet feature extraction makes real-time implementation a possibility. In future, we'll extend this work to other modalities and actions.

The performance of the proposed fall detection approach using various fusion rules is shown in Table 10.8. It can be observed that our temporal template gives better results for fall detection when the sliding window is at the beginning of the selected region than at the end. The use of frame difference, computed from colour, in the computation of the temporal templates and the short duration of fall events in UP Fall, compared to SDU Fall, are the possible factors, leading to the negative impact of considering the ending of the event, on the overall performance. The potential use of the proposed temporal template without region selection, for fall detection is demonstrated on UR Fall dataset.

From the computation of temporal templates elaborated in Algorithm 1, if n is the number of pixels in a frame of the input video, then the space and time complexity of computing this temporal template representation is $O(n)$. For observations of UP-Fall

Table 10.8 Evaluation of the fall-detection framework on SDU-Fall and UP-Fall datasets

Dataset	Modality	Event-begin	Event-end	Early-fusion	Late-fusion
SDU Fall	Depth	88.92	79.25	**91.58**	88.83
UP-Fall	RGB	**68.24**	29.41	29.41	41.18

and SDU Fall datasets, it took less than a quarter second on a CPU-only machine. Hence, the study suggests that this work can be extended to design a real-time fall detection system even with a moderate non-GPU hardware.

10.5 Compliance with Ethical Standards

This work was carried out without any funding. The corresponding author on behalf of all the authors declares that there are no conflict of interest. This article does not contain any studies with human or animal subjects by any of the authors. The manuscript has been approved by all authors and has never been published, or under consideration for publication elsewhere.

References

1. M. Cheffena, Fall detection using smartphone audio features. IEEE J. Biomed. Health Inform. **20**(4), 1073–1080 (2016)
2. L.J. Kau, C.S. Chen, A smart phone-based pocket fall accident detection, positioning, and rescue system. IEEE J. Biomed. Health Inform. **19**(1), 44–56 (2015)
3. W.C. Cheng, D. M. Jhan, Triaxial accelerometer-based fall detection method using a self-constructing cascade-adaboost-svm classifier. IEEE J. Biomed. Health Inform. **17**(2), 411–419 (2013)
4. J. Cheng, X. Chen, and M. Shen, A framework for daily activity monitoring and fall detection based on surface electromyography and accelerometer signals. IEEE J. Biomed. Health Inform. **17**(1), 38–45 (2013)
5. C. Garripoli, M. Mercuri, P. Karsmakers, P. J. Soh, G. Crupi, G. A. E. Vandenbosch, C. Pace, P. Leroux, and D. Schreurs, Embedded dsp-based telehealth radar system for remote in-door fall detection. IEEE J. Biomed. Health Inform. **19**(1), 92–101 (2015)
6. P.A.C. Aguilar, J. Boudy, D. Istrate, B. Dorizzi, J.C.M. Mota, A dynamic evidential network for fall detection. IEEE J. Biomed. Health Inform. **18**(4), 1103–1113 (2014)
7. C.F. Lai, M. Chen, J.S. Pan, C.H. Youn, H.C. Chao, A collaborative computing framework of cloud network and wbsn applied to fall detection and 3-d motion reconstruction. IEEE J. Biomed. Health Inform. **18**(2), 457–466 (2014)
8. G. Lo, S. González-Valenzuela, V.C.M. Leung, Wireless body area network node localization using small-scale spatial information. IEEE J. Biomed. Health Inform. **17**(3), 715–726 (2013)
9. M.A. Mousse, C. Motamed, E.C. Ezin, Percentage of human-occupied areas for fall detection from two views. Vis. Comput. **33**(12), 1529–1540 (2017)
10. E.E. Stone, M. Skubic, Fall detection in homes of older adults using the microsoft kinect. IEEE J. Biomed. Health Inform. **19**(1), 290–301 (2015)
11. C. Chen, K. Liu, N. Kehtarnavaz, Real-time human action recognition based on depth motion maps. J. Real-Time Image Process. **12**(1), pp 155–163 (2016)
12. H. Chen, J. Chen, C. Chen, R. Hu, Action recognition with gradient boundary convolutional network, in *Proceedings of IEEE International Conference on Image Processing (ICIP)*, Beijing, pp. 1047–1051 (2017)
13. G. Wang, Y. Zhao, D. Wang, A protein secondary structure prediction framework based on the extreme learning machine. Neurocomputing **72**(1), 262–268 (2018)
14. X.-G. Zhao, G. Wang, X. Bi, P. Gong, Y. Zhao, XML document classification based on ELM. Neurocomputing **74**(16), 2444–2451 (2011)

15. Yongjiao Sun, Ye Yuan, and Guoren Wang, An OS-ELM based distributed ensemble classification framework in P2P networks, in Neurocomputing, Volume 74, no. 16, Sep 2011, pp 2438–2443

16. Y. Lecun, L. Bottou, Y. Bengio, and P. Haffner, Gradient-based learning applied to document recognition, Proc. of IEEE, vol. 86, no. 11, pp. 2278–2324, Nov 1998

17. Extreme learning machine: theory and applications. Neurocomputing **70**(1–3), 489–501 (2006)

18. M.D. McDonnell, T. Vladusich, Enhanced image classification with a fast-learning shallow convolutional neural network. CoRR abs/1503.04596 (2015)

19. G. B. Huang, H. Zhou, X. Ding, and R. Zhang, Extreme learning machine for regression and multiclass classification, IEEE Transactions on Systems, Man, and Cybernetics, Part B (Cybernetics), vol. 42, no. 2, pp. 513–529, April 2012

20. X. Ma, H. Wang, B. Xue, M. Zhou, B. Ji, and Y. Li, Depth-based human fall detection via shape features and improved extreme learning machine, IEEE Journal of Biomedical and Health Informatics, vol. 18, no. 6, pp. 1915–1922, Nov 2014

21. L. Martínez-Villaseñor, H. Ponce, J. Brieva, E. Moya-Albor, J. ñez-Martínez, C. Peñafort-Asturiano, UP-fall detection dataset: a multimodal approach. Sensors (Basel) **19**, 1–28 (2019)

22. Bogdan Kwolek, and Michal Kepski, Human fall detection on embedded platform using depth maps and wireless accelerometer, in Computer Methods and Programs in Biomedicine, Volume 117, Issue 3, Dec 2014, pp 489–501

23. V. Carletti, P. Foggia, G. Percannella, A. Saggese, M. Vento, *Recognition of Human Actions from RGB-D Videos Using a Reject Option*. Berlin, Heidelberg: Springer Berlin Heidelberg, pp. 436–445 (2013)

24. P. Foggia, G. Percannella, A. Saggese, M. Vento, Recognizing human actions by a bag of visual words, in *Proceedings IEEE International Conference on Systems, Man, and Cybernetics*, pp. 2910–2915 (2013)

25. L. van der Maaten and G. Hinton, Visualizing data using t-SNE, Journal of Machine Learning Research, vol. 9, pp. 2579–2605, 2008

26. M. Aslan, A. Sengur, Y. Xiao, H. Wang, M. C. Ince, and X. Ma, Shape feature encoding via fisher vector for efficient fall detection in depth-videos, Applied Soft Computing, vol. 37, pp. 1023–1028, 2015

27. M. Aslan, . F. Alçin, A. Şengür, M. C. İnce, Fall detection with depth-videos, in *Proceedings IEEE Signal Processing and Communications Applications Conference (SIU)*, May 2015, pp. 443–446 (2015)

28. E. Akagunduz, M. Aslan, A. Sengur, H. Wang, and M. Ince, Silhouette orientation volumes for efficient fall detection in depth videos, IEEE Journal of Biomedical and Health Informatics, no. 99, pp. 1–8, 2016

29. G.-B. Huang, Z. Bai, L.L.C. Kasun, C.M. Vong, Local receptive fields based extreme learning machine. IEEE Comput. Intelli. Mag. **10**(2), 18–29 (2015)

30. C. Chen, M. Liu, H. Liu, B. Zhang, J. Han, N. Kehtarnavaz, Multi-temporal depth motion maps-based local binary patterns for 3D human action recognition. IEEE Access **5** 22590–22604 (2017)

31. N. Zerrouki, and A.Houacine, Combined curvelets and hidden Markov models for human fall detection, in Multimedia Tools and Applications, Volume 77, Issue 5, March 2018, pp 6405–6424

32. C. Ge, I. Yu-Hua Gu, J. Yang , Human fall detection using segment-level cnn features and sparse dictionary learning, in *Proceedings of IEEE International Workshop on Machine Learning for Signal Processing (MLSP)*, pp. 1–6 (2017)

33. L. Brun, G. Percannella, A. Saggese, M. Vento, Hack: a system for the recognition of human actions by kernels of visual strings, in *Proceedings IEEE International Conference on Advanced Video and Signal Based Surveillance (AVSS)*, pp. 142–147 (2014)

34. P. Foggia, A. Saggese, N. Strisciuglio, M. Vento, Exploiting the deep learning paradigm for recognizing human actions, in *Proceedings IEEE International Conference on Advanced Video and Signal Based Surveillance (AVSS)*, pp. 93–98 (2014)

35. L. Brun, P. Foggia, A. Saggese, M. Vento, Recognition of human actions using edit distance on aclet strings, in *Proceedings International Conference on Computer Vision Theory and Applications (VISAPP)*, pp. 97–103 (2015)

Chapter 11
Challenges in Assistive Living Based on Tech Synergies: The Cooperation of a Wheelchair and A Wearable Device

Nikolaos G. Bourbakis

Abstract The direct medical costs for falls only—what patients and insurance companies pay—was $34 billion [*Center for Disease Control and Prevention, 2013*]. The necessity for continuous monitoring and assisting of persons in risk, and the increasing prevalence of chronic conditions among our aging population present great challenges to our healthcare system. These costs are also expected to surpass $54 billion by 2020 (http://www.cdc.gov/ncipc/factsheets/fallcost.htm, in 2007). In addition, the larger healthcare providers offer low quality services. Thus, making a robotic machine "lifting-up" elderly persons or persons with disabilities from a sitting position and transferring them to the bathroom is a very challenging scenario that urgently require innovated solutions. Thus, advanced technological achievements based on robotic nurses and wearable health monitoring devices may represent the up-coming hopeful and desirable solutions. Thus, the challenges here are based on the importance of IT-Engineering solutions to healthcare needs. In addition these challenges offer a paradigm on the already growing personalized/person-centered healthcare and the role of 'smart homes' and 'smart beds' for independent living and self-management of chronic conditions. While the importance of such care is recognized, not many efforts have been pursued to develop low-cost (and inconspicuous) technologies. Thus, the chapter presents issues for the development of a novel synergistic interactive intelligent framework-model between intelligent wearable health-monitoring devices (WS) and intelligent robotic wheelchairs (IRC) for assisting people (HS) in risk at smart homes, improving safety and quality of life, and reducing healthcare cost. In particular, it will focus on two challenging scenarios (lifting-transferring) reported above providing concepts but at the same time maintaining the generality of this synergistic model. The challenges are related with a new generation of intelligent synergistic models of inexpensive, unobtrusive intelligent wearable sensors, monitoring devices, intelligent robotic assistants, and surveillance methodologies. In addition, the healthcare advanced research technologies and educational training is expected to transform the way physicians and nurses provide care and the way of people at risk will safely stay at their homes. Concurrently, they will offer a ground for new research directions for engineers to contribute in

N. G. Bourbakis (✉)
Wright State University, Dayton, OH, USA
e-mail: Nikolaos.Bourbakis@wright.edu

© The Author(s), under exclusive license to Springer Nature Switzerland AG 2022
G. A. Tsihrintzis et al. (eds.), *Advances in Assistive Technologies*, Learning and Analytics in Intelligent Systems 28, https://doi.org/10.1007/978-3-030-87132-1_11

healthcare applications. In addition, commercialization of these devices will certainly transform our healthcare monitoring systems for a large population, and transform the living conditions of people at need.

11.1 Overall Description of the Challenges

Those in risk and the elderly often require diligent monitoring to detect healthcare problems, when they are most-easily treated. Unfortunately, members of those populations can be either unable or disinclined to detect (or communicate) the existence of critical changes in their own health. A common solution for human healthcare professionals is to closely monitor patients directly or via crude patent's data collection devices. Of course, that solution does not scale economically. In addition, due to the high cost of medical care for groups (elderly, paraplegic and in general people in risk) and at the same time the lack of large number of qualified healthcare providers, the robotic nurses and health monitoring devices represent the most effective and desirable solution. Moreover, the continuous monitoring for people-in-risk and elderly offers safety. These people are vulnerable to crimes (theft, homicide, mistreatments, etc.). The continuous remote monitoring offers them a "protection" that may drive the suspects away. Thus, we envision a future in which people at risk are provided with inexpensive (in long term), intelligent robotic nurses and unobtrusive wearable devices in a smart infrastructure that automatically monitor and detect critical changes to their health and to offer first response help.

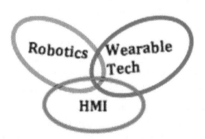

In response to this vision, here we present a synergistic intelligent framework, which will offer safe and efficient service to a Human Subject (HS) for one well focused healthcare challenging scenario that involves the intelligent interaction (HMI) between two systems (a robot-chair (IRC) and a wearable health monitoring device (WS)). This scenario is: *lifting a HS from a sitting position, transferring him/her to the bathroom.* To achieve this objective (scenario) we will address some of these innovative challenges:

Challenge-1: It will transform an existing power-wheelchair (which has a laptop, a screen, two cameras, a microphone, two speakers, range sensors, GPS, two robotic-arms) into *a novel and intelligent robotic-chair (IRC) by embedding methods with intelligent capabilities in order to make decisions and learn when and how to assist people (cognizant).* How: *These intelligent capabilities will come from challenging novel subtasks performed by IRC like lifting people, detecting, tracking, grasping, extracting, detecting body positions, and interacting with humans* [1–16]. *Based on the existing knowledge there is no such an IRC with the intelligent capabilities;*

Challenge-2: There are several wearable sensor-based devices, *but no one has intelligence prognosis capabilities and voice interaction (IVI) between HS and Wearable devices* [17–19]. *The wearable device with its IVI will reliably monitor (reflective) the HS body bio-signals for detecting not only sensor-based measurable symptoms after de-noising, but also a variety of important non-measurable symptoms (pain, cough, nausea, etc.).* An existing wearable device can be transformed *into a novel intelligent device to learn HS' health-history and interactively respond to a range of commands relevant to HS health condition.* How: *The novelty of WS is coming from the learning capabilities to interactively understand sentences incompletely spoken by HS, due to noise or to HS difficulty of speech to express his/her non-measurable symptoms.*

Challenge-3: The integration of WS, the IRC and the HS into *an interactive framework-model (reflective, protective and knowledge-rich) is a challenging issue and has to be tested on realistic scenarios not only for proving the concepts (exchange health information, decision-making, assist the human user, monitoring bio-signals, extract body signatures (patterns), but also to evaluate the process by offering safety and monitoring healthcare for people in need;* The novelty of this task is based on *the synergistic interaction of two intelligent assistive devices for effectively assisting human-users at need.*

11.2 Background and Significance

The demographic imperative for the US and many other industrialized countries is that the number of elderly people-at-risk will increase dramatically over the next 50 years. Many older adults will become chronically ill and frail, and over 50% predicted to have at least some cognitive impairment. There are two major challenging reasons to keep them safe at home for as long as possible: (1) home is where most people want to be, and (2) institutionalization is extremely expensive. As an example approximately 33% of persons over 65 and 50% of persons over 85 experience a fall each year [20–22]. The injuries associated with these falls can have serious consequences. Especially, following hip fracture, 50% are unable to live independently, 25% will die within six months, and 33% die within one year [20, 22]. In one study [23], hospital admissions due to fall-related injuries (of any type) carried the highest risk for disability. The percentages are as follows: 79.4% of the falls having some disability, 45.2% led to chronic disability, and 58.8% are related with nursing home admission. The costs associated with falls amongst the elderly are staggering. Expenditures associated with hip fractures alone exceed $10 billion annually [21].

Overall, this cost to treat fall injuries in people age 65 or older in 1994 was $27.3 billion (in 1996 dollars) and, by 2020, the cost is expected to reach $43.8 billion (in 1996 dollars) [24, 25].

Wearable Health Monitoring Devices (WS)

Technology solutions to people at risk, lifelong or acquired with age, have demonstrated numerous advantages. Among them, wheelchairs with sensory capabilities have provided mobility for millions of people with physical impairments. However, the long-term reliance on the upper limbs for mobility and in performing daily activities has led to increased repetitive strain injuries (RSI) and chronic pain. Upper limb RSI pain is indeed very common in manual wheelchair users. Specifically, carpal tunnel syndrome occurs in 49 to 73% of them [26, 27]. Also, rotator cuff tendinopathy and shoulder pain occurs in 31 to 73% of wheelchair users [28, 29]. Consequently, advances in sensor communication and IT have enabled health care providers to monitor and manage chronic diseases and to detect potentially urgent or emergent conditions [30]. Therefore, we are now in a position to reliably detect and prevent RSI injuries. This can be accomplished by monitoring the patient in one of two possible ways [31]: (a) Employ ambulatory monitors that relay on wearable sensors and devices which record physiological signals; (b) Sensors embedded with the home to collect behavioral and physiological data unobtrusively. The acceptance and positive psychological impact of these monitoring technologies have been confirmed in our studies that have already been included for people with dementia and other chronic conditions [3, 32–49].

Autonomous Wheelchairs and Robotic Nurses [50–79]

It is well known that the cost of human providers in serving people with disabilities or elderly is steep. Indeed, several reports such as the "2010 Computerized Nursing Facility Cost Report "and the "SPRC-Cost-of Providing Specialist Disability Services", [41–88] verify these high costs. In addition, human support degrades over time and is subject of distraction and abuse. Thus, researchers, engineers and practitioners have the recent years focused on autonomous wheelchairs with intelligent capabilities. Even, robotic-nurses with capabilities to lift or assist the people-at-risk have been considered. Several research groups at MIT, CMU, UPitt, UG, UBC, UTA, Purdue U, CART-WSU have actually developed autonomous or semi-autonomous wheelchairs for assisting people with disabilities. These wheelchairs already carry laptops, cameras, join-sticks, touch-screens, range sensors (laser, sonar, other) and sophisticated algorithms for path planning, navigation, human-device interaction (voice communication) map generation, etc. At the same time several companies and Universities in Europe (Germany, Italy) and Japan (Honda, Toyota) and S. Korea have already developed robotic nurses to assist the needs of people with disabilities and elderly.

11.3 The Associated Research Challenges

As it stated above, the effort here aims to address of a smart, synergistic and inter-active model between the HS and the Assistive Devices. In particular, this model is composed of three components: an IRC, a WS and a HS. To prove the importance of this synergistic model, we present only one novel scenario, where the synergy of IRC and WS is important and necessary for the wellbeing of the HS. This scenario emphasizes the novel synergistic interaction and not the improvement or perfection of each robotic sub-tasks:

Scenario: *"A human user is laying on his(her) bed or sitting on a comfortable chair and he(she) needs to visit the bathroom. We assume here that the human user needs help to stand-up, sit on the IRC, which will transfer him/her towards the bathroom".*

The scenario is composed by 3 major challenging tasks. Thus, we present the important components (or subtasks) of each task needed for its feasibility and successful implementation. Also, we show how these components (or subtasks) work together in a synergistic way under a complex scenario to achieve the goal. Some of these subtasks are unique and some others play an important complementary role. On the overall, the synergy is novel and based on challenging decision-making and unique interfacing of these subtasks in order successfully and safely to accomplish the goal(s).

11.3.1 Main Innovative Tasks

In this section we briefly present the main challenging tasks and their association for the successful implementation of the project, Fig. 11.1. For each of these tasks we also provide experimental results to show feasibility for their implementation.

Fig. 11.1 The three main tasks and their associations for the implementation of research project here

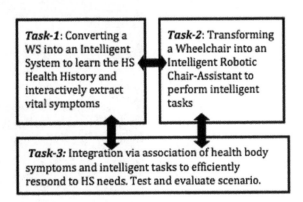

11.3.1.1 Challenge-1: A Wheelchair as an Intelligent Robotic Chair (Irc)

A. *The Robotic Wheelchair as a Testbed*

There are several research efforts involving the improvement of power-wheelchairs [75, 76]. Here the unique challenge is not on the power wheelchair (laptop, two cameras, a microphone, an ear-speaker, range sensors, a computer screen, a GPS, two robotic arms) as shown in Fig. 11.2, but on the synergy of wearable sensors, robotic parts and cutting-edge-intelligent methods to advance a wheelchair into a robot-assistant to people-at-risk. In particular, in this challenge-1 an existing power-wheelchair (Fig. 11.2) will transformed into an intelligent robotic assistant by developing *intelligent algorithms (computational component)*.

There are also researches using robotic arms to assist humans with special needs. In this case, although there is available expertise in robotic-arms design it is better for someone to buy two arm-hands from the market to eliminate the time of building them (Fig. 11.3). Thus, the main challenging research issue here is the intelligent algorithms to efficiently synchronize the IRC components and actions to safely service the HS. Another challenging point is the amount (in pounds or kilograms) that the IRC arms can lift or assist during standing-up and walking stages. From experience,

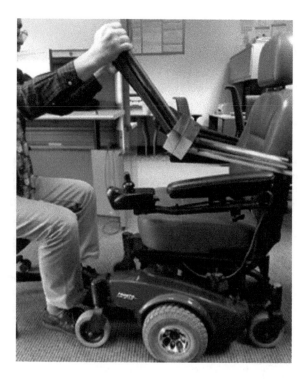

Fig. 11.2 The available IRC with a laptop placed under the seat. The view of a HS holding the "robotic-arms"

Fig. 11.3 The graphical view of the IRA with two arms and two cameras

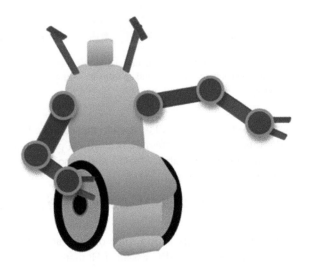

the lifting will be near 80 lb without the IRC loosing its balance. For assistive modes this amount is manageable for pooling, lifting and assist standing stages [89, 90].

B. *The IRC Innovative and Intelligent Capabilities*

Under the challenge-1, existing interface intelligent methods (sub-tasks) have to be modified or new to be developed in order to achieve compatibility among the IRC intelligent capabilities (like *autonomous Navigation, Tracking, Grasping, Body Recognition/Tracking, Learning Patterns of Behavior*) and effective Human-Machine-Interaction (HMI) (like *health monitoring, assistive lifting,* etc.). An additional challenge is the interactions between Human-Subject and IRC based on Voice or simple Gestures. Although there is a long list of research subtasks to be achieved, most of the research labs have already in house these standalone methodologies. Thus, the challenge here is to appropriately interface these methods to work as a sequence achieving goals. These sub-tasks will give to IRC intelligent capabilities needed for the successful completion of the scenarios presented here. Below we describe these sub-tasks for this challenge "Standing-up, Turning around, Sitting-down on the IRC":

B1. The Robotic arm-hand gentle Grasping (**Gr**)

In the robotic field researcher have developed various robot arm-hands [91, 92]. The sophistication of these robotic arm-hands has reached an acceptable level of sensitive grasping. For example there are robotic arm-hands that can handle eggs without break them. Also there are robotic arms used in cooking in smart kitchens. Thus, for the robotic-hand grasping here a neural net model will be used to control the robotic arm-hand to grab certain objects [90, 93]. For this grasping subtask, two robotic-arms can be bought from the market by reducing the time for design and developing new ones. Thus, the expertise on grasping methods with neural nets and fuzzy algorithms will be used to modify the existing robotic arms to be suitable for gentle grasping interactively coordinated by the HS, [90, 93]. For the robot arms

Fig. 11.4 It graphically shows the initial and intermediate states of the IRC positions to assist a human subject (orange color) to stand up from the sitting position (chair, green color), to turn around and sit down

grasping, touch sensors will be used to measure the force (pressure) that HS applies on the robotic arm. The pressure is a critical parameter for the IRC to judge if it is safe to lift the human subject from a sitting position or assist him/her to sit back to the place that was sitting.

B2. Novel Assistive Lifting (AL) for Standing (St) and Walking (Wa) (major novel sub-task)

Figure 11.4 graphically shows some of the states (body and IRC positions) of the robotic arms, when the IRC assists a Human user to stand-up from a chair. More specifically, the IRC will go close to the human user following either a command from HS ("help me to stand up" or "help me to walk", or "help to go to the bathroom" or for rehab exercises") or making a decision under urgent conditions. In the scenario (stand-up), the IRC will use the cameras to create stereo-vision images, detect the HS (where most of its time IRC will track HS movements) and determine the distance to travel in order to reach a position from which it will extend its robotic-arms to safely reach the HS. The HS has to grab the rob-arm and synchronize his/her effort with the robotic-arm movement upwards (a neural net can be used to learn the human subject's responses and the robot assistant adjustments). This will result the HS to reach the standing position. Thus, for the walking to the bathroom, the HS will hold and follow the IRC and command it to go there.

B3. Novel Learning Patterns of Body Signatures, BP) [2, 4, 94, 95]

In this case presents extraction, tracking and representation of a human performing an action by using local global graphs and stochastic Petri-nets (SPNs). In Fig. 11.5 presents the extraction of the human body and the association of different positions that describe an activity. Here 3 non-consecutive image-frames are presented for demo only, (Fig. 11.6). In these image-frames, the segmented human is colored with different colors to demonstrate the main parts of the body (head, (blue), arm (red), hand (green), and legs (yellow)).

From each image-frame the main parts of the segmented human body are extracted, represented and interrelated using the L-G graph. The Stochastic Petri-net (SPN) graph can be used to associate these body-parts in these frames to determine the changes of the states that took place. Between the first and the second frame, the token activates or causes the activation of the head to "look at" a certain direction, at the same time the arm is moving up and the legs change position. The SPN graph graphically presents and connects these changes. The same happens between the second and the third frames, where the SPN graph connects the body-parts using the affect of the previous token. The important issue here is that the SPN graph provides

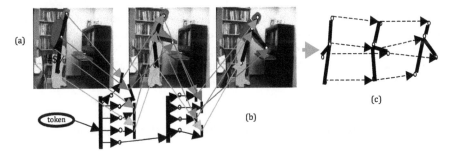

Fig. 11.5 A sequence of body positions. **a** The detection and extraction of the body posion using a skeleton approach; **b** The L--G graphs are represented by with thick and thin lines respectively on the segmented image frames (left) and on the right a sequence of L--G graph-patterns extracted and associated. The color lines the same body-regions in different positions in time; c) the SPN sequence of body positions

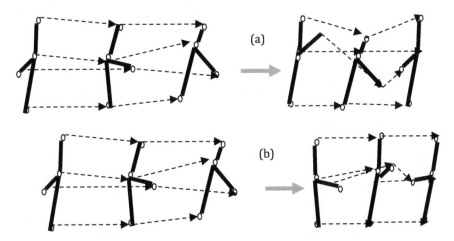

Fig. 11.6 In brief, **a** A sequence of states for drinking (or behavioral pattern); **b** a sequence of carrying a cup. Note here that the uses of three body L-G graphs is only for demonstration of the idea. In reality these are dozens or hundreds of image-frames and so the same number of L-G graphs for body-patterns

the ability of synchronizing the actions performed by the body-parts and creates the body behavioral patterns or signatures in time.

Predicting (Anticipating)and Associating Body Positions: Here in Fig. 11.6 also presents an extension of the previous example by anticipating (predicting) the next states of behavioral patterns. In particular, Fig. 11.6a shows a starting behavioral pattern and the possible next states of the pattern (or signature) for the scenario *"the man drinks water"*. Figure 11.6b shows the same starting behavioral pattern but with different next states of a pattern that lead to the scenario, *"the man takes the cup and returns to his desk"*. The prediction in a pattern will be determined by the higher probability of occurrence of the states that compose a pattern, according to the frequency of repetitions of that pattern in certain time (or according to an external *"token"* that might trigger an appropriate sequence of states). Note here that these or similar sequences of patterns (in more detailed representation) will be associated with heath conditions.

11.3.1.2   Challenge-2: The Intelligent Wearable Health Monitoring Device

A.   *The Wearable Monitoring Device*

A WS uses a variety of sensors to continuously monitor body bio-signals (**MB**) and evaluate the HS health conditions. It is known that bio-signals patterns and/or states may be presented in the various types of physiological parameters and vital signs measured by wearable biosensors, as symptoms [96, 97]. However, for a more accurate estimation of a user's health condition and the diagnosis of many diseases (if not the most), several other symptoms need to be taken into consideration [98, 99]. These symptoms, like cough or malaise, are either not measurable at all or they cannot be estimated without using invasive methods, e.g. as in the case of determining electrolyte levels in the body. Thus, these symptoms, once detected or quantified, provide important information, which together with the measured vital signs provide a more comprehensive description of what is referred to as the clinical presentation. Thus, the clinical representation under proper interpretation may lead to a specific diagnosis. However, in order to get feedback from the patient about the possible existence of these non-measurable symptoms the WS has to have speech recognition/synthesis capabilities and interactively ask the patient himself has to describe them. For this case a set of sensors can be used.

Figure 11.7 depicts the vital signal generated from our WS model [48]. Physiological biosensors constitute the front-end components of the system and will be integrated with textile communicators embedded into clothes as described above [100–105].

A Body Area Network (BAN) will be also employed for exchanging sensor data [106, 107]. In the latter case, the collected measurements will be amplified, filtered and digitized at the sensor nodes. The transmission from the sensors will be done via the BAN and collected at the central node. In the former case, there is the option where physiological signals can be transmitted in analog forms and then be digitally processed at the central node. Figure 11.8 gives a comprehensive overview of most of the physiological parameters and the most common symptoms that need to taken into consideration and properly evaluated to derive a specific diagnosis. This list is not exhaustive and it does not include findings, which can only be obtained from thorough clinical examinations and tests like MRI, CT scan, chest radiology and other medical and laboratory examinations typically performed in a hospital.

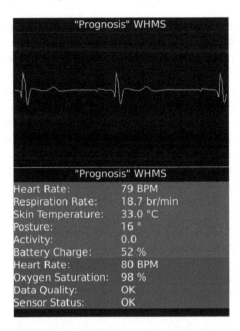

Fig. 11.7 WS sensors output sent to the physician for final approval

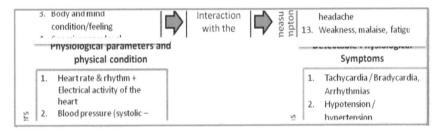

Fig. 11.8 It shows physiological parameters, health symptoms & biosensors for WS devices

B. *WS Transformation (Learning and Novel Interaction)*

A key subtask is to intelligently upscale a wearable device is the installation of (a) "Speech Recognition/Synthesis" software; (b) A "health-history database"; and (c) "Training/learning scheme" Fig. 11.10. That is, the WS central node responsible for several possible subtasks [Prognosis]: (1) collecting various types of physiological data from the biosensors (see Fig. 11.8) applying further DSP on the signals (e.g. for feature extraction), (3) comparing the extracted features from the body-signals with the ones in the "Healthy History Database" using to provide a valuable decision support, (4) generation of alarm signals for the user, (5) displaying the estimated health status and/or collected data on the node's screen, (6) transmitting medical data to a remote base station (e.g. hospital or cell phone of a supervising physician) or even a dispatched ambulance and (7) generating sensor's control signals for initializing measurements or setting up parameters such as sampling interval and A/D frequency).

A novel feature of a WS system is the ability to receive feedback from the patient through voice (or through writing on the central node's keypad). Once implemented, this functionality could enable the user to provide system-feedback regarding *symptoms (coughing, nausea, malaise, back or chest pains etc.) that cannot be measured through standard non-invasive biosensors.* The feedback regarding these symptoms could enable the detection of a wide variety of health conditions.

Alarm signals, measured physiological data and feedback from the patient can be exchanged with the IRC and securely transmitted through the cellular network or the Internet to the medical center or to dispatch ambulance. As the healthcare center keeps the long-term detailed medical history of the patient, the received data and patient symptoms can be further evaluated to derive a more accurate condition estimate or even verify the detected health risk.

The Health History Learning Scheme (HHLS) is depicted in Fig. 11.9. Here, we show the steps needed to extract health history and save it in a database under the physician's approval [49, 108].

Here, the learning is based on a fuzzy-based neural net, a personalized filter and the Prognosis formal method that leads to possible "diagnosis" forwarded to the physician for approval or correction.

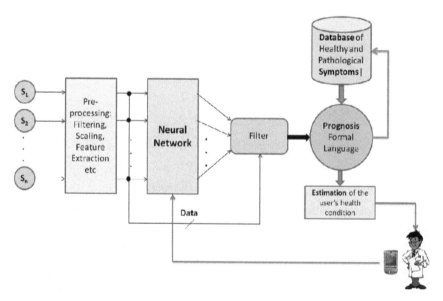

Fig. 11.9 The Health-history learning scheme for the WS

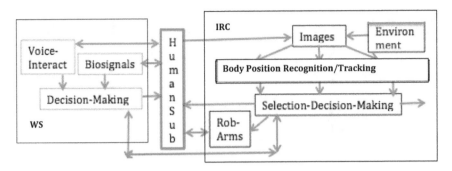

Fig. 11.10 The interaction among HS, IRC and WS

11.3.1.3 Challenge-3: The Novel Synergistic-Interactive Framework-Model

Under this task deals with to integration of the methodologies associated with the IRC and WS devices and at the same time to create a framework model of collaboration with the IRC, WS devices and the Human Subject (HS) [40, 86, 109]. The integration of the HS, IRC, WS and their operational subtasks are based on the development of an efficient interactive communication scheme. Figure 11.10 graphically shows the operational subtasks of the WS and IRC and the way that are connected or interacted internally and with each other. More specifically, the WS interacts with the HS to extract not only bio-signals (measurable symptoms), but also verbal symptoms as indicated in Figs. 11.7 and 11.8.

Then, it makes decisions (or prognoses) for the health condition of the HS. At the same time, IRC captures images from the environment and the HS and process them to recognize its own position in the 3D environment and the body position of the HS. Then it associates these pieces of information to determine its new position or actions, like to extract the navigation path (strategy) to be followed. In addition, IRC makes decisions to assist the HS by activating the robotic arms as shown in Fig. 11.4. A set of requirements has to be developed for the interactions between assistive devices and three different HMI schemes. The first scheme is for the HS + IRC interaction, the second scheme is for the HS + WS interaction and the last is for the efficient communication and information exchange between IRC + WS. Note that the WS and IRC will use the intelligent schemes mentioned in the tasks above for decision making, tracking and learning.

To facilitate the HS-IRC-WS interaction, the WS must have a verbal interactive scheme by giving to HS the option to ask (command) the operation of the robotic arms, when needed. In particular, the HS will have the option to ask the WS-IRC simple questions regarding to his/her conditions. As a response to these commands the WS passes them to IRC, which will operate the robotic arms in a slow but steady motion to assist the HS. For instance, the HS commands the IRC-WS **"help me to stand-up"**. Thus, the WS uses its voice recognition system to understand this command. Then it will pass this "command" in the form of sequence of subtasks to

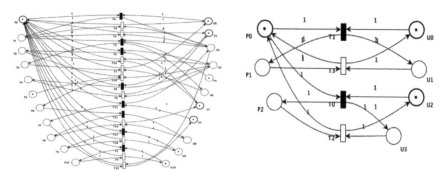

Fig. 11.11 The SPN Human-Device -Interaction scheme (left) available in house, [86], and its detailed representation of a small portion (right)

IRC. The IRC recognizes that sequence, checks the HS position again and gradually extends the robotic-arms to reach the HS. At this point the IRC uses its cameras to continuously track and monitor the HS body and hands positions to appropriately reach the pressure sensors at the end of the robotic-arms. The HS grabs the robotic arms at the position that the pressure sensors are. The IRC continuously monitors the pressure on the sensors and position of the HS hands and slowly raises its robotic-arms until the HS says "**stop**" or "**ok**", or the WS-IRC will ask the HS "**is it ok**". The explanations of such a subtask are graphically provided in the SPN scheme of Fig. 11.11. In a similar way we will study the synergistic collaboration (interaction) among HS-WS-IRC for other commands, like "**help me to turn around**" or "**help me to sit-down**", etc. Risky conditions, like, "**I cannot stand-up**" or the HS accidently releases the pressure on the sensors and IRC has to immediately react by grabbing the hands of the HS and assisting HS to sit-down by avoiding a possible collapse. Finally, such a synergy must be tested and evaluated in a collaborative framework of a small group of volunteers to secure its correct operation without harming the HS. Through this process. The model has to be trained to efficiently learn and adjust its responses according to the human subject's reactions.

A. *Requirements for Interactions (RI)*

For this subtask a set of requirements of interactions are provided for the efficient communication of the HS with the IRC and WS systems. Some of these requirements are shown in Table 11.1.

B. *Scenarios for Proof of Concept (Human Device Interaction)*

In Fig. 11.11 we show an SPN Human-Device Interaction model based on voice commands [40, 86] to satisfy the needs of certain scenarios that require communication between Human user and WS and IRC devices.

In the left side there are the states of various request (commands) and in the right side there are the machine interactions according to its status and capabilities. In the middle are the transitions that fire when the conditions of interactions are satisfied. This means that the same model will be appropriately modified to satisfy both HS +

Table 11.1 Interaction requirements

(1) The HS has always the highest priority for communication (commands) with the IRC and WS systems;

(2) Both IRC and WS have to periodically "ask" the HS for his/her health status. The HS and/or the physician will set the duration of that period;

(3) Both IRC and WS have to periodically exchange information regarding the health status of the HS;

(4) Both IRC and WS have to continuously check and exchange their status for ensuring their highest performance;

(5) Both IRC and WS have access and communication with the physician or a medical center providing the status of the HS, if needed;

(6) If any of the these two devices has a faulty operation, the other has to inform the medical center and the HS;

(7) In case of emergency, both devices have to collaborate to offer first response help to HS (if possible after MD "approval") and inform the medical center and the physician;

IRC and HS + WS interactions. Note that SPN stands for Stochastic Petri-net. The SPN model is well known in the literature, thus we provide no more information about it. A HMI–SPN model has been developed [86, 109] and appropriate modifications will be done to satisfy the **RI** of this project. Here, an example-scenario is offered for the HS + WS interaction for better understanding about the information extraction via questions (Table 11.2). WS will use this piece of information for a more accurate decision-making regarding the HS health status.

C. Scenario for Testing and Integration

Here, the idea is to use as a proof of concept the conduct of experiments in one innovative scenario described at the beginning of this chapter. Here we do not intend to study high risk and dangerous scenarios, as high-risk studies can be studied in a

Table 11.2 Example scenario

WS: < Detects a slight chest discomfort and some of his vitals start to go out of range>

WS: *Some of your vitals are out of regular range. Do you need help?* **HS**: *No*

WS: *Do you have any other symptoms?* **HS** *Yes*

WS: *Please describe your symptoms.* **HS**: *I have slight chest discomfort.*

WS: < the system does not acknowledge the user's phrase > *Please state your symptom again in a brief manner.* **HS**: *chest pain*

WS: < Recognizes the symptom > *Is the symptom intermittent or continuous?* **HS** *continuous*

WS: *Is the symptom instantaneous or persistent?* **HS** *persistent*

WS: *Is the pain radiating to your arms?* **HS**: *Yes*

WS: *Is the pain aggravated by movement or breathing?* **HS**: *No*

WS: *Are you experiencing any other symptoms, hemoptysis or sputum?* **HS** *No*

WS: *Describe your pain level on a scale from one until ten.* **HS**: *five*

WS: < decides that according to the user's answers and his recent vital signs the user could be in an alarming health condition > < notifies the medical-center by sending an alarm message via GPRS>

next phase. Thus, for the simple scenario presented here, various sequences of sub-tasks have to be performed, as they have been described above to show feasibility of this project:

Scenario: "*A human user is laying on his/her bed or sitting on a comfortable chair and he/she needs to visit the bathroom. The hypothesis here is that the human user needs help to stand-up, turn around and sit down on the IRC to go to the bathroom*".

For this particular scenario, the IRC can provide support (using its robotic arms) to lift or assist the subject to stand-up. The IRC can also assist the HS to walk in the bathroom. Now, if a human provider is present, there is a quick response to this scenario. However, if there is no human provider or is busy with something else, a single device (WS) or a system (IRC) cannot provide a satisfactory quick response. In particular, for this scenario, the WS device cannot lift the HS and assist (assistive mode) him/her towards to the bathroom. The presence of the IRC therefore is necessary for this case. This scenario shows the need for both devices in a smart home environment. For people at risk the synergy of IRC and WS can improve and save lives, and at the same time reduce cost by reducing the time needed for a human provider to be present in such a group home. Table 11.3 itemizes the tasks required of the IRC.

Table 11.3 Required tasks of the IRC

1. *IRC recognizes the HS voice request to "go to the bathroom"* (**Voice Recognition**);
2. *IRC searches the room to locate the HS (if needed), although IRC will track the HS around the room if needed to "know" the HS location* (**Body Recognition/Tracking**);
3. *IRC moves towards to the HS location* (**Robot Navigation/Path Planning**);
4. *The IRC uses its synthetic-voice to inform the HS to be ready for assistance* (**Assistive Lifting**);
5. *IRC interactively (it talks to HS regarding the position of the robotic arms) extends its robotic-arms on the "right" position for the HS to grab them* (**Assistive Lifting**);
6. *IRC asks the HS if (s)he is ready to stand-up* (**Assistive Standing**);
7. *IRC interactively assists the HS to turn around* (**Assistive Turn Around**);
8. *IRC interactively assists the HS to sit on IRC seat* (*Assistive Sitting*);
9. *IRC asks the HS if (s)he is ready to move towards to the bathroom* (**Assistive Walking/Predicting Body walking behavior and Patterns**);
10. *IRC continuously inspects (monitors) the free navigation space for safety and collision avoidance* (**Navigation/Planning**);
11. *IRC assists the HS to take the "correct" position near the toilette* (**Tracking/Assistive Lifting**);
12. *IRC lows its robotic-arms interactively to assist the HS to seat on the toilette* (**Assistive Lifting/Tracking**);
13. *During these steps, WS continuously checks the HS health signs (especially the blood pressure) for alarming the IRC for a possible incident; Also, IRC continuously checks the HS body and her/his face for possible indications of uncomforted situation;*

11.4 Discussion

This chapter offers ideas for a challenging project with a profound social impact, but at the same time deals with the human machine efficient interaction that is also a challenging issue for the machine to respond in the same way that humans do. A brief description was given for a set of tasks to be performed under safe conditions (although risk issues can be evaluated) using only one novel scenario in an indoor environment. Thus, the success of such project will lead to the development of more efficient HMI in this very important area for the people in need.

The impact of these challenges presented are profound at many levels. It will advance discovery by promoting new technologies and models for quick first response of medical conditions for assisting people in need at their homes. This will be done through the interactive-collaborative synergy of monitoring sensors via wearable devices and intelligent robotic assistants based on power wheelchairs. It is expected that many research sub-projects and new directions will grow out of this idea in an area of great need for our society. Examples of this kind are numerous and include: (1) significant overall healthcare cost reduction for people at risk by improvements in early prognosis; (2) significant device cost reduction by developing a new generation of low-cost "Robotic Assistants" and sensors to intelligently monitor & detect health conditions, (3) patient safety improvements by reducing risk for falls, and other injuries and in general improving quality of life; (4) enhanced knowledge of human behavioral/health patterns in complex environments to guide development of future medical devices and treatments.

Finally a bibliography, although not complete, is presented for those with interest to contribute to these very important social issues and projects.

References

1. N. Bourbakis, P. Kakumanu, P. Yuan, 3D Object Recognition using synthesis of views. J. Artif. Intell. Tools **18** (2009)
2. N. Bourbakis, A. Esposito, D. Kavraki, A case study of invariant meta-features for learning and understanding disable people's emotional behavior related to their health conditions, *IEEE International Symposium on BIBE-06*, Oct. 2006, WDC, pp. 357–369
3. N. Bourbakis, S. Tzafestas, Bioengineering for people with disabilities. IEEE Robot. Autom. **8** (2002)
4. A. Tsitsoulis, N. Bourbakis, A survey on human activity recognition methodologies. Int. Journal on AI Tools **22** (2013)
5. L. Zhang, Z. Wang, J.L. Volakis, Textiles Antennas and sensors for body-Worn applications. IEEE Antennas Wireless Propag. Lett. **11**, 1690–1693 (2012)
6. J. Aleotti et al., Perception and grasping of object parts from active robot exploration. J. Intell. Robot. Syst. **76**(3–4) (2014)
7. A. Cherian et al., Mixing body-part sequences for human Pose estimation. *CVPR 2014 - IEEE Conference on Computer Vision & Pattern Recognition*, Jun 2014, Columbus
8. M. Rohrbach et al. A database for fine grained activity detection of cooking activities, in *IEEE Conference CVPR* (2012)

9. B. Sapp, B. Taskar, MODEC: multimodal decomposable models for human pose estimation, in *IEEE Conference CVPR* (2013)
10. B. Sapp et al. Parsing human motion with stretchable models, in *IEEE Conference CVPR* (2011)
11. M. Dantone, J. Gall, C. Leistner, L. van Gool. Human pose estimation using body parts dependent joint regressors, in *IEEE Conference CVPR* (2013)
12. V. Morariu, D. Harwood, L. Davis, Tracking people's hands and feet using mixed network and/or search. IEEE T-PAMI **35**(5), 1248–1262 (2013)
13. K. Fragkiadaki et al., Pose from flow and flow from pose, in *IEEE Conference CVPR* (2013)
14. Y. Yang, D. Ramanan, Articulated human detection with flexible mixtures-of-parts. IEEE T-PAMI (2012)
15. R. Tokola, W. Choi, S. Savarese, Breaking the chain: liberation from the temporal Markov assumption for tracking human poses, in *ICCV* (2013)
16. J. Shotton et al., Real-time human pose recognition in parts from single depth images, *IEEE Conference on CVPR*, pp. 1297–1304 (2011)
17. A. Pantelopoulos, N. Bourbakis, A Survey on wearable systems for monitoring and early diagnosis for the elderly. IEEE Trans. SMC-C **40**(1), 1–12 (2010)
18. P. Bonato, Wearable sensors and systems. IEEE Eng. Med. Biol. **29**(3), 25–36 (2010)
19. S. Patel et al., A review of wearable sensors and systems with application in rehabilitation www.jneuroengrehab.com/content/9/1/21
20. J. West, J. Hippisley-Cox, C.A. Coupland, G.M. Price, L.M. Groom, D. Kendrick et al., Do rates of hospital admission for falls and hip fracture in elderly people vary by socioeconomic status? Public Health **118**(8), 576–581 (2004)
21. B.S. Robinson, J.M. Gordon, S.W. Wallentine, M. Visio, Relationship between lower extremity joint torque and the risk for falls in a group of community dwelling older adults. Physiother. Theory Pract. **20**, 155–173 (2004)
22. I. Melzer, N. Benjuya, J. Kaplanski, Postural stability in the elderly: a comparison between fallers and non-fallers. Age Ageing **33**(6), 602–607 (2004)
23. T.M. Gill, H.G. Allore, T.R. Holford, Z. Guo, Hospitalization, restricted activity, and the development of disability among older persons. J. Am. Med. Assoc. **292**(17), 2115–2124 (2004)
24. F. Englander, T.J. Hodson, R.A. Terregrossa, Economic dimensions of slip and fall injuries. J. Forensic Sci. **41**(5), 733–746 (1996)
25. I. Ktistakis, N. Bourbakis, *A survey on Smart Handed Wheelchairs,* IEEE Conference NAECON, Dayton, June 2015
26. I.H. Sie, R.L. Waters, R.H. Adkins, H. Gellman, Upper extremity pain in the postrehabilitation spinal cord injured patient. Arch. Phys. Med. Rehabil. **73**, 44–48 (1992)
27. H. Gellman, D.R. Chandler, J. Petrasek, I. Sie, R. Adkins, R.L. Waters, Carpal tunnel syndrome in paraplegic patients. J. Bone Joint Surg. [Am]. **70**, 517–519 (1988)
28. J.C. Bayley, T.P. Cochran, C.B. Sledge, The weight-bearing shoulder. The impingement syndrome in paraplegics. J. Bone Joint Surg. [Am], **69**:676–678 (1987)
29. Preserving Upper Limb Function in Spinal Cord Injury: A Clinical Practice Guideline for Health-Care Professionals: *Consortium for Spinal Cord Medicine, Spinal Cord Medicine; 2005.*
30. B.G. Celler, N.H. Lovell, D.K. Chan, The potential impact of home telecare on clinical practice. MJA **171**, 518–521 (1999)
31. I. Korhonen, J. Parrka, M. Van Gils, Health monitoring in the home of the future. IEEE Eng. Med. Biol. Mag. **22**(3), 66–73 (2003)
32. M. Alwan, S. Dala, D. Mack et al., Impact of monitoring technology in assisted living. IEEE Trans. Inform. Technol. Biomed. **10**(1), 192–198 (2006)
33. N. Bourbakis, A. Pantelopoulos Modeling a health prognosis generic wearable system using SPNs, *Proceedings IEEE Conference Tools with AI-09*, NY, Nov. 2009
34. A. Pantelopoulos, N. Bourbakis, A comparative survey on wearable biosensor systems for health monitoring, *Proceedings IEEE EMBS Workshop on Wearable Systems and Sensors*, Aug. 20–22, Vancouver, Ca., 2008

35. N. Bourbakis, A. Pantelopoulos, R. Kannavara, iMASS Project: The LG graph model for detecting and associating signatures in NRF signals, *Proceedings IEEE International Conference on TAI-08*, Dayton, Nov. 2008

36. N. Bourbakis, Prognosis: a language for monitoring and prevention, AIIS Inc. Technical Report, TR-06-2007, 4 p

37. N. Bourbakis, J. Gallagher, *DIAGNOSIS: A Synergistic Co-Operative Framework of New Generation Health Diagnosis Systems for People with Disabilities and the Elderly* (International Conference Circuit. Computer, Crete, Greece July, 2008), pp. 20–22

38. R. Kannavara, N. Bourbakis, An FPGA Implementation of the L-G Graph based Voice Biometric Authentication, *IEEE International Symposium on FCCM*, April. 2009

39. N. Bourbakis, A. Esposito, R. Kannavara, An Iris authentication FPGA scheme using 1D LG graphs, *International Workshop on MMI-09*, Vietri, Italy, 2010

40. A. Esposito, N. Bourbakis, N. Avouris, Y. Hatziligeroudis, (eds) Verbal and Non- Verbal H-H & M-H Interaction, *Springer* (2008)

41. R.C. Simpson, P. Daniel, F. Baxter, The Hephaestus Smart Wheelchair System. IEEE Trans. Neural Syst. Rehab. Eng. **10**(2), 118–122 (2002)

42. Q. Zeng, C.L. Teo, B. Rebsamen, E. Burdet, A collaborative wheelchair system. IEEE Trans. Neural Syst. Rehab. Eng. **16**(2), 161–170 (2008)

43. R.C. Simpson, Smart wheelchair a literature review. J. Rehabil. Res. Dev. **42**(4), 423–438 (2005)

44. R.L. Kirby, D.J. Dupuis, A.H. MacPhee et al., The Wheelchair Skills Test (version 2.4): Measurement properties. Arch. Phys. Med. Rehab. **85**, 794–804, 2004

45. V. Zalzal, R. Gava, S. Kelouwani, P. Cohen, Acropolis: a fast Protoyping robotic application. Int. J. Adv. Rob. Syst. **6**(1), 1–6 (2009)

46. A. Atrash, R. Kaplow, J. Villemure, R. West, H. Yamani, J. Pineau, Development and validation of a robust interface for improved human-robot interaction. Int. J. Soc. Robot. **1**(4), 345–356 (2009)

47. A. Pantelopoulos, E. Saldivar, M. Roham, A wireless modular multi-modal multi-node patch platform for robust Biosignal monitoring, *in Proceedings 33rd Annual International IEEE EMBC*, pp. 6919–6922 (2011)

48. A. Pantelopoulos, N. Bourbakis, A wearable health monitoring system for people at risk: methodology and modeling. IEEE Trans. ITB **14**(3), 613–621 (2010)

49. A. Pantelopoulos, N. Bourbakis, Fuzzy petri nets based decision making in human-device interaction with wearable health-monitoring system, *Proceedings IEEE International Conference Bodynets*, Corfu, Greece, Sept. 2010

50. Recently published articles from Computer Vision and Image Understanding, www.journals. elsevier.com/computer-vision-and-image-understanding/recent-articles/

51. C. De La Cruz et al., A robust navigation system for robotic wheelchairs. Control. Eng. Pract. **19**, 575–590 (2011)

52. P.S. Gajwani et al., Eye motion tracking for wheelchair control. Int. J. Inform. Technol. Knowl. Manag. **2**(2), 185–187 (2010)

53. H. Yanco, *Integrating Robotic Research: A Survey of Robotic Wheelchair Development*, MIT, AI Lab, holly@ai.mit.edu

54. H. Yanco, *Shared User-Computer Control of a Robotic Wheelchair System*, Ph.D. 2000, MIT

55. M. Hillman et al., *The Weston Wheelchair Mounted Assistive Robot-The design, Robotica* (2002) volume 20, pp. 125–132. © Cambridge University Press

56. C. Stanger et al., Devices for assisting manipulation: a summary of user task priorities. IEEE Trans, Rehab, Engr. **2**(4) (1994)

57. K. Kawamura et al., Trends in service robots for the disabled and the Elderly, VU, Center of Intelligent Systems, TN, kawamura@vuse.vanderbilt.edu

58. P. Trahanias et al., Navigational support for robotic wheelchair platforms: an approach that combines vision and range sensors, *IEEE International Conference on Robotics and Automation Albuquerque*, New Mexico - April 1997

59. W. Song et al., KARES: intelligent Rehabilitation robotic system for the disabled and the elderly, *Proceedings of the 20thAnnual International Conference IEEE EMBS* Vol. 20(5) (1998)

60. S. Tzafestas, *Research on Autonomous Robotic Wheelchairs in Europe*, IEEE RAM, March 2001

61. E. Prassler et al., *A robotic Wheelchair for Crowded Public Environments*, IEEE RAM, March 2001, pp.38–45

62. W.-K. Song, Visual servoing for a user's mouth with effective intention reading in a wheelchair-based robotic, *Arm, IEEE Conf. Robotics and Automation Seoul*, Korea, May 21–26, 2001

63. Z. Bien et al., Development of a Wheelchair-based Rehabilitation robotic system (KARES 11) with various human-robot interaction interfaces for the disabled, *IEEWASME International Conference on Advanced Intelligent Mechatronics (AIM2003)*

64. H. Seki et al., Minimum Jerk control of power assisting tobot based on human arm behavior characteristic, *2004 IEEE International Conference on Systems, Man and Cybernetics*

65. R.M. Alqasemi et al., Wheelchair-mounted robotic Arms: analysis, evaluation and development, *IEEE/ASME International Conference Advanced Intelligent Mechatronics Monterey*, California, USA, 24–28 July, 2005

66. C. Balaguer et al., Live experimentation of the service robot applications for elderly people care in home environments, IST-2001, *balaguer@ing.uc3m.es*

67. C. Balaguer et al., The MATS robot, IEEE RAM, March 2006, pp. 51–58

68. K. Edwards et al., Design, construction and testing of a wheelchair-mounted robotic armopment, *EEE International Conference Robotics and Automation Orlando , Florida*, May 2006

69. C.-H. Kuo et al., *Human-Oriented Design of Autonomous Navigation Assisted Robotic Wheelchair for Indoor Environments*, 1-424–9713–4/06/©2006IEEE, pp.230235

70. M. Palankar et al., Control of a 9-DoF wheelchair-mounted robotic arm system ua P300 brain computer interface: initial experiments, *IEEE International Conference on Robotics and Biomimetics Bangkok*, Thailand, February 21 - 26, 2009

71. D.-J. Kim et al., Eye-in-hand Stereo visual servoing of an assistive robot arm in unstructured environments, *2009 IEEE International Conference Robotics and Automation*, Kobe, Japan, May 12–17, 2009

72. F.A. Cheein et al., Autonomous assistance navigation for robotic Wheelchairs in confined spaces, *32nd Annual International Conference IEEE EMBS Buenos Aires*, Argentina, Aug.31– Sept.4, 2010

73. P. Schrock et al., Design, simulation and testing of a new modular wheelchair mounted robotic arm to perform activities of daily living, *2009 IEEE 11th International Conference on Rehabilitation Robotics Kyoto International Conference Center*, Japan, June 23–26, 2009

74. T. Mukai et al., Development of a nursing-care assistant robot RIBA that can lift a human in its arms, *2010 IEEE/RSJ Int. Conference on Intelligent Robots and Systems* October 18–22, 2010, Taipei, Taiwan

75. V. Maheu et.al., Evaluation of the JACO robotic arm, *2011 IEEE International Conference Rehabilitation Robotics Rehab Week Zurich*, ETH Zurich Science City, Switzerland, June 29 - July 1, 2011

76. Z. Bien et al., Integration of a rehabilitation robotic system (KARES II) with human-friendly man-machine interaction units. J. Autonom. Robot. **16**, 165–191 (2004)

77. N. Bourbakis, I. Papadakis Ktistakis, M. Alamaniotis, L. Tsoukalas, An autonomous intelligent wheelchair for assisting people at need in smart homes: a case study, *Proceedings IEEE International Conference Information, Intelligence, Systems & Applications (IISA)*, July 2015, Corfu Greece

78. Costs of Falls Among Older Adults, CDC, Center for Disease Control and Prevention, http://www.cdc.gov/homeandrecreationalsafety/falls/fallcost.html

79. M. Oish et.al., Modeling and control of a powered wheelchair, *Proceedings Conference CANCAM*, 2011

80. R. Cooper, *Research and Development of Intelligent Wheelchairs (2000–2011), School of Rehabilitation*, Univ of Pitt, Pittsburgh, PA
81. N.I. Katevas et al., The autonomous mobile robot SENARIO: a sensor aided intelligent navigation system for powered wheelchairs. IEEE Robot. Autom. Mag. **4**(4), 60–70 (Dec 1997)
82. Persons with severe disabilities: a clinical survey. J. Rehab. Res. Develop. **37**(3), 353–360
83. Y. Murakami, Y. Kuno, N. Shimada, Y. Shirai, Intelligent wheelchair moving among people based on their observations. IEEE T-Neural Syst. Rehab. Eng., 466–1471 (2000)
84. R.C. Simpson, S.P. Levine, Voice control of a powered wheelchair. IEEE T-Neural Syst. Rehab. Eng. **10**(2), 122–125 (2002)
85. A. Pantelopoulos a wearable platform with off-the-shelf components for quality analysis of physiological data. *IEEE International Conference on Bodynets,* 2010, Corfu, Greece
86. R. Keefer, Lin, N. Bourbakis, The development and evaluation of an eyes-free interaction model for mobile reading devices. *IEEE Trans SMC: Human-Mach. Syst.* **43**(1), 76–91 (2013)
87. S. Tzafestas, *Advanced intelligent Systems* (Kluwer Publisher, 2001)
88. A Wheelchair steered through voice commands and assisted by a reactive fuzzy-logic controller. J. Intell. Robot. Syst. **34**(3), 301–314
89. N. Bourbakis, Kydonas—an autonomous walking robot. IEEE Robot. Autom. **5**(2), 52–59 (1998)
90. A.Tascillo, N.Bourbakis, A neural-fuzzy control of a robotic hand. IEEE Trans. Syst. Man Cybern. **29**(5), .636–642 (1999)
91. A. Tsitsoulis, N. Bourbakis, A methodology-model for extracting human bodies from single images. IEEE Trans. Humans Mach. Syst. (2015)
92. S. Tzafestas, *Introduction to Mobile Robot Control* (Elsevier Pub. 2013)
93. N. Bourbakis, A. Tascillo, An SPN-neuro planning methodology for coordination of multiple robotic arms with constrained placement. Int. J. Intell. Syst. Robot. **17**, 321–337 (1997)
94. N. Bourbakis, J. Gattiker, G. Bebis, Representing and interpreting human activity and events from video. Int. J. AI Tools **12**, 1 (2003)
95. N. Bourbakis, G. Bebis, A method for associating patterns of motion in events from video, *IEEE Conference on TAI-06*, WDC, Nov. 2006, pp. 187–196
96. A. Lymperis, A. Dittmar, advanced wearable health systems and applications, research and development efforts in the European Union, *IEEE Engineering in Medicine and Biology Magazine*, May/June 2007
97. G. Tröster, The agenda of wearable healthcare, *Wearable Computing at the Electronics Lab ETH* Zürich, Switzerland, Review Proposal in IMIA Yearbook of Medical Informatics, 2005
98. D.A. Rund, R.M. Barkin, P. Rosen, G.L. Sternbach, *Essentials of Emergency Medicine*, , 2nd Edition, Mosby-Year Book Inc (1996)
99. L.M. Tierney Jr., S.J. McPhee, M.A. Papadakis, *CURRENT Medical Diagnosis & Treatment*, 45th Edition, McGraw-Hill (2006)
100. J. Habetha, The MyHeart Project – fighting cardiovascular diseases by prevention and early diagnosis, 28th *Annual Internatonal Conference, IEEE Engineering in Medicine and Biology Society (EMBS)*, August 30 – Sept. 3, 2006, NY City, USA
101. J.L. Weber, F. Porotte, MEdical Remote MOnitoring with cloTHes, *PHealth, Luzerne,* 31 Jan, 2006
102. K.J. Heilman, S.W. Porges, Accuracy of the Lifeshirt® (Vivometrics) in the detection of cardiac rhythms. Biolog. Psych. **75**, 300–3005 (2007)
103. Sensatex, Inc., *Development of the Sensatex SmartShirt, pHealth,* (2006)
104. K. Nikita et.al., Digital Patients and their impact on healthcare: personalized glucose-Insulin Metabolism model based on self-organizing Maps for patients with type 1 diabetes Mellitus, *IEEE International Conference BIBE-13*, Nov. 10–13, 2013 Greece
105. Z. Wang, L. Zhang, J. Volakis, Textile Antennas for Wearable radio frequency applications. J. TLIST **2**(3) (2013)
106. A. Milenkovic, C. Otto, E. Jovanov, Wireless sensor networks for personal health monitoring: Issues and an implementation. Comput. Commun. **29**, 2521–2533 (2006)

107. E. Monton et al., Body area network for wireless patient monitoring, telemedicine and E-health communication systems. IET Commun. **2**, 215–222 (2008)
108. M.-A.. Fengou et. al., Unsupervised clustering of patient-centric models to cluster-centric models for ubiquitous healthcare environment, *IEEE International Conference BIBE-13*, Nov. 10–13, 2013 Greece
109. N. Bourbakis, A. Esposito, R. Keefer, D. Dakopoulos, A multi-modal scheme for HCI between blind users and AT devices, *IEEE Conference on TAI*, Dayton, Ohio, Nov. 2008
110. B.S. Duerstock, Purdue Research Project "Enhanced Control of a Wheelchair-Mounted Robotic Manipulator Using 3-D Vision and Multimodal Interaction" to be appear in CVIU Journal

Further Reading

111. C.E. Smith, N. Papanikolopoulos, Research projects on vision-guided robotic grasping: theory, issues and experiments (1995–2005), University of Minnesota, CSE Dept
112. A. Bergland, T.B. Wyller, Risk factors for serious fall related injury in elderly women living at home. Inj Prev. **10**(5), 308–313 (2004)
113. E.M. Steultjens et al., Occupational therapy for community dwelling elderly people: a systematic review. Age Ageing **33**(5), 453–460 (2004)
114. T. van Bemmel, J.P. Vandenbroucke, R.G. Westendorp, J. Gussekloo, In an observational study elderly patients had an increased risk of falling due to home hazards. J. Clin. Epidemiol. **58**(1), 63–67 (2005)
115. C.A. Dyer, G.J. Taylor, M. Reed, D.R. Robertson, R. Harrington, Falls prevention in residential care homes: a randomised controlled trial. Age Ageing **33**(6), 596–602 (2004)
116. B.L. Cho, D. Scarpace, N.B. Alexander, Tests of stepping as indicators of mobility, balance, and fall risk in balance-impaired older adults. J. Amer. Geriatric Soc. **52**(7), 1168–1173 (2004)
117. L.Z. Rubenstein, D.H. Solomon, C.P. Roth, R.T. Young, P.G. Shekelle, J.T. Chang et al., Detection and management of falls and instability in vulnerable elders by community physicians. J. Amer. Geriatric Soc. **52**(9), 1527–1531 (2004)
118. S.E. Kurrle, I.D. Cameron, S. Quine, Predictors of adherence with the recommended use of hip protectors. J. Gerontol. Med. Sci. **59**(9), M958–M961 (2004)
119. B. Clarkson, K. Mase, A. Pentland, Recognizing user context via wearable sensors, in *Proceedings of the Fourth International Symposium on Wearable Computing*, Oct. 2000, pp. 69–75
120. M. Jones, T. Martin, Z. Nakad, A Service Backplane for E-Textile applications, *Workshop Modeling, Analysis & Middleware Support for Electronic Textiles (MAMSET)*, Oct. 2002, pp. 15–22
121. S. Jung, C. Lauterbach, W. Weber, Integrated microelectronics for smart textiles, in *Workshop on Modeling, Analysis, and Middleware Support for Electronic Textiles*. MAMSET 2002, Oct. 2002
122. T. Martin, M. Jones, J. Edmison, J. Shenoy , Towards a design framework for wearable electronic textiles, *Proceedings of the Seventh International Symposium on Wearable Computers, White Plains,* NY, October 21–24, 2003, pp. 190–199
123. D. Marculescu, R. et al., Electronic textiles: a platform for pervasive computing. Proc. IEEE **91**, 1995–2018 (2003)
124. Z. Nakad, *Architectures for E-Textiles*. Ph.D. Dissertation, Dept. of ECE, Virginia Tech, 2003
125. E. Post, M. Orth, P. Russo, N. Gershenfeld, E-broidery design and fabrication of textile-based computing. IBM Syst. J. **39**(3 and 4) (2000)
126. T. Martin, E. Jovanov, D. Raskovic, Issues in wearable computing for medical monitoring applications: a case study of a wearable ECG monitoring device, *Proceedings of the 2000 International Symposium on Wearable Computers*, Atlanta, GA, October 16–17, 2000, pp. 43–50

127. D. Raskovic, T. Martin, E. Jovanov, Medical monitoring applications for wearable computing. Computer J. **47**(4), 495–504 (July 2004)
128. E. Jovanov, J. Price, D. Raskovic, K. Kavi, T. Martin, R. Adhami , Wireless personal area networks in Telemedical environment, in *Third IEEE EMBS Information Technology Applications in Biomedicine Workshop of the International Telemedical Information Society ITAB-ITIS* 2000, Arlington, Virginia, November 2000, pp. 22–27
129. K. Lyons, T. Starner, Mobile capture for wearable computer usability testing, *Proceedings of the Fifth International Symposium on Wearable Computing*, Oct. 2001, pp. 69–76
130. J. Steven, McPHEE and Maxine Papadakis, 2007 Current Medical Diagnosis and Treatment, *McGraw Hill*, 2007
131. G. Kalb, Fixing America's Hospitals. *Newsweek*
132. L. Gatzoulis, I. Iakovidis, Wearable and portable eHealth systems, technological issues and opportunities for personalized care, *IEEE Engineering in Medicine and Biology Magazine*, Sept/October 2007
133. P. Bonato, wearable sensors/systems and their impact on biomedical engineering. IEEE Eng. Med. Biol. Mag., 18–20 (2003)
134. U. Anliker, J. et.al., AMON: a wearable Multiparameter medical monitoring and alert system. IEEE Trans. Inform. Technol. Biomed. **8** (2004)
135. R. Paradiso, G. Loriga, N. Taccini, A wearable health care system based on knitted integral sensors. IEEE Trans. Inform. Technol. Biomed. **9**(3) (2005)
136. M. Sung, C. Marci, A. Pentland, Wearable feedback systems for rehabilitation. J. NeuroEng. Rehab. (2005)
137. C.D. Katsis, G. Gianatsas, D. I. Fotiadis, An integrated telemedicine platform for the assessment of affective physiological states, *Diagnostic Pathology*, 01 August, 2006
138. J. Rodriguez, A. Goni, A. Illarramendi, Real-time classification of ECGs on a PDA. IEEE Trans. Inform. Technol. Biomed. **9**(1) (2005)
139. C. Wen, M.F. et.al., Real-time telemonitoring system design with mobile phone platform. Measurement **41**, 463–470 (2008)
140. A.L. Goldberger et al., PhysioBank, PhysioToolkit, and PhysioNet: components of a new research resource for complex physiologic signals. Circulation **101**(23), e215–e220 [Circulation Electronic Pages; http://circ.ahajournals.org/cgi/content/full/101/23/e215]; 2000 (June 13)
141. PhysioNet at www.physionet.org
142. Fredrik Nilsson and Communications Axis, Intelligent Network Video: Understanding Modern Video Surveillance Systems, (*CRC Press* 2008)
143. B. Phillips, The Complete Book of Home, Site and Office Security, (*BH* 2006)
144. H. Kruegle, *CCTV Surveillance: Video Practices & Technology* (Amazon 2006) 2nd edition
145. Paolo Remagnino Sergio A Velastin, *Intelligent Distributed Video Surveillance Systems (Professional Applications of Computing)* (IEE 2006)
146. Z. Xiong, R. Radhakrishnan, A. Divakaran, Y. Rui, *A Unified Framework for Video Summarization, Browsing & Retrieval: with Applications to Consumer and Surveillance Video* (Springer 2005)
147. G.L. Foresti, P. Mähönen, C.S. Regazzoni, *Multimedia Video-Based Surveillance Systems: Requirements, Issues and Solutions* (The Springer International Series in Engineering and Computer Science, 2000)
148. O. Javed, M. Shah, *Automated Multi-Camera Surveillance: Algorithms and Practice (The International Series in Video Computing)* (Amazon 2008)
149. P. Remagnino, G.A. Jones, N. Paragios, C.S. Regazzoni, *Video-Based Surveillance Systems: Computer Vision and Distributed Processing* (Amazon 2001)
150. A. Pantelopoulos, N. Bourbakis, A formal language as a Detection and Prognosis model for a Wearable system, *Proceedings IEEE International Conference on BIBE-08, Oct. 2008*, Athens Greece
151. A. Moghaddamzadeh, N. Bourbakis, A fuzzy-like region growing approach for segmentation of colored images. PR Soc. J. Pattern Recogn. **30**(6), 867–881 (1997)

152. A. Mogadamzadeh, D. Goldman, N. Bourbakis, A fuzzy-like approach for smoothing and edge detection in color images. Int. J. Pattern Recogn. AI **12**(6), 801–816 (1998)

153. X. Yuan, A. Mogaddamzadeh, D. Goldman, N. Bourbakis, Segmentation of Colour Images with Highlights and Shadows using Fuzzy-like Reasoning. IAPR Pattern Anal. Appl. **4**(4), 272–282 (2001)

154. N. Bourbakis, P. Yuan, S. Makrogiannis, Recognizing objects using wavelets and LG graphs. PR Soc. J. Pattern Recogn. **40** (2007)

155. N. Bourbakis, P. Kakumanu, S. Makrogiannis, R. Bryll, S. Panchanathan, An ANN based approach for image chromatic adaptation for skin color detection. Int. J. Neural Syst. **17**(1), 1–12 (2007)

156. W. Li, G. Bebis, N. Bourbakis, Improving 3D recognition based on algebraic functions of views. IEEE Trans. Image Process. **17**(11), 2236–2255 (2008)

157. N. Bourbakis, Low resolution target tracking and recognition from a sequence of images. Int. J. AI Tools **11**(4) (2002)

158. T.R. Gowrishakar, N. Bourbakis, Specifications for the development of a knowledge-based image interpretation system. Int. J. Engr. Appl. Artif. Intell. **3**(1), 79–90 (1990)

159. P. Kakumanu, N. Bourbakis, Recognition of facial expressions using LG graphs, *IEEE Conference on TAI-06*, WDC, Nov. 2006, pp. 685–692

160. A. Esposito, N. Bourbakis, The role of timing on the speech perception and production processes and its effects on language impaired individuals, *Proceedings IEEE Symp. BIBE-06*, Oct. 2006, WDC, pp.348–356

161. N. Bourbakis, P. Patil, A methodology for visual detection-extraction of texture-paths in images, *IEEE Conference on TAI-07*, Greece, Patras, Oct. 2007, vol. I, pp. 504–512

162. N. Bourbakis, Automatic conversion of images into natural language text-sentences for visual impaired home interfaces, *IEEE International Conference on TAI-08*, Dayton, OH, Nov. 2008

163. J. Moore, *Computer-Controlled Power Wheelchair Navigation System* (VA Reference No. 01–021)

164. Z.B. Musa, J. Watada, Multi-camera tracking system for human motions in different areas and situations. Int. J. Innov. Comput. Inform. Control **4**(5) (2008)

165. S.M. Monk, M.T. Nanagas, J.L. Fitch, A. Stofli, A.S. Pickoff, Comparison of resident and faculty patient satisfaction surveys in a pediatric ambulatory clinic. Teaching Learn. (Medicine) **18**(4), 343–347 (2006)

166. J.L. Reynolds, A.S. Pickoff, Accelerated ventricular rhythm in children: a review and report of a case with congenital heart disease. Pediatr Cardiol. **22**(1), 23–8 (2001), Review

167. L.W. Lawhorne, Managing laboratory results in the nursing home setting. J. Amer. Med. Dir. Assoc. (2009)

168. D. Dakopoulos, N. Bourbakis, Wearable obstacle avoidance electronic travel aids for visually impaired: a Survey. IEEE Trans. Syst. Man Cybern. **40**(1), 25–35 (2010)

169. The European Pilot Project on Ambient Assistive Living, ICT based solutions for Prevention and Management of Chronic Conditions of Elderly People" pages 15, April, 2008

170. N. Bourbakis, S. Makrogiannis, D. Dakopoulos, A system-prototype representing 3D space via alternative sensation for visual impaired navigation. IEEE Sens. J. **13**(7), 2535–2547 (2013)

171. L. Rothrock, S. Kantamneni, C. Harvey, S. Narayanan, A re-configurable telerobotics system for human factors engineering education. Int. J. Model. Simul. **24**(4), 1–9 (2004)

172. N. Bourbakis, A traffic priority language for collision free navigation of autonomous mobile robots in unknown space. IEEE Trans. Syst. Man Cybern. **27**(4), 573–587 (1997)

173. S. Mertoguno, N. Bourbakis, A retina-like, low level vision processor. IEEE Trans. Syst. Man Cybern. **33**(5), 782–788 (2003)

174. A.M. Cook , J.M. Polgar, *Assistive Technologies: Principles and Practice.* St. Louis: Mosby/Elsevier (2008)

175. A.Y.C. Nee, K. Whybrew, A. Senthil kumar, *Advanced Fixture Design for FMS.* London: Springer (1995)

176. L. Segal, C. Romanescu, N. Gojinetchi, Methodologies for automated design of modular fixtures. *Proceedings of International Conference on Manufacturing Systems*, Iasi, Romania. pp. 151–156 (2001)

177. C.H. Xiong, M.Y. Wang, Y. Tang, Y.L. Xiong, On prediction of passive contact forces of Workpiece-fixture systems. Proc. Inst. Mech. Eng. Part B-J. Eng. Manuf. **219**(3), 309–324 (2005)

178. C.H. Xiong, M.Y. Wang, Y. Tang, Y.L. Xiong, compliant grasping with passive forces. J. Robot. Syst. **22**(5), 271–285 (2005)

179. S. Makrogiannis, N. Bourbakis, Motion analysis with application to assistive vision technology, *IEEE Conference TAI-2004*, FL, Nov. 2004, 344–352

180. N. Bourbakis, detecting faces and facial expressions for monitoring emotional behavior. Int. J. Monitoring Surveillance Res. Technol. **1**(2) (2013)

181. R. Patrick, N. Bourbakis, Surveillance systems for smart homes: a comparative survey, *IEEE Conference on Tools with AI-09*, Newark, NJ, Nov. 2009, pp. 248–252

182. A. Tsitsoulis, R. Patrick, N. Bourbakis, Surveillance issues in a smart home environment, *IEEE Conference on TAI-12 and International Symposium on Monitoring & Surveillance Systems*, Nov 2012, Piraeus, Greece

183. C.H. Xiong, Y. Rong, Y. Tang, Y.L. Xiong, Fixturing Model and Analysis. Int. J. Comput. Appl. Technol. **28**(1), 34–45 (2007)

184. Fundamentals of robotic grasping and fixturing © World Scientific Publishing Co. Pte. Ltd. http://www.worldscibooks.com/engineering/6610.html

185. L.S. Wong, *Robotic Grasping*, May 2008, Stanford University

186. Y. Jiang, M. Stephen, A. Saxena, Efficient Grasping from RGBD images: learning using a new rectangle representation, *International Conference on Robotics and Automation* (ICRA) (2011)

187. N. Bourbakis, *Artificial Intelligent Methods and Applications*. World Scientific Publication (1992)

188. N. Bourbakis, I. Papadakis Ktistakis, Design micro-robot structures for detecting humans under debris in disasters, *Proceedings IEEE Conference NAECON-11*, July 2011, Dayton, OH

189. Assistive Technologies Research Center, Wright State University, *Projects and Pubs* 2003-present, http://www.cs.wright.edu/atrc

190. N. Bourbakis, A. Pantelopoulos, R. Kannavara, Mobile-health: continuous monitoring, bio-signals de-noising for processing and secure information exchange, book-chapter in Biomedical Telemetry. Wiley-IEEE, ed. K. Nikita (2013)

191. Advanced Remote Patient Monitoring Systems http://www.reportlinker.com/p01172037

192. Center for Technologies and Aging, *Technologies for Remote Patient Monitoring for Older Adults*, www.techandaging.org/RPMPositionPaper.pdf

193. A. Tsitsoulis, N. Bourbakis, A. A representation of the flow of images on a multiprocessor surveillance system. Int. J. Monitoring Surveillance Res. Tech, IGI Global Publisher, **1**(1), 1–21 (2013)

194. G. Vavoulas et al., The mobifall dataset: an initial evaluation of fall detection algorithms using smartphones, *IEEE International Conference on Bioinformatics and Bioengineering (BIBE)* 2013, Nov. 10–13, 2013, Chania, Greece

195. P. Kannus et al., Prevention of falls and consequent injuries in elderly people. The Lancet **366**(9500), 1885–1893 (2005)

196. J.A. Stevens et al., The costs of fatal and non-fatal falls among older adults. Inj. Prev. **12**(5), 290–295 (2006)

197. V. Vaidehi et.al., Video based automatic fall detection in indoor environment, *in Proceedings 2011 International Conference on Recent Trends in Information Technology (ICRTIT)*, pp.1016–1020, 3–5 June 2011

198. C. Rougier et al., Robust video surveillance for fall detection based on human shape deformation. IEEE Trans. Circuit. Syst. Video Technol. **21**(5), 611–622 (2011)

199. A.K. Bourke et al., Evaluation of a threshold- based tri-axial accelerometer fall detection algorithm. Gait Posture **26**, 194–199 (2007)
200. S. Fudickar et al., Fall-detection Simulator for accelerometers with in-hardware prepro-cessing, *in Proc. 5th International Conference on Pervasive Technologies Related to Assistive Environments*, Heraklion, Crete, Greece, June 6–8, 2012.
201. F. Bagalà et al., Evaluation of accelerometer-based fall detection algorithms on real-world falls, *PLoS ONE,* vol. 7, no. 5, e37062, pp. 1–9, doi:https://doi.org/10.1371/journal.pone.003 7062 (2012)
202. F. Sposaro, G. Tyson, iFall: an Android application for fall monitoring and response, *in Proceedings 31st International Conference IEEE EMBS*, Mineapolis, Minasotta, USA, September 2–6, 2009
203. J. Dai et al., Mobile phone-based pervasive fall detection. Pers. Ubiquit. Comput. **14**, 633–643 (2010)
204. Y. He et al., Falling-incident detection and alarm by smartphone with multimedia messaging service (MMS). E-Health Telec. Syst. Netw. **1**, 1–5 (2012)
205. R.Y.W. Lee, A.J. Carlisle, Detection of falls using accelerometers and mobile phone technology. Age Ageing **40**, 690–696 (2011)
206. R. Vardasca et al., Clinical Needs and Implementation of health services in Ambient Assistive living, *2nd International Living Usability Lab Workshop on AAL Latest Solutions, Trends and Applications - AAL 2012; 01/2012 and IEEE T-SMC*
207. Medical and Biological Engineering in the Next 20 Years: the Promise and the challenges. IEEE T-Biomed. Eng. **60**(7) (2013)
208. M. Spanakis et al., Exploitation of patient avatars towards stratified medicine through the development of in silico clinical trials approaches, *IEEE International Conference BIBE-13*, Nov. 10–13, 2013 Greece
209. E. Maniadi et al., Designing a digital patient avatar in the context of the My Health Avatar project initiative, *IEEE Int. Conference BIBE-13*, Nov. 10–13, 2013 Greece
210. F. Chiarugi et. al., A cirtual individual's model based on facial expression analysis: a non-intrusive approach for wellbeing monitoring and self-management, *IEEE Int. Conference BIBE-13*, Nov. 10–13, 2013 Greece
211. X. Zhao et al., A Scalable data repository for recording self-managed longitudinal health data of digital patients, *IEEE International Conference BIBE-13*, Nov. 10–13, 2013 Greece
212. H. Wei et al., A Cross-platform approach for treatment of Amblyopia, *IEEE International Conference BIBE-13*, Nov. 10–13, 2013 Greece
213. I. Papadakis Ktistakis, N. Bourbakis, An SPN modeling of the H-IRW getting-up Task, *IEEE Conf. on Tools with AI*, Nov. 2016, S.Jose CA, USA
214. I. Papadakis Ktistakis, N. Bourbakis, A multimodal human-machine interaction scheme for an intelligent robotic nurse, *IEEE Conf. on Tools with AI*, Volos, Greece, Nov. 2018

Chapter 12
Human–Machine Requirements' Convergence for the Design of Assistive Navigation Software: The Case of Blind or Visually Impaired People

P. Theodorou and A. Meliones

Abstract Autonomous navigation is a desirable capability for various types of "smart" devices or vehicles. The development of software designed for this purpose has become a central research field for major companies, as well as in academia. This trend is accompanied by a tendency to equip moving devices with artificial intelligence (AI) features. Interestingly, however, the most (and not artificially) intelligent unit which may require assistance from digital applications in order to achieve autonomous navigation is a blind or a visually impaired person (BVI). It is argued that as the capabilities of AI are being enhanced, convergence will occur among a significant subset of the requirements concerning assistive navigation software for the BVI and AI-equipped moving devices, respectively. The corresponding requirements which have been elicited through interviews with BVI people are presented. A subset of these requirements, which exhibit direct or prospective convergence with the corresponding requirements of AI devices is then outlined, with emphasis on possible opportunities for interaction between the two research topics.

Keywords Autonomous navigation · Artificial intelligence · Blind and visually impaired people · Assistive navigation software · Requirements analysis

12.1 Introduction

Autonomous navigation has been a difficult problem for traditional vision and robotic techniques, mainly due to the noise and variability associated with real-world scenes. Autonomous navigation systems based on traditional image processing and pattern recognition techniques often perform well under certain conditions but have difficulties with others. Part of the difficulty stems from the fact that the processing carried out by these systems remains fixed in a variety of driving situations [1].

Autonomous navigation of a unit, being either a human, a robot or a vehicle relies on processing sensor and dynamic map data to derive guidance information. As far as non-human units are concerned, the necessity of employment of artificial intelligence

P. Theodorou (✉) · A. Meliones
Department of Digital Systems, University of Piraeus, 18534 Piraeus, Greece

© The Author(s), under exclusive license to Springer Nature Switzerland AG 2022   263
G. A. Tsihrintzis et al. (eds.), *Advances in Assistive Technologies*, Learning and Analytics
in Intelligent Systems 28, https://doi.org/10.1007/978-3-030-87132-1_12

(AI) was early enough identified (see [2] or [3], among others) due to the complexity of these operations and the high safety and accuracy requirements (see also [4, 5] for a recent review and survey of contemporary practices on the impact of AI on autonomous vehicle safety).

More than thirty years after Treder's [3] paper, the rate of deployment of efficient AI-equipped units which are capable of autonomous navigation is still slow. However, the benefits from the use of such devices have attracted the attention of many among the biggest technology companies worldwide during the last decade (see e.g. Amazon prime air [6], Uber's self-driving car technology [7], Tesla autopilot [8], Google, Apple, etc.), boosting the privately funded research on this topic. This research has already led to the production of cars, unmanned aerial vehicles (UAVs) and robots equipped with features that enable restricted forms of autonomous or semi-autonomous navigation.

Although the funds directed to research concerning autonomous navigation are continuously increasing, the complexity of the task along with the reasonable high safety and efficiency requirements reveal that the current status of AI and system control technology is still not adequate for the commercial deployment of units capable of multi-purpose autonomous navigation. A look, however, towards the near future, should not ignore already existing technologies that are yet to be deployed (such as 5G), which will enhance significantly critical requirements of autonomous navigation, such as interconnectivity, positioning accuracy and obstacle detection. At the same time, the capabilities of AI are expected to be continuously improving, along with the available data for applications of deep learning procedures.

Autonomous Navigation Technologies, however, does not concern only artificial units, such as cars, UAVs, field robots, etc. It is also a very significant issue for certain groups of people, such as the blind and visually impaired (BVI). In particular, this is a case where human intelligence is obliged to overcome the barrier caused by the impairment of the most significant human sensory ability concerning autonomous navigation, namely, the vision.

Navigation is the acquisition and use of spatial knowledge in order to determine a movement through a physical or virtual environment, along with the movement itself. This is a fundamental aspect of our cognitive range of perception [9]. The purpose of the navigation system is to provide users with the required and/or helpful data to reach the destination point and to monitor their position in previously modelled maps [10]. This task becomes very difficult or impossible without assistance from either sighted people, guide dogs, or technology solutions. Consequently, the ability of the BVI to use public spaces, including urban areas, transport systems and public buildings is reduced [11], except reliable, usable, safe and cost-effective technological solutions are discovered for both outdoor and indoor guidance needs.

12.2 Related Work

Given the advances in the fields of mobile technologies, software and communications during the last two decades, there is an increasing demand for technological solutions aiming to assist the BVI towards autonomous navigation. The adoption and use of such solutions would improve accessibility, self-service and autonomous living, upgrading significantly the quality of life of the BVI. Various assistive mobile apps that aim to contribute to the autonomous navigation of the BVI are already available. Apps that offer improved GPS functionality, such as Loadstone GPS [12], Mobile Geo [13] and Seeing Eye GPS [14] offer enhanced positioning accuracy in order to assist the BVI during pedestrian navigation, while apps such as BlindSquare [15] inform the BVI about points of interest during outdoor navigation.

Academic research has also been focusing on the requirements and the development of systems that assist autonomous navigation of the BVI in outdoor environments (see, e.g., [16–18], in interior spaces [19–23], etc.) or in both [24]. None of the corresponding solutions, however, is widely adopted by the BVI according to Giudice and Legge [25], and the demand for autonomous navigation systems especially designed for the BVI is increasing. This is also verified by the discussions during the interviews concerning our research with the BVI. Part of the explanation may be due to a disconnection between engineering factors and a system's perceptual and practical usefulness, which means that a product may be theoretically effective but may not work in reality for the intended consumer who adopts this navigational technology [25].

BVI now can use general navigation applications through accessibility functions, but the acceptance of mobile and location-based services is a gradual process [26]. It involves understanding of the benefits offered by these services before their acceptance, and systematic use by the majority within a target group. In general, the development of smart apps does not take into account the particular requirements of people with special needs, and especially of the BVI [27]. Even in the case that the app is designed especially for the BVI (see Csapo et al. [28], for a survey of assistive mobile apps for blind users), several features that facilitate the BVI to learn how to use the app are missing.

Assistive navigation apps for blind people are good examples where the familiarization process with the app depends on the access to specific locations (see Meliones and Sampson [23], among others). This problem directly affects the rate of technology acceptance of the BVI concerning smart app use. Interestingly, however, little or no research appears to have been undertaken on the inherent particularities in the take up of assistive navigation software by the BVI.

A necessary step in the development process of autonomous navigation software is that of the elicitation and analysis of requirements of the potential users (where, in the broader sense, the user may be an AI unit). This paper aims to draw parallels between the requirements of sensory deprived human intelligence (the BVI) and AI-equipped autonomous units with respect to autonomous navigation. This is achieved using data collected from interviews with members of the BVI community. Specifically, the answers of the BVI are classified into four main, characteristic requirements categories which are further divided into eight sub-categories (a part of the final requirements of the whole project). Then, for every elicited requirement of the BVI, it is examined whether the consideration of a corresponding requirement for an AI-equipped unit is reasonable. This procedure also allows us to identify requirements that correspond to the "human nature" of the BVI, with respect to the current status of AI robots or autonomous vehicles. Finally, the possibility of convergence of requirements that stem from human and artificial intelligence is examined.

The rest of this paper is organized as follows: Next section describes the structure of the interviews, as well as the characteristics of the BVI participants. The same section presents the classification of the requirements, as it was derived by an analysis of the answers of the BVI. The third section presents the analysis of the elicited requirements with particular focus to the requirements that seem compatible with the current status of mobile AI units. The section also includes a discussion about the possibilities of convergence between AI and human requirements in the cases where AI evolves towards a human-like intelligence or not. The last section concludes the paper.

12.3   Methodology

12.3.1   Interviews with BVI People and Requirements Classification

A user needs and requirements analysis has been conducted during the initial phase of development of two assistive mobile apps for autonomous navigation of the BVI by our research team. These assistive apps are being developed within a project entitled "MANTO" (funded by the Greek RTDI State Aid Action RESEARCH-CREATE-INNOVATE of the National Operational Programme Competitiveness, Entrepreneurship and Innovation 2014–2020 in the framework of the T1RCI-00593 contract). The first mobile app (Blind RouteVision) aims to assist the BVI during outdoor pedestrian navigation. The app's design includes enhanced GPS functionality and interconnectivity with other apps that may be useful during navigation, such as the corresponding service of Google Maps. The app is a part of an assistive navigation system which includes ultrasound sensors for real-time recognition and avoidance of obstacles along the BVI's path, synchronization with traffic lights and weather information, and utilization of information telematics of the Athens Mass Transit System (AMTS) for routes and urban transport stops. The initial version of the Blind Route-Vision system is presented in [23]. The smartphone application and its supportive external components consist of the aforementioned system for outdoor interactive autonomous navigation for BVI (see Fig. 12.1).

Fig. 12.1 Blind RouteVision outdoor navigation—advanced field navigation sensor

The second mobile app concerns autonomous blind navigation in indoor spaces. Since GPS is not reliable for indoor positioning, the app is supported by a highly accurate indoor location determination subsystem which includes accessibility mapping of indoor spaces with overlays of the positions of points of interest (POIs) (see Fig. 12.2). Moreover, Bluetooth beacons are used as proximity sensors and location indicators. The app will inform the BVI about his/her relative position of POIs and will use dynamic issuing of voice navigation instructions towards POIs considering the current position of the BVI. Blind IndoorGuide inherits the features of the Blind MuseumTourer system, a system for indoor interactive autonomous navigation for blind and visually impaired persons and groups (e.g., pupils), which has primarily addressed blind or visually impaired (BVI) accessibility and self-guided tours in museums, as they are presented in [23]. Blind IndoorGuide aims to extend the functionality of the Blind MuseumTourer beyond the case of museums. The conceptualization of the Blind IndoorGuide is discussed in [23].

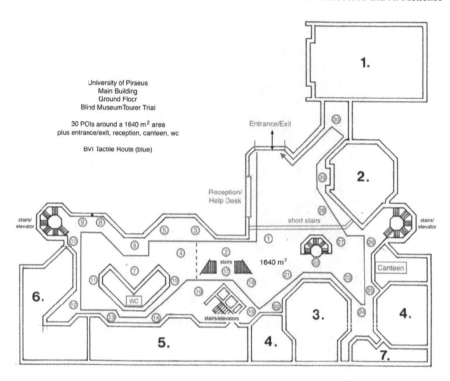

Fig. 12.2 Blind MuseumTourer indoor navigation and guidance system preliminary

12.3.2 Description of the Participants

In this section, we present the setup of the interviews with the BVI, which were conducted to determine the user needs and requirements for indoor and outdoor navigation assistive mobile apps, but also psychological characteristics and practices (preferences, habits, facts) that are related to their impaired vision (see also [29]). The interviews were conducted at the premises of the Lighthouse for the Blind of Greece, which is the main non-profit organization for education and assistance of the BVI in Athens. Thirteen male and female members of the BVI community participated in semi-structured interviews. Their vision problems ranged from severely impaired vision to complete blindness, according to self-report. Each interview lasted at least 45 min. The descriptive characteristics of the interviewees are presented in Table 12.1.

The number of participants not only appeared to be in line with the literature on qualitative research (Guest et al. [30], and Adu et al. [31], including that concerning people with visual impairments, Wolffe and Candela [32], Kane et. al. [33] and Guerreiro et. al. [34], among others) but also proved, in practice, adequate because data saturation appeared to be already reached before the last interview.

Now we turn to the methodology for deriving the classification of the BVI and the creation and analysis of the requirements. We validated this classification by

Table 12.1   Characteristics of the BVI who participated in the interviews

	Gender	Age	Degree of vision loss	Cause of vision loss	Digital sophistication
P1	Male	55	Complete	By birth	High
P2	Female	35	Severe	By birth	Average
P3	Male	36	Complete	Diabetes	High
P4	Male	40	Almost complete (95%)	By birth	Low
P5	Male	40	Almost complete (95%)	By birth	Low
P6	Female	55	Complete	Retinopathy (23 years old)	Low
P7	Male	40	Almost complete (90–95%)	By birth	Low
P8	Male	40	Complete	Cancer (7 years old)	Low
P9	Male	35	Almost complete (>95%)	Benign tumor (15 years old)	Low
P10	Male	60	Complete	By birth	High
P11	Male	30	Complete	By birth	High
P12	Male	40	Complete	By birth	High
P13	Male	38	Almost complete (90–95%)	craniocerebral injuries at 23	High

conducting a pilot study with two BVI interviewees who were specialized and experienced in the field, with whom we created the questions and the thematic axes, a large part of which are listed in Table 12.2. The whole classification and its subcategories are listed and fully described in [35]. This paper identifies a subset of these requirements, to which the corresponding requirements of AI devices may converge, as the

Table 12.2   Classification of the requirements of the BVI for assistive navigation apps

1.	Requirements concerning usefulness and capabilities	a. Obstacle detection
		b. Navigation
2.	Functionality requirements	a. Navigation
		b. Navigation
		c. Navigation
3.	Usability requirements	a. Characteristics/features of apps and devices
		b. Device handling
4.	Compatibility and parallel operation with other apps	a. Compatibility and parallel operation with other apps

capabilities of the devices increase. Indicative quotations from some of the 13 participants are provided close to the following requirements classification. Additionally,
the questions asked during the interview are given in supplementary material.

12.3.3 Requirements Classification

Next, we present the subjects which were discussed during the interviews with the
BVI. First, the participant introduced herself or himself. During this part of the
interview, BVI people were asked about characteristics such as age, degree of vision
loss, cause of vision loss and age at which this occurred. They were also asked
whether they were employed and, if this was the case, to describe their job. They
were also asked about how familiar they feel with digital apps. Then, section of
the interviews included a presentation of the features and capabilities of the apps
as they were initially conceptualized by our research team. Finally, a discussion
followed, which included specific questions concerning the BVI's indoor and outdoor
navigation habits. During this section of the interview, the BVI were asked specific
questions about their preferences and requirements that would ideally lead them to
adopt and efficiently use the apps. These questions concerned requirements about
the usefulness and capabilities, the functionality and the usability of the apps. They
also had suggestions concerning the compatibility of the apps with other apps and
services they had already been using.

The initial design of the interviews, along with the feedback from the answers and
suggestions of the interviewees led us to a classification of the subjects of interest
with respect to the design and development of assistive mobile apps for the BVI.
The interviews with the BVI were recorded on paper. Their answers, suggestions
and comments were then classified into four main categories, namely, requirements
concerning the usefulness and capabilities of the apps, functionality requirements,
usability requirements and requirements concerning the compatibility and parallel
operation of the apps with other apps and services. These categories were further
divided into subcategories as presented in Table 12.2.

The first category includes the requirements concerning the usefulness and capabilities of an assistive navigation mobile app from the BVI perspective. The structure
of the interviews allowed the interviewees to focus on the issues that they find more
important. Specifically, we observed that the main subjects of interest in BVI people
can be classified into two subcategories: (a) Obstacle detection, (b) navigation. The
second category includes the functionality requirements, as the BVI perceive them.
Again, we identified three sub-categories: (a) requirements concerning the treatment
of external stimuli, (b) requirements about the way sound (or voice) could be used
to facilitate the interaction between the BVI and the smartphone (or the app) and
(c) requirements about the accuracy of tracking and the devices that could be used
to improve it. The third category concerns the usability requirements of the apps
and interconnected devices. Apart from the first sub-category which concerns the
characteristics or features that a BVI requires from the assistive navigation apps and

devices, it is of special interest to identify the optimal way of handling the smartphone (the apps) and interconnected devices. Finally, the BVI showed particular interest in the way the assistive apps will seamlessly collaborate with the usual apps a BVI uses, for example, screen readers and web mapping services.

The left column of the classification presented in Table 12.2 is abstract enough to be adapted in a framework of an AI requirements analysis concerning autonomous navigation software. It can be observed however that subcategories 2.a, 2.b, 3.a and 3.b of the fine classification (second column) are significantly related to human–machine features for the case of the BVI. Part of the analysis of the requirements of the BVI in the following section focuses on the identification of which of them may be compatible with possible requirements of AI-equipped autonomous moving devices.

12.4 Analysis of the Elicited Requirements

This section presents in detail the elicited requirements of the BVI concerning assistive mobile navigation apps. It also identifies possible links and differences between these requirements and the ones of AI-equipped autonomous moving devices. Finally, it concludes with a discussion concerning the possible human–machine requirements convergence under the prism of two main directions of the evolution of AI in the near future.

12.4.1 Elicited Requirements of the BVI

An important qualitative advantage of a designer of navigation software for the BVI with respect to the corresponding one for autonomous vehicles or moving devices is the possibility of conducting interviews with the BVI. The material collected from these interviews can be classified to remarks, preferences and suggestions of the BVI that are related to their human substance (including psychological characteristics and ambitions) and to those that are mainly related to the specific task of autonomous navigation. This subsection mainly focuses on the latter class of requirements which exhibit direct or prospective convergence with the corresponding requirements of AI devices.

12.4.1.1 Requirements Concerning Usefulness and Capabilities

In this subsection, the requirements concerning the usefulness and capabilities of assistive navigation applications and devices are presented, as they were elicited from the interviews. These requirements have been classified into two sub-categories. These are the following: (a) object/obstacle detection, (b) navigation.

a. Object/Obstacle Detection

- The software should be capable of simultaneously processing sensor signals concerning multiple obstacles and reporting them appropriately. (P3: "It should be possible to simultaneously detect multiple obstacles and report them appropriately, guiding the BVI to manoeuvre with precision.", P8: "It is a problem to have obstacles at a different height at the same time, so there should be adequate ability to identify multiple obstacles and to properly inform the BVI user").

- Sensors should be able to detect obstacles at different heights. These obstacles may not be ground-based like signs sticking out from buildings. The sensor data should be capable of providing height information, and the software should be able to interpret this information in specific directions. (P3: "The sonar should be able to detect obstacles that are relatively high and not only ground-based in front of the BVI, such as low balconies, awnings, signs, etc.", P1: "The sonar should be capable of scanning both horizontally and vertically").

b. Navigation

- The software should be capable of identifying traffic lights and interpret their status. (P3: "The app should be able to identify the traffic lights on the streets, as well as their status").

- The software should be able to identify possible threats for the unit even when traffic lights do not allow motion. In other words, the stream of data from the sensors should be continuously processed and the correctness of the units' path and speed should be continuously validated. Safety is a prior requirement and its maintenance should not be threatened at every point of the route. (P1: "When the app notifies the BVI about the status of a traffic light, it should be able to identify the danger that may arise from a driver who does not follow the signal of the traffic light.", P1: "It must be possible to identify the danger at traffic lights even when the application has correctly stated that the blind person should proceed.", P7: "The maintenance of the software should be frequent, while adequate control systems should report any malfunction in real-time.", P11: "In order that such a system functions properly, regular maintenance of the communication systems must be provided. Moreover, appropriate control and information systems or processes should be used to promptly identify any hardware malfunction").

- The software should be swift at detecting a wrong route and correcting or adjusting it accordingly. (P2: "The app should be possible to detect a wrong route and to correct or adjust the route in case of deviation from the selected path").

- It is desirable that the software could manage multiple destinations or stops and continuously optimize the route. (P3: "It is desirable to be able to add multiple destinations or stops along a route").

12.4.1.2 Functionality Requirements

This subsection presents the elicited requirements concerning the functionality of the mobile assistive navigation apps. These requirements have been classified into three sub-categories. The first (sub-category a) concerns how the apps and the peripheral devices should react to external sounds, while the second (sub-category b) includes requirements concerning the interaction of the BVI with the device through voice commands and directions, as well as through other types of sounds. Finally, the third (sub-category c) corresponds to requirements concerning the accuracy of tracking and positioning of the navigation system (app and auxiliary devices), as well as functional requirements of the auxiliary devices.

a. External Stimuli

-   The software must be able to deal with noisy data from the sensors. As far as the BVI are concerned, this also include the case where external sounds interfere with the voice interaction with the mobile app. (P1, P3: "The ambient sounds should not be covered by the sounds produced by the app because the BVI always use their hearing in order to perceive the surrounding space. Therefore, the use of a headset that covers both ears is excluded from the implementation of the assistive navigation system. One ear should be able to hear the sounds of the surroundings").
-   Only important cell phone information should be reported phonetically in order not to cover or suppress the ambient sounds. It seems difficult to claim that for an AI moving unit a similar requirement holds. (P1: "Only important cell phone information should be reported phonetically, and ambient sounds should not be depreciated or covered").

b. Sound (or Voice) BVI-Smartphone (or App) Interaction

This sub-category corresponds to requirements that are related to the human nature of the BVI, and mainly to the fact that they substitute their vision with hearing during their interaction with the apps.

-   The app should offer voice menus of the key destination options. These options must be able to include combined pedestrian navigation and the use of other means of transport. (P1: "Voice guidance is preferable to audio", P1: "Combine pedestrian route and public transport (voice reporting of basic options).", P4: "The app should provide voice reporting of key route identification options that may combine pedestrian navigation and Mass Public Transit (MPT) use").
-   The app must provide an audio signal for the traffic lights. (P1: "It is necessary to have an audio signal on the traffic lights (the blind should not be left without audio information)").

c. Tracking and positioning accuracy and auxiliary devices

- GPS and sonar amplifiers should be discreet and not too obvious. Of course, the geometry of a machine allows easier incorporation of such amplifiers. (P1: "It takes a great deal of precision in positioning. When it comes to using a GPS precision amplifier, this should be discreet and not too obvious. There is a general problem with GPS accuracy").
- Positioning accuracy must be high (at least at the centimetre level). (P1, P6: "Positioning accuracy should be as high as possible").
- Any sonar or GPS device must refresh the information it provides at high frequency because some BVIs can move at a fast pace (the same would also be the case for AI-equipped units). (P1: "About sonar for obstacle detection: Some of the BVI move very fast, especially the elderly. Ideally, the frequency of refreshing information transmitted by the sonar should be at least once every 1 s").

12.4.1.3   Usability Requirements

This subsection presents the elicited requirements concerning the usability of mobile assistive navigation apps. These requirements have been classified into two sub-categories. The first (sub-category a) refers to the special characteristics and features of the apps and the peripheral devices that the BVI would wish to have, while the second (sub-category b) concerns device handling requirements.

a.   Characteristics/features of apps and devices

As in subcategory 2(b), some usability requirements of the BVI are not compatible with the navigation software requirements of a machine.

- The BVI showed preference to simple, easy-to-learn keyboards or pads. (P1: "The user interface on the touch screen must operate with clear and simple gestures. Keyboards are also useful").
- Because cables are often entangled, BVIs prefer to use Bluetooth headphones. (P1, P3: "Bluetooth headphones have the advantage that they do not have cables which may become entangled. On the other hand, they are more easily lost and they require to be changed often").

b.   Device Handling

- The software should be easily integrated or interconnected with the user. Concerning the BVI, this corresponds to their ability to set the destination on their own. As far as an AI machine is concerned, this would imply seamless communication and connectivity with the navigation software.
- Concerning outdoor navigation, it would be desirable that the software provides an editable list of "favourite" destinations. (P7, P12: "As far as outdoor navigation is concerned, it would be ideal for the app to include a list of 'favourites' (or preferred) destinations to be edited by the BVI").
- The BVI should have the ability to dictate the destination address. The ability of phonetical exchange of commands and information between AI-equipped

machines is currently a very interesting research topic. (P1: "The BVI should be able to set the destination on their own (the independence of the blind is very important). It is preferable for the blind to be able to dictate the destination address on the device and to have voice navigation in the search menu").

12.4.1.4 Compatibility and Parallel Operation with Other Apps

This subsection presents the elicited requirements concerning the compatibility and parallel operation of the assistive navigation mobile apps with other apps or services.

a. In general, the software should be able to run in parallel with other applications (such as screen readers, in the case of the BVI). (P2, P3: "Allow other applications (e.g. screen reader) to run in parallel with the navigation application").

b. It would be useful that the software could interconnect/collaborate with an application that describes images. (P4: "It would be very useful if the navigation apps could cooperate with image recognition apps that describe images, such as Google Lens").

12.5 Discussion

Within our initial project a detailed analysis of the BVI needs and requirements with respect to the design and development of assisting navigation systems is conducted. For this purpose, seven main categories are identified for the classification of the user needs and requirements which are elicited from extensive interviews with the BVI. To the best of our knowledge, this is the first time that the needs and requirements of the BVI are presented in such a complete and structured way that outlines a useful framework for the development of assistive navigation systems. This framework aims to provide insights concerning features and approaches that will potentially increase the use rates of the navigation systems, along with the benefits offered to them. A very interesting example of such an approach, with emphasis to integrated multipurpose assistive navigation systems is the combination of organized training offered to the BVI with a training version of the corresponding navigation app. We also identified that even the BVI who have good abilities in using mobile apps would prefer to test the navigation system in a controlled environment before trying it in conditions likely to be found in the outside world.

The previous subsection presented a part of the elicited requirements from interviews with members of the BVI community. These requirements were classified in the main categories and subcategories. For the purpose of the present study, which is to link requirements gathered from BVI individuals for wayfinding activities to those required by an AI system such as those found within autonomous vehicles, the description of the requirements was abstract enough. This was intended to highlight, when this was possible, the similarities with reasonable requirements for the navigation software of autonomous AI-equipped units. The variety of such units, however, does not support the view of "one size fits all", because these may include humanoid robots, autonomous vehicles, UAVs, or even lawnmower machines, vacuum cleaners etc.

There exists always an argument supporting a view that these machines do not actually need to possess AI capabilities when they operate in fully controlled, or remotely identified (or perceived) environments, because in this case, the machines would just need to be remotely controlled or operated (by a human or, possibly, by a remote AI). On the other hand, however, current trends support autonomous (or semi-autonomous) operation, and consequently, a requirement for autonomous (or, semi-autonomous) AI. The latter case (that concerns AI-equipped units) is the one that mainly is related to the analysis of this study.

Concerning current trends in the development of AIs for autonomous machines, two directions may be identified. The first corresponds to the effort to imitate human intelligence. Structurally, this does not only depend on how well we understand and are able to simulate the human brain, but also on how the AI is confined physically by the geometry of the unit it resides, and by the sensors that imitate human vision and hearing (although smell, taste provide information to a human, the main senses concerning communication and learning are the first two). In the case of a strict imitation of human geometry and sensory abilities, the elicited requirements of the BVI may directly correspond to the sub-modules of the AI that will have to deal with signals from the visual and sound sensory systems of the robot.

The second direction corresponds to AIs that aim to exploit every available degree of freedom in order to accelerate and enhance their learning ability, speed and efficiency. Even in this case, the appropriate treatment of sensor data will be central to autonomous navigation.

In both, the cases described above, the requirements from the interviews of the BVI concerning necessary information about the route, or safety, can be considered as variables with respect to whether the user of the software is a human or not. On the other hand, requirements concerning user interfaces mostly depend on human nature and the particular characteristics of the BVI. This, however, does not preclude the case where a humanoid robot in the future will use an interface designed to facilitate the interaction between an assistive navigation app and a BVI.

12.6   Conclusion

The purpose of this study was to examine whether requirements which are elicited in the initial phase of the design of assisting navigation apps for the BVI are comparable with reasonable requirements concerning navigation software that could be possibly used by AI-equipped units.

The case of the BVI, as far as the design of navigation software is concerned, is particularly interesting because of two main facts: The first concerns their general requirement that visual information, along with data from other sensors has to be interpreted to specific directions. This may be thought of as an analogue of the necessary treatment of the data from the visual sensory system of any machine before specific directions are derived. The second fact concerns the availability and the ability to conduct interviews with the BVI within the framework of user requirements analysis. Unfortunately, for the time being, it is not possible to conduct interviews with AI-equipped autonomous units for this purpose. This fact makes the elicited requirements of the BVI a valuable source of information concerning the design of not only stand-alone software but also of specific modules that will aim to treat sensor data in order to provide efficient and accurate navigation directions to autonomous units. Conclusively, it is important to investigate the human–machine requirements convergence for the design of assistive navigation software, because this research could benefit from getting implications for designing AI-equipped autonomous units and the reverse.

Appendix A

QUESTIONNAIRE/QUESTIONS TO DISCUSS.

We are developing two systems that aim to assist blind people to navigate.

The first system concerns outdoor navigation and autonomous and safe pedestrian travel to predetermined destinations.

1. *Description*: The app is intended to be used by Android smartphones.

   Questions:

   – Are there any comparative advantages of the iPhone over the Android smartphones? If so, what are they?
   – Do you know smartphones specially designed for blind people (for example SmartVision 2)? What is the preferred operating system by blind people?
   – Do they opt for Apple iPhone? Will it be easy to switch to Android smartphones or to get a second smartphone that will use Android?

2. *Description*: Our system utilizes Google maps for voice-guided navigation.

   Question:

   – Are the voice capabilities of smartphones used by blind people and to what extent? Do you find Google Maps easy and functional to use?

3. *Description*: There will be used headphones that do not isolate both eardrums. We recommend the use of bone conduction headphones, or of a single ear headphone so that the ambient sounds are not dampened.

   Question:

– Do BVI people use headsets connected to their smartphones? Is it easy for a blind person to simultaneously recognize sounds from different sound sources by each ear? Do you find this specification reasonable? Do you have anything else to recommend? What kind of handset do you prefer, Bluetooth or wired?

4. *Description*: A simple keypad will be used for the blind person to easily interact with the application, to select routes and other available functions.

   Question:

   – Do you think the keypad should have any specifications regarding its functionality, ease of use and usability? (that is, how good and easy it is to use).

5. *Description*: The app will use voice commands to inform the BVI for obstacles in the direction of their movement.

   Question:

   – Questions: How do you think obstacles should be reported and what instructions would be given to them along their route? Increasing continuous sound, interrupted sound, or vibration with increasing frequency as the obstacle approaches? Simultaneous or only voice reporting? How do you think the warning of the obstacle will be more user-friendly or practical?

6. *Description*: There will be a configuration activity that allows the user to create an extensive list of destinations to be selected from the keyboard.

   Question:

   – Do you find this easy for the BVI? Are there any examples of navigation in the options menu? Are options menu widely used? (e.g. smartphone Smartvision has an audio description function for the menu)

7. *Description*: It will be possible to synchronize the application with traffic lights. We recommend that the system will be implemented centrally through the traffic management system so that the BVI does not depend on whether or not each traffic light is equipped with a sound broadcasting system.

   Questions:

   – Do you have any suggestions concerning these features?
   – Is it sufficient that the mobile phone can produce a sound similar to that of traffic lights equipped with sound broadcast features for the blind people? Do you have any suggestions for improvements?

8. *Description*: Weather information will be provided so that the blind person can dress appropriately for the pedestrian route.

   Question:

- How do you keep up to date with current weather?

9. *Description*: The app will send notifications to selected persons about the current position of the BVI in case of need

   Question:

   - Who should be informed about this relative, police, ambulance)? In the case of automatic activation, is there the nearest person who can receive a message? (It is logical that if no one answers the phone, there will be a hierarchy of options that will be called automatically). Note that the SmartVision 2 commercial smartphone has a SOS button that can send GPS coordinates (SmartVision 2, 2019).

10. *Description*: The app will use real-time information from the OASA telematics system for routes and stops for the development of complex routes that may include urban transport, etc.

   Question:

   - How does a BVI now choose the means of transport?

11. *Description:* The app will be connected to an external wearable subsystem, which could be fitted to e.g., a hat to ensure clearer reception of the GPS receiver and a sonar.

   Questions:

   - Is there an original to evaluate the ease of use?] Do you think a wearable device can easily be adopted by a BVI? What could be the type of wearable device that should be used/worn by the BVI to improve GPS accuracy (eg vest, hat, or embedded in a cane)?

12. *Description:* The application will be able to extract semantic information (along the way) which will be communicated to the BVI.

   Question:

   What do you think are the objects of interest that a BVI would want to identify along the way (toilets, pedestrians, obstructed vehicles, shops/species identification)?—list completion.

The second application is a blind navigation system in public interior spaces with a pilot application to provide autonomous tours of museums.

1. *Description*: It is designed for Android smartphones

   Questions:

   - The question serves both versions of apps (indoor and outdoor).

2. *Description*: It will use voice guidance.

   Questions:

- The question serves both versions of apps (indoor and outdoor).

3. *Description*: Guidance will be provided along the tour route.

    Question:

    - Which one of the following is preferred by a BVI? Voice guidance (speech), audio or a combination of both?

4. *Description*: The app will provide audio information about the exhibits the BVI has approached. It will also notify the user about whether it is allowed to touch the exhibit.

    Questions:

    - Do you have any suggestions for additional specifications?
    - The app will have the ability to give accurate voice guidance at any time on how to access the help desk, the exit, the WC or the restaurant. Are there any other similar points of interest?

5. *Description*: Ability to request assistance from the Museum staff at any time.

    Question:

    - The question serves both versions of apps (indoor and outdoor).

6. *Description*: Emergency call option.

    Question:

    - The question serves both versions of apps (indoor and outdoor).

7. *Description*: Design and implementation of an appropriate simple user interface on the touch screen of the smartphone.

    Question:

    - How do you propose to split the screen of the smartphone so that the blind can choose commands?

8. *Description*: Screen reading will support special reading functionality for the BVI.

    Question:

    - What is applicable today?

9. *Description*: The app will be used at the Tactual Museum - Lighthouse for the Blind of Greece, and after the completion of the project at the National Archaeological Museum and the Acropolis Museum.

    Question:

    - Is there another indoor destination you would like to visit?

10. *Description*: The implementation of a training version of the applications and the development of educational and training methods and strategies will be of high importance.

    Questions:

    – Do you find what those apps have to offer appealing?
    – Will we be able to have the BVI's massive participation in training on these apps, on the use of equipment, and by what means?
    – How long is a BVI going to spend on training in these applications?
    – What is the process and training that will allow the BVI to gain the confidence to navigate on his/her own?

    Questions:

    – Do you believe that a training version of the apps, which could be easily parametrized and applied by a sighted trainer to your routes or locations, would increase the rate of acceptance and involvement of the apps?
    – Do you think this could successfully replace the need for training in real conditions (for example, in a museum)?

    Questions:

    – Now that the main features of the assistive navigation apps have been described to you, do you believe that the presence of a sighted escort along your trip (outdoors or indoors) is still necessary, given that you have learned how to use the apps? To what extent can these apps support autonomous navigation of a BVI person?

11. *Description*: The operation of navigation aids for BVI depends on the external systems (such as beacons, signal recognition, NFC sensors, etc.) whose space must be equipped to support the operation of the application. In essence, these external systems determine how the application interacts with the environment. It cannot, therefore, be learned in places that are not equipped with such systems.

    Questions:

    – Would you be reluctant to use the app for the first time during your visit to the museum? What do you think about it?

    Questions:

    – Does learning the app detract from the main goal of the visitor, to enjoy and learn from the exhibits?
    – Does this happen also when navigating to an external destination which becomes complicated by learning the corresponding auxiliary application-utility?

Questions:

– Do you think an educational version of each application would be beneficial which could be tried with the help of an instructor, in familiar places first, before being used in conditions likely to be found in the outside world?

References

1. D.A. Pomerleau, in *Alvinn, an Autonomous Land Vehicle in a Neural Network*. Carnegie Mellon University (1990)
2. J. McCarthy, T. Binford, D. Luckham, et al., in *Basic Research in Artificial Intelligence and Foundations of Programming*. Defense Technical Information Center (1980)
3. A.J. Treder, Autonomous navigation-when will we have it? Navigation **34**, 93–114 (1987)
4. A.M. Nascimento, A.Y. Hata, L.F. Vismari, et al., A systematic literature review about the impact of artificial intelligence on autonomous vehicle safety. IEEE Trans. Intell. Transp. Syst., 1–19 (2019)
5. Y. Ma, Z. Wang, H. Yang, L. Yang, Artificial intelligence applications in the development of autonomous vehicles: a survey. IEEE/CAA J Autom Sinica **7**, 315–329 (2020)
6. www.amazon.com (2018) Amazon prime air: frequently asked questions. https://www.amazon.com/Amazon-Prime-Air/b?ie=UTF8&node=8037720011. Accessed 29 Jan 2020
7. www.uber.com (n.d.) Uber self-driving car technology. https://www.uber.com/us/en/atg/technology/. Accessed 29 Jan 2020.
8. Tesla.com, Autopilot (2019). https://www.tesla.com/autopilot. Accessed 29 Jan 2020
9. H. Segond, D. Weiss, E. Sampaio, Human spatial navigation via a visuo-tactile sensory substitution system. Perception **34**, 1231–1249 (2005)
10. A. Real, Navigation systems for the blind and visually impaired: past work, challenges, and open problems. Sensors **19**, 3404 (2019)
11. P. Strumillo, Electronic interfaces aiding the visually impaired in environmental access, mobility and navigation. In: IEEE 3rd International Conference on Human System Interaction (2010)
12. www.loadstone-gps.com. (n.d.). Loadstone GPS - Home. www.loadstone-gps.com. Accessed 2 Jan. 2020
13. www.senderogroup.com. (n.d.). Sendero Online Store: Mobile Geo http://www.senderogroup.com/products/shopmgeo.html#upg. Accessed 2 Jan. 2020
14. www.senderogroup.com (n.d.) Sendero Group: The Seeing Eye GPSTM App for cell-enabled iOS devices. http://www.senderogroup.com/products/SeeingEyeGPS/index.html. Accessed 2 Jan. 2020
15. BlindSquare.com (n.d.) http://blindsquare.com. Accessed 2 Jan. 2020
16. J.M. Loomis, R.G. Golledge, R.L. Klatzky, Navigation system for the blind: auditory display modes and guidance. Presence **7**, 193–203 (1998)
17. M. Bousbia-Salah, M. Fezari, A navigation tool for blind people, in *Innovations and Advanced Techniques in Computer and Information Sciences and Engineering* (Springer, Netherlands), pp. 333–337
18. S. Koley, R. Mishra, Voice operated outdoor navigation system for visually impaired persons. Int. J. Eng. Trends Technol **3**, 153–157 (2012)
19. G. Ghiani, B. Leporini, F. Paternò, C. Santoro, Exploiting RFIDs and tilt-based interaction for mobile museum guides accessible to vision-impaired users. In: Miesenberger K. et al. (eds.) ICCHP 2008, LNCS 5105, pp. 1070–1077 (2008)
20. M. Miao, M. Spindler, G. Weber, Requirements of Indoor Navigation System from Blind Users, in *Lecture Notes in Computer Science*. (Springer, Berlin Heidelberg, 2011), pp. 673–679

21. T. Gallagher, E. Wise, H.C. Yam et al., Indoor navigation for people who are blind or vision impaired: where are we and where are we going? J. Locat. Based Serv. **8**, 54–73 (2014)
22. M. Dias, Indoor navigation challenges for visually impaired people, in *Indoor Wayfinding and Navigation* (CRC Press, 2015), pp. 141–164
23. A. Meliones, D. Sampson, Blind MuseumTourer: A System for Self-Guided Tours in Museums and Blind Indoor Navigation. Technologies **6**, 4 (2018)
24. L. Ran, S. Helal, S. Moore, Drishti: an integrated indoor/outdoor blind navigation system and service, in Proceedings of the Second IEEE Annual Conference on Pervasive Computing and Communications (2004)
25. N.A. Giudice, G.E. Legge, Blind Navigation and the Role of Technology, in *The Engineering Handbook of Smart Technology for Aging, Disability, and Independence* (Wiley, 2008), pp. 479–500
26. Z. Osman, M. Maguire, M. Tarkiainen, Older Users' Requirements for Location Based Services and Mobile Phones, in *Lecture Notes in Computer Science* (Springer, Berlin Heidelberg, 2003), pp. 352–357
27. E. Ghidini, W.D.L. Almeida, I.H. Manssour, M.S. Silveira, Developing apps for visually impaired people: lessons learned from practice, in IEEE 2016 49th Hawaii International Conference on System Sciences (HICSS) (2016)
28. Á. Csapó, G. Wersényi, H. Nagy, T. Stockman, A survey of assistive technologies and applications for blind users on mobile platforms: a review and foundation for research. J. Multimodal User Interf. **9**(4), 275–286 (2015)
29. P. Theodorou, A. Meliones, Towards a training framework for improved assistive mobile app acceptance and use rates by blind and visually impaired people. Educ. Sci. **10**, 58 (2020)
30. G. Guest, A. Bunce, L. Johnson, How Many Interviews Are Enough? Field Methods **18**, 59–82 (2006)
31. M.D. Adu, U.H. Malabu, A.E.O. Malau-Aduli, B.S. Malau-Aduli, Users' preferences and design recommendations to promote engagements with mobile apps for diabetes self-management: Multi-national perspectives. PLoS ONE **13**, e0208942 (2018)
32. K.E. Wolffe, A.R. Candela, A qualitative analysis of employers' experiences with visually impaired workers. J. Visual Impairm. Blindness **96**, 622–634 (2002)
33. S.K. Kane, J.P. Bigham, J.O. Wobbrock, Slide rule, in Proceedings of the 10th international ACM SIGACCESS conference on Computers and accessibility—Assets '08 (ACM Press, 2008)
34. J. Guerreiro, D. Ahmetovic, D. Sato, et al., Airport accessibility and navigation assistance for people with visual impairments, in Proceedings of the 2019 CHI Conference on Human Factors in Computing Systems - CHI '19 (ACM Press, 2019)
35. P. Theodorou, A. Meliones, Gaining Insight for the Design, Development, Deployment and Distribution of Assistive Navigation Systems for Blind and Visually Impaired People through a Detailed User Requirements Elicitation. Unpublished manuscript (2021)

Part IV
Advances in Privacy and Explainability in Assistive Technologies

Chapter 13
Privacy-Preserving Mechanisms with Explainability in Assistive AI Technologies

Z. Müftüoğlu ⓘ, M. A. Kızrak ⓘ, and T. Yıldırım ⓘ

Abstract With the developing technology and increasing amount of data, Artificial Intelligence (AI) shows the effect in almost every field of our lives. Thanks to AI systems that are revolutionizing the technology world, things only humans could do before can be automated such as visual perception, decision-making, speech recognition, and translating between languages. Even though AI brings to mind the perception of a futuristic future, it also plays a big role in facilitating human life. This role is vital especially for people with disabilities trying to adapt to daily life and be fulfilled with the Assistive Technologies domain. In Model ICT Accessibility Policy Report (Available: https://www.itu.int/en/ITU-D/Digital-Inclus ion/Persons-with-Disabilities/Documents/ICT%20Accessibility%20Policy%20R eport.pdf. Accessed: 31 Jan 202 [1]), the Assistive Technologies are described as any information and communications technology used to protect, increase, or advance the functional abilities of individuals with particular needs or disabilities. But this brings the potential for simultaneous privacy issues such as spying, exploitation, and data breaches due to context-awareness interfaces and penetrable information everywhere. According to WHO (World Health Organization)'s report on disability, approximately 15% of the world's population suffers from some form of disability (Summary World Report on Disability, Available: https://apps. who.int/iris/bitstream/handle/10665/70670/WHO_NMH_VIP_11.01_eng.pdf;jse ssionid=50C7F4199A25E26711B5A903759B35C6?sequence=1. Accessed: 31 Jan 2021 [2]). Considering that people with disabilities as well as many people in need to use these technologies, we can say that the number of potential users is much higher. Since they are designed to be used by vulnerable individuals with physical and cognitive disabilities, ensuring data privacy is of greater importance.

Z. Müftüoğlu (✉) · M. A. Kızrak
The Presidency of Republic of Turkey, The Digital Transformation Office, Ankara, Turkey
e-mail: zumrut.muftuoglu@cbddo.gov.tr

M. A. Kızrak
e-mail: ayyuce.kizrak@cbddo.gov.tr

T. Yıldırım
Yıldız Technical University Istanbul, Istanbul, Turkey
e-mail: tulay@yildiz.edu.tr

© The Author(s), under exclusive license to Springer Nature Switzerland AG 2022   287
G. A. Tsihrintzis et al. (eds.), *Advances in Assistive Technologies*, Learning and Analytics
in Intelligent Systems 28, https://doi.org/10.1007/978-3-030-87132-1_13

While sensitive data needs to be protected, on the other hand, the use of this data is very critical for the functionality of technology. Besides all these, when decisions of AI systems affect humans' lives, the need for figuring out how such decisions are taken comes out (Goodman B, Flaxman S, European union regulations on algorithmic decision-making and a right to explanation. AI Magazine 38 (3) 50–57 [3]). Especially with the development of deep neural networks which are considered as black-box models in recent years (Castelvecchi in Nature 538:20–23 [4]), the importance of explainability in Artificial Intelligence systems has come to the fore. In the literature, model explainability and privacy/security by design come together under the notion of Responsible AI (Arrietaa et al. in Information Fusion 58:82–115 [5]). In the scope of this chapter, privacy-preserving solutions for assistive technologies and the relationship between explainability and privacy will be tackled.

Keywords Data privacy · Privacy-preserving · Secure assistive technologies · Private AI intelligence

13.1 Introduction

User-friendly and innovative solutions brought by emerging technologies have been easing the life of individuals who have constraints to meet their own needs. These solutions, called assistive technologies, have become widespread with the entry of AI into our lives. According to U.S. Census Bureau data; while assistive technology provides 77% of people with disabilities to look after themselves, it also enables nearly 75% of them to take upon the role of economic heads of households [6]. Statistics say that someone in the world develops dementia every 3 s. There were an estimated 46.8 million people worldwide living with dementia in 2015 and this number is believed to reach 131.5 doubling every 20 years almost in 2050 [7] (Fig. 13.1).

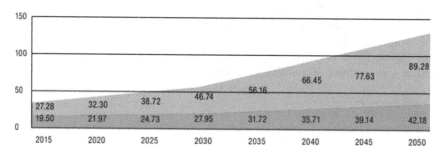

Fig. 13.1 The growth in numbers of people with dementia (millions) in high-income and low and middle-income countries [7]

In the ever-changing world, so many advancements in technology are done day by day. One of these advancements is solutions developed with AI intelligence. Assistive technology has a helping hand role in the lives of people with disabilities. Thanks to a touch of AI, a person who has a disability can live without feeling the lack of constraint.

However, this supportive role also brings ethical and privacy concerns since they require collecting sensitive data to perform their function. Considering these technologies are designed to be used by vulnerable individuals, these concerns are becoming more important. At this point data ethics, data privacy, and data security terms gain importance to define a framework for data protection. It shouldn't be forgotten that data protection is related to ethics which deals with judgments and truths that can include all human practices involving respect for individuals and their rights regarding privacy.

13.1.1 Data Ethics

Ethics is defined as a specific set of principles and rules regarding duty with dictionary meaning. Depending on personal data usage, privacy and ethics are among the sine qua non for trusted technologies. Ethics and Data Privacy issues were placed amongst the top 10 strategic technology trends for 2019 in Gartner's Report. These two notations protect their place with the title of Transparency and Traceability for 2020 as well [8]. Dubois and Miley define ethics under two subtitles as micro ethics and macro ethics [9]. While micro ethics direct principles for individual implementation, macro ethics tackle organizational regulations. Considering the focus point for both, the medical ethics framework might be valid for assistive technologies as well [10].

The newly emerging field of machine ethics is concerned with the ethical dimension of this human–machine interaction. While computer ethics focuses on ethical issues covering humans' usage of machines, machine ethics focuses on maintaining whether if the behavior of machines against humans and other machines is ethical [11]. James Moor describes a concept called ethics impact agents which provides to evaluate computing technologies in terms of ethical norms: implicit ethical agents, explicit ethical agents, full ethical agents [12] (Fig. 13.2).

13.1.2 Data Privacy

Although the history of data privacy concerns is very old, discussions on privacy started shortly after the Second World War [14]. The data, which is considered as the most valuable mine of today, also makes the person strong who owns it. This situation creates an unfair competitive environment for the sake of owning the data and brings the chain of violations causing victimization in people within the twentieth

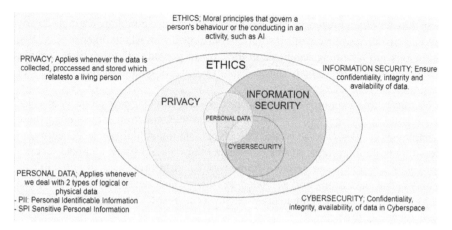

Fig. 13.2 Relationship between data privacy, data protection and ethics [13]

century, privacy violations have increased due to the rapid technological progress. With the release of the General Data Protection Regulation (GDPR) which includes European countries, most non-European countries prepared their policies. It is aimed to protect citizens from these violations, define a framework on individuals' rights, and raise awareness of providers to implement privacy policies with these regulations (Fig. 13.3).

13.1.3 Data Security

Even if it seems that data security and data privacy can be used interchangeably, in fact, there are distinct differences. While data privacy governs how the data is collected and shared, data security protects the data from internal and external attacks. When talking about security, three concepts come to the fore: confidentiality, integrity, and availability. The basic goal of data security is protecting the data which is collected, received, stored, or transmitted. Data security covers a set of techniques such as physical security, administrative controls, logical controls, and other safeguarding techniques that avoid unauthorized access.

13.2 AI Applications in Assistive Technologies

AI, which is defined as a sub-subject of computer science, has settled in almost every point of daily life with the progress of brain-neuroscience research and the development of technology. Mobile phones can recognize people by using fingerprints and facial images. In order to carry out financial transactions, they encounter smart

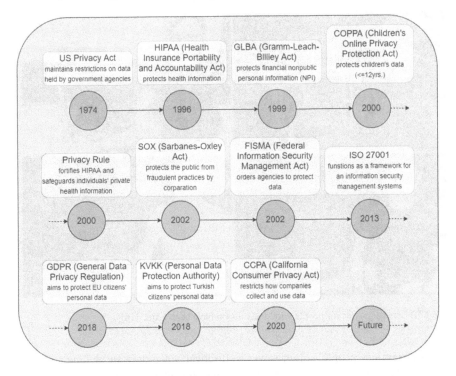

Fig. 13.3 Data privacy timeline [15]

chatbots in banks and they can recognize us by our voices. Thanks to wearable technologies, health data can be processed in the cloud in real-time, thus enabling doctors to detect diseases early by making predictions about the future health conditions of living things. Shopping made using the internet, movies watched and music listened to make life faster and easier by providing advice on areas of interest. These are just a few examples of AI products used today. With the processing of the produced data, personalized products and environments can be designed and personalized training services are also provided.

- The person using the Seeing AI [16] application directs his phone to the person he wants to know. Thus, he hears the physical characteristics of the other person such as hair color, eye color, age, gender, mood, as sound. Thanks to this product developed by Microsoft for visually impaired individuals, the person who goes grocery shopping can see the expiry dates and contents of the products.
- AI-supported smart glasses also have the feature of helping in many areas. Neuroscientists and computer vision experts developed it in 2014 at Oxford University to have verbal vision and avoid obstacles [17, 18]. Thus, it was ensured that people easily overcome obstacles such as sidewalks and cars (Fig. 13.4).
- Numerous examples of assistive technologies can be cited in hearing. One of these was developed by Columbia University researchers in 2017. The device

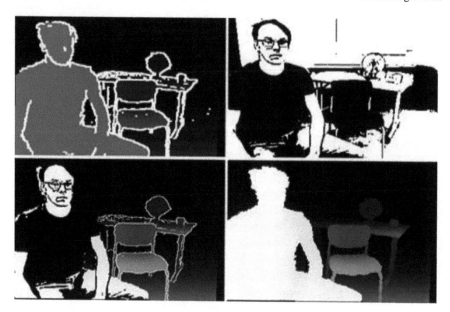

Fig. 13.4 Smart glasses for people with weak vision have been tested [18]

also examines people's brain waves to discover when they are willing to listen more. The device not only helps the person hear sounds but also adjusts the brain waves to control what the person wants to listen to. It also ensures that if the user has more than one sound, music, and speech, it is focused on only one of them [19, 20] (Fig. 13.5).

- Efforts are ongoing to create a device that can convert sign language to text or sound to help people with hearing impairments communicate effectively. For example, the software application developed by scientists in Aberdeen can be

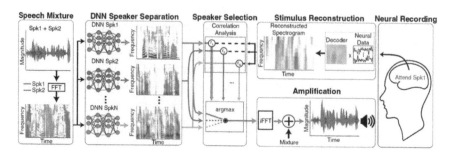

Fig. 13.5 The two speakers, Spk1 (red) and Spk2 (blue) are mixed together in a single acoustic channel. A spectrogram of the corresponding mix is fed to several DNNs, each trained to separate one of the particular speakers. As the user joins one of the speakers (here Spk1; red), the neural recordings can be used to reconstruct a spectrogram of that speaker that is compared to the outputs of each DNN before being amplified by the hearing aid [19]

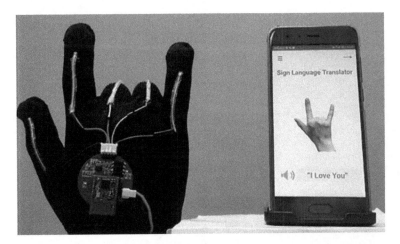

Fig. 13.6   The signs would be read out through a smartphone connected to the glove [21]

used on portable devices. In addition, users can customize the sign language according to their needs. A product, also known as AcceleGlove, works as a one-way translator. It uses built-in accelerometers to detect signs created by the user's hand and express them in written text or spoken words. Some of the applications that convert sign language to text and sound also use AI intelligence. Aberdeen is just one of them. They also offer users the option to customize sign language according to their needs. AcceleGlove has also been developed for use as a one-way translator [21, 22] (Fig. 13.6).

- Similar studies are also being carried out in Turkey. Moreover, the Sesgoritma application, which was developed by high school students, is also offered for the hearing impaired. In a study conducted for hearing-impaired children to learn sign language by interacting with humanoid robots, the robot perceives the image shown by children allows the child to repeat it with sign language, and it is aimed to enable the child to confirm or learn by repeating sign language movements [23]. It is obvious that human–robot interaction will increasingly support the health sector in the field of lifelong care and development for children, elderly, and/or disabled individuals [24].
- It is also possible for paralyzed individuals, who have to spend their lives in a wheelchair, to walk on their own. For this, they gain the ability to move intelligently with a robotic exoskeleton design and learning the walking characteristics of people [25] (Fig. 13.7).
- A built-in cleaner can now be used to help better clean living spaces. The most popular example is undoubtedly robotic vacuum cleaners that follow their patterns on the floor to suck dust, pet hair, and dirt. When you have an injury, these robotic vacuum cleaners become the greatest helpers [26].
- An example of helping and AI-supported technology can be added to make life easier not only for people with disabilities but also for everyone. The most popular

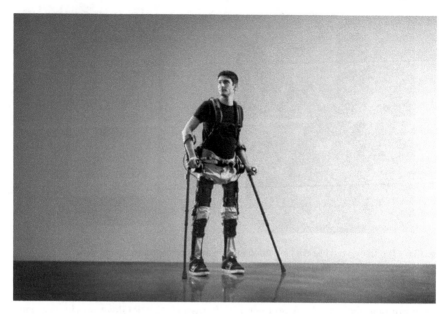

Fig. 13.7 To gain the ability to walk and move with a robotic inner or outer skeleton [25]

and the first to come to mind are cleaning assist robots. Robot vacuum cleaners that follow patterns and obstacles on the floor are always needed to improve your quality of life, not only when you have an injury.

- The personal health robot Mabu symbolizes a new era in inpatient care. Mabu can make eye contact with patients, and can also make private interviews with patients. Thus, it made a positive contribution to the well-being rate of the patients. These robots, which provide not only physical but social interaction, receive investments from global companies [27, 28].

- A wearable biofeedback product has been developed at the University of Houston for Parkinson's patients, which are defined as neurodegenerative brain disorders that cause the person to experience balance disorders. This product will aid rehabilitation. Through the data collected, it is designed to give confidence to the person and to improve posture stability in daily life. It also helps guide exercises that contribute to the recovery.

- In addition to the lack of skills such as speaking, moving, and hearing at some point in human life, he also needs help to cope with problems in the mind. One of them is the fear of staying indoors. It is also known as being a locked-in-place syndrome. With the help of such technology, the questions posed can easily be answered as no or yes with thoughts alone. The machine consists of some smart learning tools and electrodes that can provide an accuracy of over 80%.

As in other industries, an important issue for AI applications in the health sector is based on how well the privacy and security of data can be ensured. Considering the most common attacks and data breach problems, it is predicted that it is not possible to use algorithms that risk revealing the details of the patient's medical history. Personalized applications come to the fore in the use of AI intelligence as assistive technology. For this, it is necessary to emphasize that personal data should be collected, transferred, stored, and processed in the cloud.

A healthy and equitable future for everyone should be the ultimate goal with accessible, fast, precise, successful, low-cost AI-powered assistive technologies.

13.2.1 Explainable AI (XAI)

By AI, a revolutionary technology, it has become possible to overcome many limitations. However, the decision-making processes of these models are generally not interpreted by users. In various fields such as health, finance, or law, there is an increasing expectation to know the reasons behind a decision made by an AI-based system. Therefore, studies to explain neural models have been started recently.

In situations involving highly complex algorithms, the concept of reliability may become a more useful regulatory principle than explainability. It focuses on the reliability of the algorithmic system needed to reduce possible risks. For example, algorithms such as decision trees are less efficient in performance than more recent AI solutions, but they are more preferred due to the explanation of their relatively transparent and intuitive structures [29].

The widespread use of AI-powered technologies has led to an increasing impact on individuals, organizations, and societies. The decision of such a system must be understood and it must be clear who is responsible for risky situations (Fig. 13.8).

When we consider the aforementioned concerns of people regarding AI, adopting the principles of the concept of Trustworthy AI can produce beneficial results for society.

At the international level, there is currently a consensus on 6 dimensions of Trustworthy AI:

- Justice,
- Accountability,
- Value,
- Robustness,
- Repeatability,
- Explainability.

While justice, accountability, and value embody our social responsibility; robustness, repeatability, and explainability pose enormous technical challenges for us [31].

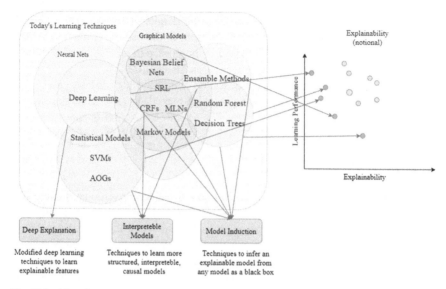

Fig. 13.8 AI performance versus explainability [30]

13.3 Data Privacy and Ethical Challenges for Assistive Technologies

Almost every innovative technology comes into our lives with some concerns. For example, keeping the phone numbers in our memory has become a thing of the past with mobile phones, the development of social media technologies has affected physical communication negatively. At the beginning of the twenty-first century, systems were converted into electronic records from paper-based records [32]. But it did not take long for humanity to adapt to these and similar negativities by developing some countermeasures.

In the digital age, AI intelligence and assistive technologies have an important effect on human life. While it seems that this effect will increase in the near future, these developments raise privacy concerns. When we come to the term intelligent assistive technology, it is a general definition of technologies that ease life for the older or disabled population making use of robotics and machine intelligence for assistive purposes. People have worries about raising privacy challenges, especially when these technologies involve AI, due to collecting sensitive data.

Boise et al. showed in their study about unobtrusive monitoring through motion sensors and computers that 60% of people who participate in their study reported their worries related to privacy while over 72% of participants reported acceptance of willingness to share the data with their doctor or family members [33].

Risks in terms of ethics for AI technologies are divided into four: Reputational Risk, Legal Risk, Environmental Risk, and Social Risk [34].

Biased systems can cause reputational damage. Buolamwini and Raji investigated the commercial impact of Gender Shades to analyze the effect of publicly naming and revealing performance results of biased AI systems [35].

Legal risk covers situations where a system is seen as causing an anti-competitive environment. It can be given the example of Microsoft being fined by the European Commission because of anti-competitive behavior [36].

Environmental risk is related to the ethical responsibilities of technology to the natural environment [37]. A leakage, which damages nature, is an example of environmental risk.

Social risk encompasses the cases which include society and communities. These risks may include issues related to human morale or traffic problem. When we look in terms of assistive technologies; increased technology-driven social isolation, inequality, local community matters covering the acceptable uses of technology and AI [34].

13.3.1 Data Collection and Data Sharing

Current technological advances significantly increase digital data generation, leading to large amounts of data in different disciplines from finance to medicine. We especially see an increase in data generation in the field of biomedical and health-care. Assistive technologies collect data including sensitivity for their functionality. Massive data collecting raises significant ethical concerns, while it affects the performance of AI systems positively. AI-driven assistive technologies collect a great amount of data about their users.

13.3.2 Secure and Responsible Data Sharing Framework

Data sharing has key importance in ensuring continued progress in the effective functionality of AI-driven technologies. It is inevitable to need more sensitive data to develop AI Assistive Devices, which provides assist human successfully. To design private systems, all parties should be considered such as regulators and technological measures. It is obvious that only legal measures or only technical measures are not enough to provide private systems. Knopper's study [38] provides a general overview of responsible and secure data sharing (Fig. 13.9).

Each of the elements seen in Fig. 13.4 constitutes a ring for the design of a private system. Some of them are provided with legal measures. But for some of them, only legal measures are not enough to create a proper private system. In recent years, the literature offers new approaches to this issue.

With the development of deep neural networks which are considered as black-box models in recent years [4], the importance of explainability in Artificial Intelligence systems has come to the fore. In the literature, model explainability and

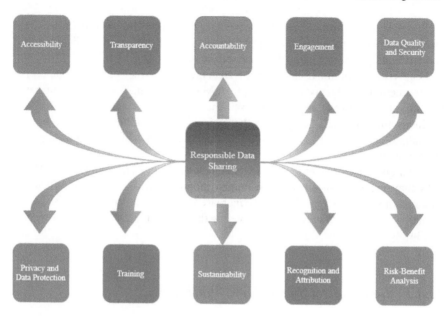

Fig. 13.9 Responsible data sharing [39]

privacy/security by design come together under the notion of Responsible AI [5]. Responsible AI determines that fairness, accountability, and privacy should also be noted when implementing AI models in real environments. Explainability in ML models is important to assess privacy. Because it may cause a privacy breach not being able to understand what has been captured by the model [4].

Arrieta et al. address model interpretability jointly with requirements and constraints related to data privacy, model confidentiality, fairness, and accountability (see Fig. 13.10). They also emphasize in their study that a responsible implementa-

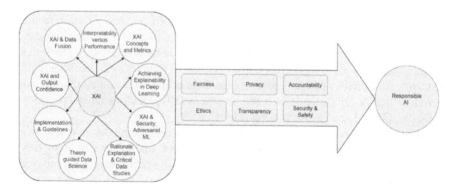

Fig. 13.10 Relationship between XAI and privacy under the concept of responsible AI [5]

tion of AI methods in organizations and institutions will be only guaranteed in case all these AI principles are studied jointly [5].

13.4 AI Assistive Technologies with Privacy Enhancing

Some of the potential attacks such as Denial of Service attacks (DoS), cloud malware injection attacks, man-in-the-middle attacks [40], spoofing [41], and collusion attacks [42] implicate information disclosure.

Hoepman introduced privacy design strategies in his study [43]. Privacy-enhancing by design can be achieved by applying particular design strategies (see Fig. 13.11). In this chapter, technological solutions will be addressed. Ferrer and Justicia introduce privacy and data protection in terms of design and privacy-enhancing techniques [44]. They enumerate the design strategies as minimize, hide, separate, aggregate, inform, control, enforce, and demonstrate. Hide strategy infers to ensure secrecy for collected data. This can be provided by encrypting, pseudonymizing or anonymizing data. According to a separate strategy, personal data should be stored and processed in a decentralized way. Regulations, policies on privacy also should be established and enforced. And control strategy implies that individuals should be able to modify and delete the information of themselves when necessary.

13.4.1 Privacy-Preserving Mechanisms for AI Assistive Technologies

Considering that AI is based on data, privacy and AI are in conflict, progress in one must come at the expense of the other. AI-powered assistive technologies help to ease

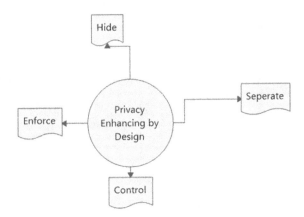

Fig. 13.11 Basic strategies for privacy enhancing by design [43]

life for people who have vision, hearing, mobility, cognition, and learning disabilities. Personal data, against unauthorized access; must be stored securely building confidentiality (privacy), data integrity, availability, and accountability. Machine learning and cryptography come together to provide privacy. The technology presents solutions with promising privacy-preserving technologies such as differential privacy, federated learning, and homomorphic encryption.

13.4.1.1   Differential Privacy

Differential privacy is a strict mathematical definition of privacy that shares information about a dataset publicly by defining the patterns of groups while keeping individual data in the dataset. The inspiration of differential privacy is that one cannot tell whether if any individual's data is placed in the original dataset or not. So, the behavior of the algorithm hardly changes in case an individual is replaced in the dataset. This can be achieved by using randomized mechanisms.

As come to the formal definition, suppose that there is a D dataset with n records and an attacker has background information in the dataset except nth record. When the attacker query on the data set to achieve aggregated information about n records, s/he can obtain the information of n by comparing the difference between query results. Differential privacy resists this kind of attack scenarios. In Fig. 13.12, a basic differential privacy model is depicted. Neighboring datasets (D and D') are created that differ with only one record in differential privacy mindset. The attacker cannot identify whether if n^{th} record in D or D', it is because the result of running query is same with high probability. To wrap up, we can say that an algorithm is

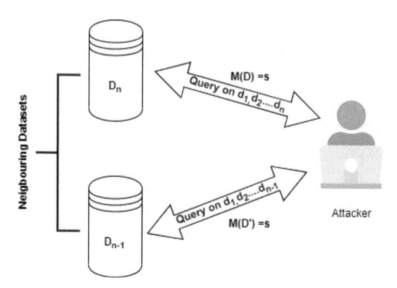

Fig. 13.12 Differential privacy model [45]

differentially private if its output doesn't tell any information about the individual used in the computation. The differential privacy masks the difference of query between neighboring datasets by perturbing the result with random noise generated according to specific distribution [46].

Differential privacy is a probabilistic privacy mechanism that ensures an information-theoretic security warranty. The formal definition of differential privacy given by Dwork as follows [47].

Differential Privacy ($\mathcal{E},\delta$): A randomized mechanism M preserves ($\mathcal{E},\delta$) differential privacy for each set of outputs S, and any neighboring datasets of D and D' differing by at most one record, if M satisfies [48]:

$$Pr[M(D) \in S] \leq \exp(\varepsilon).Pr\big[M\big(D'\big) \in S\big] + \delta$$

where $\mathcal{E}$ is the privacy budget and δ is failure probability. The ratio on two probabilities is restrained by e^ε for a certain output. The randomized mechanism M grants ε differential privacy by its strictest definition if $\delta = 0$. When $\delta = 0$, a strictly stronger notion of ε-differential privacy is achieved. ($\mathcal{E},\delta$)-differential privacy maintains latitude to break rigid $\mathcal{E}$-differential privacy for some low probability events [49].

The *privacy budget* provides to control the privacy guarantee degree of mechanism M. A smaller $\mathcal{E}$ stands for more powerful privacy [50].

Sensitivity defines how much complexity is required in the mechanism. The sensitivity calibrates the required volume of noise for $M(D)$. Two types of sensitivity are placed in the literature: global sensitivity and local sensitivity.

There are two kinds of differential privacy approaches as local differential privacy and global differential privacy. In Table 13.1, studies of differentially private deep learning methods in the literature are given.

Table 13.1 Differentially private deep learning methods [45]

Related work	Adversarial setting	System setting	Privacy guarantee method
Shokri et al. [40]	Additional capabilities	Distributed system	Differentially private Stochastic Gradient Descent (SGD) algorithm with convex objective functions
Adabi et al. [41]	Additional capabilities	Centralized system	Differentially private SGD algorithm with non-convex objective functions
Phah et al. [42]	General capabilities	Centralized system	Objective function perturbation of deep auto-encoder

13.4.1.2   Homomorphic Encryption

Homomorphic encryption enables performing computation on encrypted data without decrypting it. The output is encrypted while the decrypted output is the same as the output of computation on unencrypted data. Homomorphic encryption is an approach used computing on outsourced storage that allows processing data in cloud environments. While differential privacy adds randomness into data to prevent de-anonymization strategies from succeeding, homomorphic encryption adds a security layer by providing machine learning algorithms to run on data without decrypting it. We can meet with Homomorphic encryption at most applications from healthcare to machine learning as a service (MLaaS) (see Figs. 13.12, 13.13).

This approach which was firstly proposed by Rivest et al. [52] was applied to the convolutional neural networks by Bachrach et al. [53]. It is based on "privacy homomorphisms which is a structure-preserving transformation and allows for an arbitrary number of addition and multiplication operations on the encrypted data. Given a public-key encryption scheme with two ciphertexts $(\hat{x}_1, \hat{x}_2)$ [54].

$$\hat{x}_1 := E(p_k, x_1)$$

$$\hat{x}_2 := E(p_k, x_2)$$

There is a public-key operation $\oplus$ such that;

$$E(p_k, x_1 + x_2) \leftarrow \hat{x}_1 \oplus \hat{x}_2$$

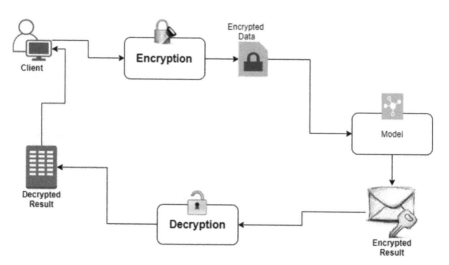

Fig. 13.13 Homomorphic encryption-use case for machine learning as a service (MLaaS) [51]

Paillier's encryption [55] and ElGamal encryption [56] can be given as an example of this kind of scheme which is referred to as homomorphic encryption. Subtraction ($\ominus$) which is the inverse of addition is supported by homomorphic encryption as well. Adding or multiplying a ciphertext by a constant also supported:

$$E(p_k, a + x) \leftarrow a \oplus \hat{x}$$

$$E(p_k, a.x) \leftarrow a \odot \hat{x}_1$$

For addition and multiplication between two ciphertexts, fully homomorphic encryption (FHE) or leveled homomorphic encryption (LHE) is required. Fully homomorphic encryption (FHE) is the most powerful concept of homomorphic encryption which provides the assessment of arbitrary circuits of unbounded depth.

13.4.1.3 Federated Learning

The notion of federated learning was proposed by Google [57–59] to build secure machine learning models running on distributed data. The data to be trained doesn't aggregate but the model is sent to the data owner. After the training completes, the updated weights are sent back to be averaged the final model. Thanks to this approach, the data remains at the original owner in a secure and trusted way without a compromise in model performance.

Even there is a little computational cost and limited neural network's ability to develop federated learning models, there is great potential. Nowadays, consumer devices such as mobile phones generate a huge amount of data. This data may contain personal data or data that gives an idea about user habits. At this point, federated learning enables us to build a personalized model without compromising privacy. This is the motivation for AI researchers to work in this field.

As seen in Fig. 13.14, Federated learning is a machine learning technique that trains an algorithm across multiple decentralized data, without exchanging them. Hence, it enables multiple databases to build a common, robust machine learning model without sharing data [60, 61].

13.4.1.4 Secure Multi-party Computation

Secure multi-party computation (MPC) provides to perform computation without disclosing any participant's private inputs. The participants can use an MPC protocol to compute the function together, which they agree on without compromising privacy.

It was proposed by Andrew Yao in 1980 [62] with Millionaires Problem. The problem is: "Two millionaires wish to know who is richer; however, they do not want to find out inadvertently any additional information about each other's wealth." The solution is obtained by computing the Boolean result of $x_1 \leq x_2$ where x_1 is

Fig. 13.14 Federated
learning [60]

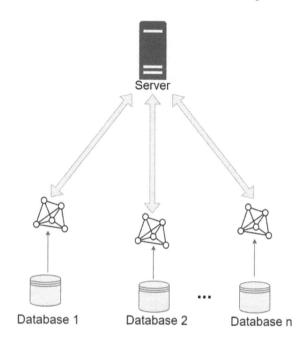

private input of the first party and x_2 is private input of the second party. Secure
multi-party computation aims to provide independent data owners who do not trust
each other to make computation that depends on their private inputs securely [63].
Malkhi et al. implemented the first secure computation system which is a full-fledged
system that implements generic secure function evaluation [64] (Fig. 13.15).

13.5 Discussions

In the chapter, examples of the use of AI technologies by individuals who are limited
in meeting their needs are emphasized. At these and similar points of use, data
privacy and data security issues that need to be considered in the collection, storage,
and processing of health and other sensitive data are addressed. Requirements for
ethics, data security, data privacy, and explainable AI technologies are summarized.
Particular emphasis is placed on recent privacy protection approaches.

Many of the principles discussed in various AI ethics statements such as transparency, explainability, and privacy can be viewed as aspects of respect for autonomy.
Decision algorithms that are glossed over for health in deep learning and related
AI approaches do not provide transparency or interpretability as decision trees.
Providing explanations for the "why" question that will accompany black-box prediction algorithms is an ongoing area of research. Techniques such as differential privacy,
homomorphic encryption, and federated learning are produced for AI-supported
assistant systems that are secure against cyber-attacks, which facilitate the lives

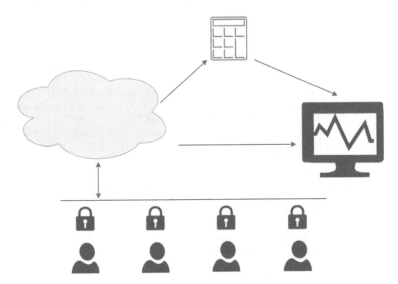

Fig. 13.15 Secure multi-party computation [65]

of individuals and protect their privacy while doing so. In this way, successful AI applications without compromising security will be able to find their place in daily life.

While developing AI technologies, international organizations publish regular reports for the use of AI technologies and the applicability of the process, which is also emphasized in this chapter. At the end of October 2020, the European Data Protection Board published the report of Data Protection by Design and by Default. The report is a guide on how to apply the same data protection principles effectively [66].

A healthy and equitable future for everyone should be the ultimate goal with accessible, fast, precise, successful, low-cost AI-powered assistive technologies.

References

1. Model ICT Accessibility Policy Report, Available: https://www.itu.int/en/ITU-D/Digital-Inc lusion/Persons-with-Disabilities/Documents/ICT%20Accessibility%20Policy%20Report.pdf (2014). Accessed 31 Jan 2021
2. Summary World Report on Disability, Available: https://apps.who.int/iris/bitstream/handle/ 10665/70670/WHO_NMH_VIP_11.01_eng.pdf;jsessionid=50C7F4199A25E26711B5A903 759B35C6?sequence=1 (2011). Accessed 31 Jan 2021
3. B. Goodman, S. Flaxman, European union regulations on algorithmic decision-making and a right to explanation. AI Magazine **38**(3) 50–57 (2017)
4. D. Castelvecchi, Can we open the black box of AI? Nature **538**(7623), 20–23 (2016). https:// doi.org/10.1038/538020a
5. A.B. Arrietaa, N. Diaz-Rodriguez, J. Del Sera, A. Bennetot, S. Tabik, A. Barbado, S. Garcia, S. Gil-Lopez, et. Al. , Explainable Artificial Intelligence (XAI): concepts, taxonomies, oppor-tunities and challenges toward responsible AI. Inform. Fus. **58**, 82–115 (2019). https://doi.org/

10.1016/j.inffus.2019.12.012

6. ScienceDirect, Assistive Technology. Available: https://www.sciencedirect.com/topics/com
   puter-science/assistive-technology. Accessed 31 Jan 2021
7. Alzheimer's Disease International, Dementia Statistics. Available: https://www.alz.co.uk/res
   earch/statistics. Accessed 31 Jan 2021
8. Gartner, Gartner identifies the top 10 strategic technology trends for 2019 (2018). Available:
   https://www.gartner.com/en/newsroom/press-releases/2018-10-15-gartner-identifies-the-top-
   10-strategic-technology-trends-for-2019. Accessed 31 Jan 2021
9. B. DuBois, K. K. Miley, Social work: an empowering profession. Available: https://www.pea
   rsonhighered.com/assets/preface/0/1/3/4/0134747399.pdf. Accessed 31 Jan 2021 (2018)
10. S. Martin, B.E. Johan, R.M. Dröes, in *Supporting People with Dementia Using Pervasive
    Health Technologies. Advanced Information and Knowledge Processing*, ed. by M. Mulvenna,
    C. Nugent. Assistive technologies and issues relating to privacy, ethics and security (Springer,
    London, 2010). https://doi.org/10.1007/978-1-84882-551-2_5
11. M. Anderson, S. Anderson, Guest editors' introduction: machine ethics. IEEE Intell. Syst.
    21(4), 10–11 (2006). https://doi.org/10.1109/MIS.2006.70
12. J. Moor, The nature, importance, and difficulty of machine ethics. IEEE Intell. Syst. **21**(4),
    18–21 (2006). https://doi.org/10.1109/MIS.2006.80
13. Commercial UAV News, How has GDPR reshaped the way drone stakeholders should approach
    data privacy? (2019). Available: https://www.commercialuavnews.com/europe/gdpr-drone-
    data-privacy. Accessed 31 Jan 2021
14. J. Holvast, in *The Future of Identity in the Information Society. Privacy and Identity 2008.
    IFIP Advances in Information and Communication Technology*, ed. by V. Matyáš, S. Fischer-
    Hübner, D. Cvrček, P. Švenda. History of Privacy, vol 298 (Springer, Berlin, Heidelberg, 2009).
    https://doi.org/10.1007/978-3-642-03315-5_2
15. Varonis, Data Privacy Guide: Definitions, Explanations and Legislation (2020). Available:
    https://www.varonis.com/blog/data-privacy/. Accessed 31 Jan 2021
16. Microsoft I AI. Seeing AI in new languages. Available: https://www.microsoft.com/en-us/ai/
    seeing-ai. Accessed 31 Jan 2021
17. University of Oxford, News&Events. Smart glasses for people with poor vision being tested in
    Oxford. Available: https://www.ox.ac.uk/news/2014-06-17-smart-glasses-people-poor-vision-
    being-tested-oxford. Accessed 31 Jan 2021
18. Oxsight, Available: https://www.oxsight.co.uk/. Accessed 31 Jan 2021
19. eeNews Europe, Cognitive hearing aid puts DNNs to work with the wearer's brain,
    eenewseuropa (2017). Available: https://www.eenewseurope.com/news/cognitive-hearing-aid-
    puts-dnns-work-wearers-brain. Accessed 31 Jan 2021
20. Neural Acoustic Processing Lab, Brain-controlled hearing aid. Available: http://naplab.ee.col
    umbia.edu/nnaad.html. Accessed 31 Jan 2021
21. CNN Health, This new high-tech glove translates sign language into speech in real time
    (2020). Available: https://edition.cnn.com/2020/06/30/health/sign-language-glove-ucla-scn-
    scli-intl/index.html. Accessed 31 Jan 2021
22. University of Aberdeen News, Technology that translates sign language into text aims
    to empower sign language users (2012). Available: https://www.abdn.ac.uk/news/4294/.
    Accessed 31 Jan 2021
23. Sesgoritma, Apple App Store, Available: https://apps.apple.com/tr/app/sesgoritma/id1369560
    353?l=tr. Accessed 31 Jan 2021
24. H. Kose et al., in *Intelligent Assistive Robots. Springer Tracts in Advanced Robotics*, ed. by
    S. Mohammed, J. Moreno, K. Kong, Y. Amirat. iSign: an architecture for humanoid assisted
    sign language tutoring, vol 106 (Springer, Cham, 2015). https://doi.org/10.1007/978-3-319-
    12922-8_6
25. IEEE Spectrum, To Build the Best Robotic Exoskeleton, Make It on the Cheap
    (2016). Available:https://spectrum.ieee.org/the-human-os/biomedical/bionics/to-build-the-fan
    ciest-robotic-exoskeleton-make-it-on-the-cheap. Accessed 31 Jan 2021

26. D.C. Patel, H.S. Patil, Design and development of low cost artificial intelligence vacuum cleaner. Int. J. Recent Trends Eng. Res. (IJRTER) **03**(11) (2017). https://doi.org/10.23883/IJR TER.2017.3494.S78ZF.

27. IEEE Spectrum, Caregiver Robots (2010). Available: https://spectrum.ieee.org/robotics/hum anoids/caregiver-robots. Accessed 31 Jan 2021

28. IEEE Robots, Mabu (2015). Available: https://robots.ieee.org/robots/mabu/. Accessed 31 Jan 2021

29. O.-M. Camburu, Explaining deep neural networks. Ph.D. Thesis, University of Oxford, ArXiv, abs/2010.01496 (2020)

30. M. Turak, DARPA: Explainable Artificial Intelligence (XAI) (2016). https://www.darpa.mil/ program/explainable-artificial-intelligence. Accessed 31 Jan 2021

31. European Comission, JRC Technical Report. Robustness and Explainability of Artificial Intelligence (2020). Available: https://publications.jrc.ec.europa.eu/repository/bitstream/JRC119 336/dpad_report.pdf. Accessed 31 Jan 2021

32. N. Dong, H. Jonker, J. Pang, in *Foundations of Health Informatics Engineering and Systems. FHIES 2011. Lecture Notes in Computer Science*, vol 7151, ed. by Z. Liu, A. Wassyng. Challenges in eHealth: From Enabling to Enforcing Privacy (Springer, Berlin, Heidelberg, 2012). https://doi.org/10.1007/978-3-642-32355-3_12

33. L. Boise, K. Wild, N. Mattek, M. Ruhl, H.H. Dodge, J. Kaye, Willingness of older adults to share data and privacy concerns after exposure to unobtrusive in-home monitoring. Gerontechnol. Int. J. Fundament Aspects Technol Serve Ageing Soc. **11**(3), 428–435 (2013). https://doi.org/ 10.4017/gt.2013.11.3.001.00

34. C. Bartneck, C. Lütge, A. Wagner, S. Welsh, *An Introduction to Ethics in Robotics and AI* (Springer, 2021). https://doi.org/10.1007/978-3-030-51110-4

35. I.D. Raji, B. Buolamwini, in *Proceedings of the 2019 AAAI/ACM Conference on AI, Ethics, and Society (AIES '19). Association for Computing Machinery*. Actionable Auditing: Investigating the Impact of Publicly Naming Biased Performance Results of Commercial AI Products (New York, NY, 2019), pp. 429–435. https://doi.org/10.1145/3306618.3314244

36. Microsoft, A History of Anticompetitive Behavior and Consumer Harm (2009). Available: http://www.ecis.eu/documents/Finalversion_Consumerchoicepaper.pdf. Accessed 31 Jan 2021

37. A.J. Thomson, Artificial intelligence and environmental ethics (1997). Available: https:// www.researchgate.net/publication/324825733_AI_Intelligence_and_Environmental_Ethics. Accessed 31 Jan 2021

38. B.M. Knoppers, Framework for responsible sharing of genomic and health-related data. Hugo J. **8**(1), 3 (2014). https://doi.org/10.1186/s11568-014-0003-1

39. V.C. Pezoulas, T.P. Exarchos, D.I. Fotiadis, Medical Data Sharing, Harmonization, and Analytics (2020)

40. N. Asokan, V. Niemi, K. Nyberg, in *Proceedings International Workshop Security, Protocols*. Man-in-the-middle in tunnelled authentication protocols (New York, NY, Springer, 2003), pp. 28–41

41. Y. Chen, W. Trappe, R.P. Martin, Detecting and Localizing Wireless Spoofing Attacks. 4th Annual IEEE Communications Society Conference on Sensor, Mesh and Ad Hoc Communications and Networks (San Diego, CA, 2007), pp. 193–202. https://doi.org/10.1109/SAHCN. 2007.4292831

42. C. Meadows, R. Poovendran, D. Pavlovic, L. Chang, P. Syverson, Distance bounding protocols: Authentication logic analysis and collusion attacks, in *Secure Localization and Time Synchronization for Wireless Sensor and Ad Hoc Networks* (Springer, New York, 2007), pp. 279–298

43. J.-H. Hoepman, Privacy design strategies. IFIP international information security conference (Springer, 2014), pp. 446–459

44. J. Domingo-Ferrer, A. Blanco-Justicia, Privacy-Preserving Technologies, in *The Ethics of Cybersecurity. The International Library of Ethics, Law and Technology*, vol 21, ed. by M. Christen, B. Gordijn, M. Loi (Springer, Cham, 2020). https://doi.org/10.1007/978-3-030-29053-5_14

45. T. Zhu, G. Li et al., *Differential Privacy and Applications* (Springer, 2017)
46. C. Dwork, F. McSherry, K. Nissim, A. Smith, Calibrating Noise to Sensitivity in Private Data Analysis. In Theory of Cryptography Conference (TCC), (Springer, 2006). https://doi.org/10.1007/11681878_14. J. Privacy Confident. **7**(3), 17–51. https://doi.org/10.29012/jpc.v7i3.405
47. C. Dwork, in *Theory and Applications of Models of Computation. TAMC 2008. Lecture Notes in Computer Science*, vol 4978, ed. by M. Agrawal, D. Du, Z. Duan, A. Li. Differential privacy: a survey of results (Springer, Berlin, Heidelberg, 2008). https://doi.org/10.1007/978-3-540-792 28-4_1
48. C. Dwork, A firm foundation for private data analysis. Communicationas of ACM **54**(1), 86–95 (2011). https://doi.org/10.1145/1866739.1866758
49. A. Beimel, K. Nissim, U. Stemmer, in *Approximation, Randomization, and Combinatorial Optimization. Algorithms and Techniques. APPROX 2013, RANDOM 2013. Lecture Notes in Computer Science*, vol 8096, ed. by P. Raghavendra, S. Raskhodnikova, K. Jansen, J.D.P. Rolim. Private Learning and Sanitization: Pure vs. Approximate Differential Privacy (Springer, Berlin, Heidelberg, 2013). https://doi.org/10.1007/978-3-642-40328-6_26
50. A. Haeberlen, B. Pierce, A. Narayan, Differential Privacy Under Fire. USENIX Security Symposium (2011)
51. OpenMined Blog, What is Homomorphic Encryption? (2020). Available: https://blog.openmi ned.org/what-is-homomorphic-encryption/. Accessed 31 Jan 2021
52. R.L. Rivest, A. Len, M.L. Dertouzos, On data banks and privacy homomorphisms. Foundations of secure computation **4**(11), 169–180 (1978)
53. R. Gilad-Bachrach, N. Dowlin, K. Laine, K. Lauter, M. Naehrig, J. Wernsing, Cryptonets: Applying neural networks to encrypted data with high throughput and accuracy, in Proceedings of ICML'16 (2016), pp. 201–210
54. J. Liu, M. Juuti, Y. Lu, N. Asokan, Oblivious Neural Network Predictions via MiniONN Transformations. CCS '17: Proceedings of the 2017 ACM SIGSAC Conference on Computer and Communications Security October (2017), pp. 619–631. https://doi.org/10.1145/3133956.3134056
55. P. Paillier, Public-Key Cryptosystems Based on Composite Degree Residuosity Classes. In EUROCRYPT (LNCS), Jacques Stern (Ed.), Vol. 1592. (Springer, 1999), pp. 223–238. https://doi.org/10.1007/3-540-48910-X_16
56. T. ElGamal, A Public Key Cryptosystem and a Signature Scheme Based on Discrete Logarithms. In CRYPTO (LNCS), Vol. 196. Springer, (1985) pp. 10–18. doi: https://doi.org/10.1007/3-540-39568-7_2
57. J. Konecný, H.B. McMahan, D. Ramage, P. Richtárik, Federated Optimization: Distributed Machine Learning for On-Device Intelligence (2016). Available: https://arxiv.org/abs/1610.02527. Accessed 31 Jan 2021
58. J. Konecný, H.B. McMahan, F.X. Yu, P. Richtárik, A.T. Suresh, D. Bacon, Federated Learning: Strategies for Improving Communication Efficiency (2016). Available: https://arxiv.org/abs/1610.05492. Accessed 31 Jan 2021
59. H.B. McMahan, E. Moore, D. Ramage, B.A. Arcas y, Communication-Efficient Learning of Deep Networks from Decentralized Data (2016). Available: https://arxiv.org/abs/1602.05629. Accessed 31 Jan 2021
60. Q. Yang, Y. Liu, T. Chen, Y. Tong, Federated Machine Learning: Concept and Applications (2019). Available: https://arxiv.org/pdf/1902.04885.pdf. Accessed 31 Jan 2021
61. Wikipedia, Federated Learning. Available: https://en.wikipedia.org/wiki/Federated_learning. Accessed 31 Jan 2021
62. A.C. Yao, Protocols for secure computations. 23rd Annual Symposium on Foundations of Computer Science (sfcs 1982) (Chicago, IL, 1982) pp. 160–164. https://doi.org/10.1109/SFCS.1982.38
63. D. Evans, V. Kolesnikov, M. Rosulek, A Pragmatic Introduction to Secure Multi-Party Computation (2020). Available: https://www.cs.virginia.edu/~evans/pragmaticmpc/pragmatic mpc.pdf. Accessed 31 Jan 2021

64. D. Malkhi, N. Nisan, B. Pinkas, Y. Sella, Fairplay-A Secure Two-Party Computation System, in USENIX Security Symposium (2004). Available: http://www.pinkas.net/PAPERS/MNPS. pdf. Accessed 31 Jan 2021

65. Hackernoon Medium Blog, What is Secure Multi Party Computation? (2019). Available: https:// medium.com/hackernoon/what-is-secure-multi-party-computation-232caef900b9. Accessed 31 Jan 2021

66. European Commission, What does data protection 'by design' and 'by default' mean?. Available: https://ec.europa.eu/info/law/law-topic/data-protection/reform/rules-business-and-organisations/obligations/what-does-data-protection-design-and-default-mean_en. Accessed 31 Jan 2021